AF807316

Advances in Karst Science

Series Editor

James LaMoreaux, Tuscaloosa, AL, USA

This book series covers advances in the field of karst from a variety of perspectives to facilitate knowledge and promote interaction thereby building stepping stones in this process. Methodologies, monitoring, data analysis and interpretation are addressed throughout the wide range of climatic, geological and hydrogeological contexts in which karst occurs. Case studies are presented to provide examples of advancement of the science.

Issues to be addressed include water supply, contamination, and land use management. These issues although occurring on a local basis share many of the same features on the global stage. This book series is a critical resource to the scientific community allowing them to compare problems, results, and solutions. The presented information can be utilized by decision makers in making decisions on development in karst regions. Contributions presented may be used in the classroom and to work with stakeholders, scientists, and engineers to determine practical applications of these advances to common problems worldwide.

The series aims at building a varied library of reference works, textbooks, proceedings, and monographs, by describing the current understanding of selected themes. The books in the series are prepared by leading experts actively working in the relevant field. The book series *Advances in Karst Science* includes single and multi-authored books as well as edited volumes. The Series Editor, Dr. James W. LaMoreaux, is currently accepting proposals and a proposal document can be obtained from the Publisher.

Xiaozhen Jiang · Mingtang Lei · Wanfang Zhou ·
Fubiao Zhou · Xiao Ma · Jianling Dai · Zongyuan Pan

Monitoring and Early Warning Technologies on Karst Lands

Surface Collapse and Groundwater Contamination

Springer

Xiaozhen Jiang
Institute of Karst Geology
Chinese Academy of Geological Sciences
Guilin, China

Wanfang Zhou
ZeoEnvironmental, LLC
Knoxville, TN, USA

Xiao Ma
Institute of Karst Geology
Chinese Academy of Geological Sciences
Guilin, China

Zongyuan Pan
Institute of Karst Geology
Chinese Academy of Geological Sciences
Guilin, China

Mingtang Lei
Institute of Karst Geology
Chinese Academy of Geological Sciences
Guilin, China

Fubiao Zhou
Institute of Karst Geology
Chinese Academy of Geological Sciences
Guilin, China

Jianling Dai
Institute of Karst Geology
Chinese Academy of Geological Sciences
Guilin, China

ISSN 2511-2066 ISSN 2511-2082 (electronic)
Advances in Karst Science
ISBN 978-3-031-59044-3 ISBN 978-3-031-59045-0 (eBook)
https://doi.org/10.1007/978-3-031-59045-0

© The Editor(s) (if applicable) and The Author(s), under exclusive license to Springer Nature Switzerland AG 2024

This work is subject to copyright. All rights are solely and exclusively licensed by the Publisher, whether the whole or part of the material is concerned, specifically the rights of translation, reprinting, reuse of illustrations, recitation, broadcasting, reproduction on microfilms or in any other physical way, and transmission or information storage and retrieval, electronic adaptation, computer software, or by similar or dissimilar methodology now known or hereafter developed.
The use of general descriptive names, registered names, trademarks, service marks, etc. in this publication does not imply, even in the absence of a specific statement, that such names are exempt from the relevant protective laws and regulations and therefore free for general use.
The publisher, the authors and the editors are safe to assume that the advice and information in this book are believed to be true and accurate at the date of publication. Neither the publisher nor the authors or the editors give a warranty, expressed or implied, with respect to the material contained herein or for any errors or omissions that may have been made. The publisher remains neutral with regard to jurisdictional claims in published maps and institutional affiliations.

This Springer imprint is published by the registered company Springer Nature Switzerland AG
The registered company address is: Gewerbestrasse 11, 6330 Cham, Switzerland

If disposing of this product, please recycle the paper.

Foreword

The phrase "expect the unexpected" is often associated with engineering construction and environmental remediation in karst terranes. This is because karst does not follow the typical rules that we learnt from most landscapes that form, as the rain, rivers, wind, waves, and even glaciers break down and wear away the rocks. This isn't the case with karst terranes. Karst developed on such rocks as limestone, dolomite, gypsum, or halite that simply dissolve in contact with acidic water, leaving little behind except insoluble leftovers. The solution process occurs both on the ground surface and in the subsurface, resulting in karst features characterized with depressions on the bedrock surface, a highly irregular bedrock surface, an underground tributary network of solution-widened conduits and small tubules that may merge into small and large caves. These karst features tend to establish an underground circulation in which the water sinks into the ground rather than flowing away in rivers and discharges through karst springs or other outlets.

The bedrock solution process can lead to cave collapses on a geological timescale. But on a human timescale, the most intimately tied to karst terranes are soil collapses in areas where the bedrock is buried beneath an overburden made up of alluvial, glacial, or marine deposits or the insoluble leftovers from the bedrock dissolution. The downward drainage into voids in the bedrock carries the overburden sediment down with it. The sediment is eroded from the bottom upwards giving no sign of the ongoing erosion, or of the impending collapse at the ground surface. When the ground sinks suddenly, small, or not-so-small depressions, often referred to as "karst collapses", suddenly appear in your construction site. They sink down 1, 2, 3, or more meters, and they wreak havoc with your plans for a foundation. This same erosion process can occur slowly, even imperceptibly, just a couple of centimeters or less per year. A noticeable shallow depression can eventually form on the ground surface in 25 years. If this process develops under the foundation of a building, major cracks occur, and the building becomes unlivable.

The scale and rate of this karst collapse process is variable. The same process that produced those small karst collapses on your site, under the right circumstances, and on a grander scale, can bring about gigantic collapses that damage roads, houses, or buildings. When the karst is covered, karst identification during subsurface exploration programs is challenging, and it is difficult to anticipate when and where these collapses will occur.

Also difficult is the monitoring of groundwater flow and the associated contaminant transport in a network of discrete channels hidden underground. So, unless the monitoring well actually intercepts a conduit in the network, it will not receive drainage and detect contamination. However, the main question is, how do we hit a conduit in the first place? Since this drainage network is under the ground, nature does not provide any "as builts" for us to review. If the water is flowing rapidly through a discrete conduit in the rock, possibly in the order of a foot wide, the chances for the monitoring well to intercept it is small.

Monitoring of springs has been considered a viable option since the entire drainage network eventually discharges at springs. The spring discharge will integrate flow from the entire basin, including any contamination that is occurring. Unfortunately, it is still not that easy; it never is in karst. Contaminant generally moves through the drainage network in flood pulses. Samples taken at low flow may be clean, while a precipitation-generated peak discharge may be highly contaminated. So, simply sampling at the spring on a quarterly basis will not do it. And it is equally important to be sure to prove it is the right spring, usually by dye tracing. Underground drainage networks have been known to cross each other and even to flow under surface streams without any interconnection.

Fortunately, there are some solutions to these challenges. We now understand that karst follows its own rules. Data on the karst collapse mechanisms and groundwater dynamics have been collected in recent years to understand these rules of karst formation. In addition to specialized techniques that have been developed for site characterization, innovative methods, and tools have been evaluated and validated for monitoring and early warning of the geohazards. I have known the lead authors of this book for three decades, and their expertise is on karst collapse formation, prevention, and mitigation. I am honored to write the foreword for this book, which summarize these innovative monitoring and early warning technologies, and to present case studies on their applications to solving practical problems for either karst collapses or groundwater contamination.

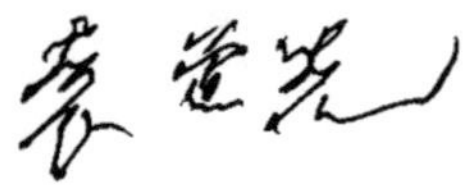

Daoxian Yuan
Professor at the Institute of Karst Geology of China
Academician of the Chinese Academy of Sciences
Guilin, China

Acknowledgements

Writing and publishing a book demands dedication, effort, and support from families, colleagues, and organizations. We are in debt to all of them, especially those who participated in the projects included in this book. The following organizations are acknowledged for their support in publishing this important book:

Institute of Karst Geology, China Geological Survey

International Research Center on Karst under the Auspices of UNESCO

Some of the materials used in the book are collected through research and investigation projects that are sponsored by Ministry of Natural Resources of China (Grant No. 201211083), China Geological Survey (Grant No. DD20160254 and DD20190266), China Natural Science Foundation (Grant No. 42077273, 41877300, 41428202, and 41272355), and Guangxi Science and Technology Planning Project (Grant No. 21196001). In addition, we would like to express our gratitude to Dr. James LaMoreaux, Dr. Annett Buettner, Yogesh Padmanaban, and Qiao Shu for the support, encouragement, and constructive comments in the preparation of the book. Last but not least, we are honored to have Prof. Daoxian Yuan, Academician of the Chinese Academy of Sciences, write the Foreword for this book.

Contents

About the Lead Author

Dr. Xiaozhen Jiang is a professor at the Institute of Karst Geology, Chinese Academy of Geological Sciences, and a senior researcher at the UNESCO International Karst Research Center. She serves as the deputy director of the Karst Exploration Technology Professional Committee of the Chinese Society for Rock Mechanics and Engineering and on the National Natural Resources and Territory Spatial Planning Committee of China (SAC/TC93/SC2) and the Karst Professional Committee of the Geological Society of China. She is an appointed technical expert by the Ministry of Natural Resources on geological disaster prevention and control.

Dr. Jiang has long been committed to the research of karst collapse prevention and control technology. As the principal investigator, she has successfully completed more than 40 projects sponsored by the National Natural Science Foundation, the Ministry of Natural Resources, and industrial enterprises. Based on over 20 years' field studies and extensive research with large-scale physical analog models, she proposed the sudden changes in water and gas pressure as the collapse mechanism in karst systems, developed criteria determination methods, and established a technical support system for karst collapse monitoring and early warning, early identification, and dynamic risk assessment. She pioneered the application of distributed optical fiber sensing to karst collapse monitoring and developed a cloud-based information management platform by integrating data from dynamic monitoring and ground penetration radar.

She has published more than 40 papers and 10 invention patents. She serves as the editor-in-chief for the industry standards on "Karst Collapse Monitoring Specifications" and "Karst Ground Collapse Prevention and Control Engineering Exploration Specifications" and is the lead author of the monographs "Karst Collapse Monitoring Technology" and "Atlas of Karst Collapses". The newly developed technologies have been applied to karst collapse monitoring along several railways including the Wuhan-Guangzhou High-speed Railway and Guiyang-Guangzhou High-speed Railway as wells as to collapse-prone provinces of China such as Guangdong, Guangxi, Hunan, Hubei, Anhui, Guizhou, Chongqing, and Shandong.

1 Vulnerability of Karst Terranes to Ground Collapses and Groundwater Contamination

1.1 Hydrodynamics of Karst Terranes

Recharge and groundwater flow characteristics are the primary factors that make karst terranes vulnerable to surface collapse and contamination. One-dimensional channels, two-dimensional planar partings such as fractures, joints, and faults, and three-dimensional inter-granular matrix pores constitute porosities for groundwater storage and flow in a karst aquifer. While rapid and possibly turbulent flow occurs in large channels, commonly termed conduits, or caves if they are accessible by people, slow and laminar flow occurs in the rock matrices and tight fractures. While the majority of the undergroundwater flows through the conduits in many karst aquifers, the conduits actually contain only a small fraction of the total volume of water in the aquifer. The chance for a randomly selected well to intercept a major conduit is small. Because groundwater levels in monitoring wells installed in different porosities respond differently to recharge events, it should be cautious to interpret the groundwater flow direction from water level measurements in monitoring wells, especially on local scales. Specialized techniques such as tracer tests or geophysics may be needed to determine the groundwater flow paths. In non-karstic terranes, there is a clean distinction between surface water and groundwater, and between surface drainage basins and groundwater aquifers. Drainage basins, which determine the catchment and runoff characteristics of surface water are defined by drainage divides set by local topography. Surface water is often only loosely coupled to groundwater systems because of the slow rate of infiltration. These conceptual distinctions between basins and aquifers are blurred in karst areas because of the integrated system of solution-enlarged fractures or conduits that carry water through the subsurface. Localized recharge to the conduit system occurs where surface streams flow underground and where sinkholes collect overland flow and direct to the subsurface. The discharge of water from the conduit system often takes place at one or several large springs that form the headwaters of a perennial surface stream. There is in karst a much more intimate relationship between groundwater and local surface water than in other types of aquifers.

1.1.1 Recharge Types

In karst terranes, the recharge includes inputs from surface catchments on non-karstic rocks. These non-karstic, but hydrologically connected rocks are referred to as borderland. Surface runoff from the borderlands drains through surface streams that flow onto the karst area, where some sink underground at the margin of the karst land. These are the sinking stream inputs to the karst groundwater system. Little runoff is present on the karst land, which is typical in well-developed karst terranes. Overland flow during storms disappears into sinkholes to enter the groundwater system as internal runoff. Depending on where the water has come from, the recharge sources to a karst aquifer can be divided into two categories—allogenic recharge and autogenic recharge. An allogenic recharge derives its water from a neighboring non-karst borderland, whereas an autogenic recharge derives its water from meteoric precipitation falling on the karst land.

The recharge processes are affected by the overlying soil (Lloyd et al. 1981) and the epikarst. As shown in Fig. 1.1, the epikarst consists of the upper most portion of the unsaturated rock where significant fracturing, solutional enlargement, and storage may occur. The network of dissolved fractures and bedding planes is normally filled with sediment, but it is typically more permeable than the rock matrix. Through this permeable zone, the drainage flows to a solution pipe, which directs the collected water down into the limestone. The cutters provide a myriad of entrance points into the epikarst zone and the solution pipes function as drains for the water in the perched epikarst zone. Because water flow from the surrounding epikarst zone converges toward these pipes, more limestone will be dissolved

© The Author(s), under exclusive license to Springer Nature Switzerland AG 2024

X. Jiang et al., *Monitoring and Early Warning Technologies on Karst Lands*, Advances in Karst Science, https://doi.org/10.1007/978-3-031-59045-0_1

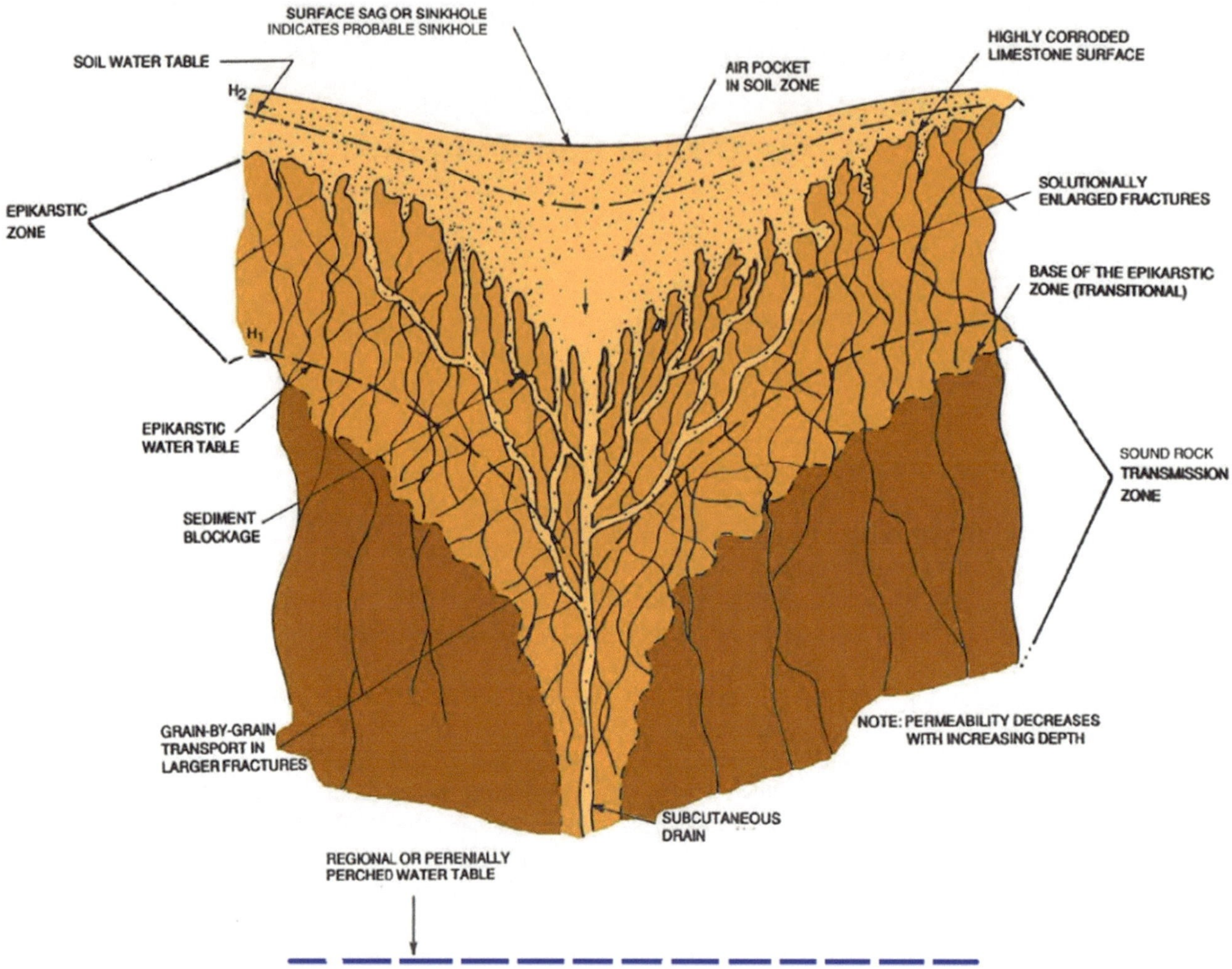

Fig. 1.1 Recharge passageways in epikarst (base map from Williams 1985) (Perching may occur in both the epikarst zone (H_1) and in the soil zone (H_2), but the flow is always directed toward a subcutaneous drain)

and removed around the pipes, thus continuing the subsurface erosion processes and the development of the broad depressions or sinkholes. Following infiltration through topsoil and internal runoff through sinkholes on the karst land, water can be collected and temporarily stored in the epikarst zone, until it subsequently joins the saturated zone. The channeling of soil water to the subcutaneous drains eventually contributes to a catastrophic collapse of the overburden (Williams, 1985).

1.1.2 Flow Characteristics

Because multiple types of porosity such as caves, conduits, fractures, and matrix pores are potentially present in a karst aquifer, the hydraulic conductivity is highly heterogeneous on the scale hierarchy. Figure 1.2 presents an example of groundwater flow patterns in a karst aquifer and compares the characteristics with a non-karst inter-granular aquifer. Groundwater flow patterns can be characterized by the co-existence of steady and unsteady flows, laminar and turbulent flows, isolated groundwater flow, and continuous groundwater flow. The groundwater flow field can be hydraulically connected to surface water and be sensitive to water recharge or extraction. Conduit flow aquifers behave hydraulically as a system of pipes with flow velocities similar to those of surface water streams, whereas diffuse flow aquifers contain solution-enlarged fractures that are relatively small and generally transmit water under laminar flow conditions.

What is generally described as the characteristics of a karst system depends very much on the type of investigation method used and on the scale of the flow domain. A detection method with a large averaging volume will produce parameter fields that appear to be almost homogeneous, whereas from small-scale measurements, the same aquifer may appear highly heterogeneous. An investigation method may be more selective in the analysis of the

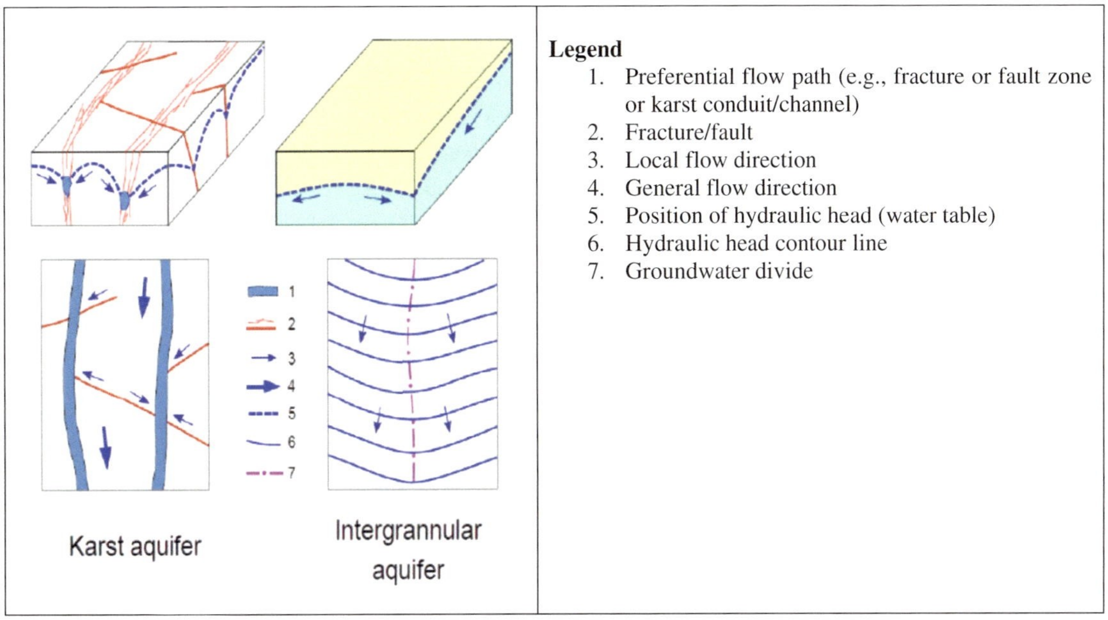

Fig. 1.2 Select characteristics of karst aquifers

high or low hydraulic zones. Therefore, the application of a singular method to karst investigation is not adequate to characterize a karst system. Only the combination of different methods at the core sample, borehole, intermediate scale, and catchment scale may provide a somewhat complete view of the relevant processes. Techniques such as tracer tests, cave mapping, and geophysical surveys provide important information on the conduit flow portion of a karst aquifer, while water table mapping relates to the diffuse flow portion of the karst aquifer. If conduit flow analysis is combined with a water table map of the area, anisotropic flow (the flow lines will not be orthogonal to the equipotential lines) will become evident. In fact, it is very possible that some flow lines may even run parallel to the equipotential lines. This concept appears to be contrary to conventional wisdom as to be completely ignored in many groundwater investigations in karst terranes, which can lead to devastating effects from chemical contaminant releases because the flow route is incorrectly predicted from data collected in monitoring wells. The methodology of using dye-tracing studies allows the more accurate definition of the true flow paths and connections between potential sources and receptors of contaminants (Zhou et al. 2002).

1.2 Vulnerability to Ground Collapses

Karst collapses, also known as sinkholes or dropouts, are a global geohazard and have occurred in 23 countries, including China, the United States of America (USA), Canada, South Africa, Italy, France, the United Kingdom, Germany, Russia, and Turkey. The soluble rock in China, for example, encompasses approximately 3.44 million km^2, consisting of 2.06 million km^2 of exposed and covered karst and 1.38 million km^2 of buried karst, accounting for more than a third of China's land area. China is one of few countries possessing every type of karst, whereas karst collapse is a unique geological feature. Based on the karst inventory conducted in 2020, there were 3800 reported karst collapses and high-risk collapse areas totaling 1.495 million km^2. The high-risk area included the cities of Guangxi, Guizhou, Hunan, Jiangxi, Sichuan, Yunnan, Hubei, and Chongqing in south China and Hebei, Shandong, Liaoning, and other provinces in northern China. The northeast China also experienced serious karst collapse disasters. Of 337 prefecture-level cities in China, 105 cities were in high-risk karst collapse area. Approximately 2000 km of high-speed railways were constructed in collapse-prone areas, and 1800 km of high-speed railways will be constructed in collapse-prone areas. Karst collapses mainly occur in covered karst where the topography is relatively flat. With the development of urbanization, concealment and suddenness of karst collapses make them less predictable. In China, human engineering activities induced more than 75% of karst collapses. Karst collapse has become a major geological problem faced by the cities, infrastructure constructions, and natural resources exploitation in karst areas. In-depth analysis of karst collapse mechanisms and the induced geohazards is the basis and pre-requisite for monitoring and early warning of karst collapses, and for establishing geohazard control and mitigation measures in karst areas.

1.2.1 General Processes of Karst Collapses

Ground collapses result from sinkhole development processes. There is no universal theory of sinkhole development. The recommended approach is to understand the processes of sinkhole development and apply them to site-specific conditions. Although the hydrogeologic settings vary from site to site, the basic processes remain the same. Sinkholes result from two different processes: either the transport of unconsolidated overburden materials downward along solution-enlarged channels or the collapse of bedrock roofs over cavities due to progressive enlargement by solutioning. Sinkholes that involve the collapse of cover sediment are appropriately termed cover-collapse sinkholes, while sinkholes resulting from roof collapse are termed cave collapse sinkholes. Figure 1.3 shows the various karst collapse hazards.

Pre-existing dissolved voids in the underlying bedrock or soil are the most critical pre-requisite for karst collapses. Therefore, carbonate rocks such as limestone and dolomite are the major candidate geologic formations of karst collapses. Karst collapses also occur in evaporites such as gypsum and halite. Under some circumstances, collapses even occur in sandstones. The most damaging sinkholes are cover-collapse sinkholes. Figure 1.4 shows a schematic diagram of the cover-collapse sinkhole development, which includes at least three basic elements: development of karst features in the underlying bedrock, overburden layer of a certain thickness, and groundwater activity. Karst collapses often result from hydrogeologic circulations in karst water systems. Circulation of groundwater through karst is significantly different from water circulation through non-karst settings. Recharge of karst aquifers is direct through sinkholes and swallets, and by percolation of rainwater through a network of joints. Moreover, karst groundwater systems can transport sediment and contaminants virtually unimpeded into an aquifer, cave, or spring system due to the rapid recharge rate and lack of filtering. The sinkhole collapses result from internal erosions including removal of dissolved limestone by dissolution, downward transport of unconsolidated sediments, or collapse of bedrock into deeper voids, where water (precipitation, surface water, or groundwater) is the driving force in karst and sinkhole formation. Understanding water dynamics in a karst system is essential in understanding sinkhole formation mechanisms. All too often we only consider exploratory borings and geophysical methods to help characterize a project site. These methods provide valuable snapshots in time of subsurface conditions, which is important, but they provide little insight into groundwater fluctuations. Engineering over karst needs to consider groundwater monitoring programs to understand both hydraulic head fluctuations as well as

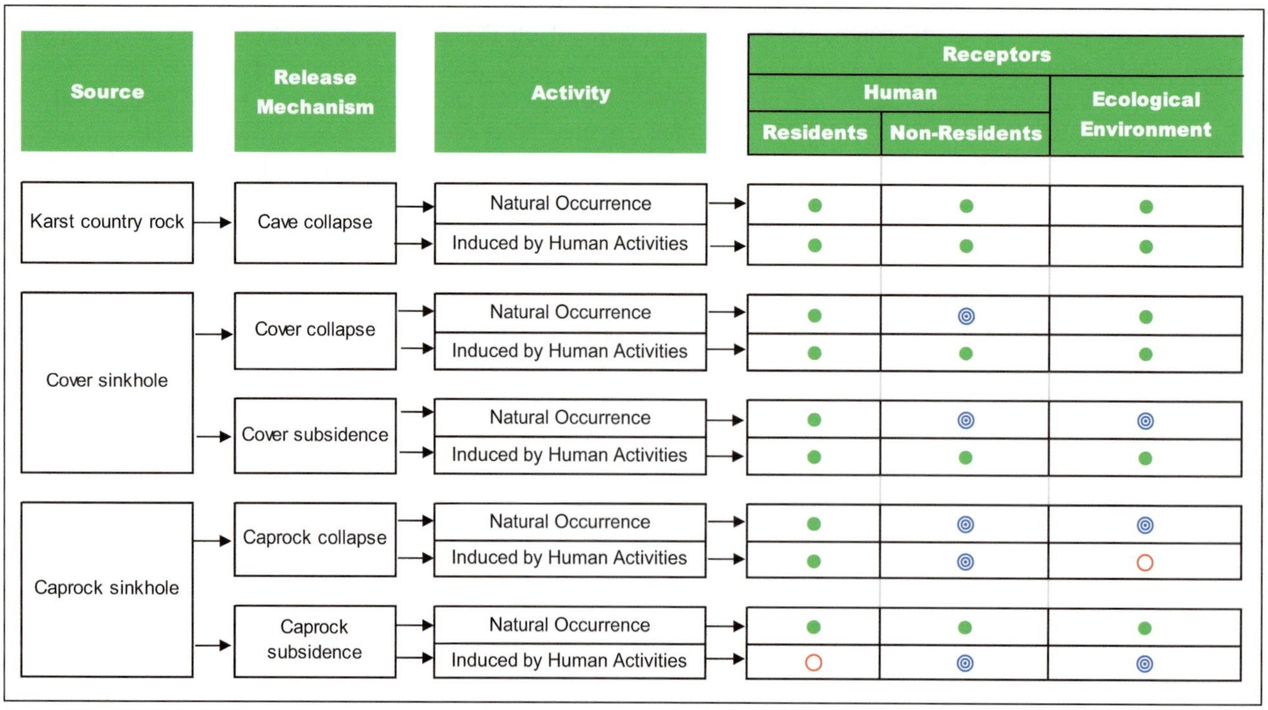

Fig. 1.3 Karst collapses and associated hazards (Green circles represent complete pathways for damages caused by collapses; blue double-circles represent potentially complete pathways for damages caused by collapses; red circles represent incomplete pathways for damages caused by collapses. These symbols shown here are for demonstration purposes only)

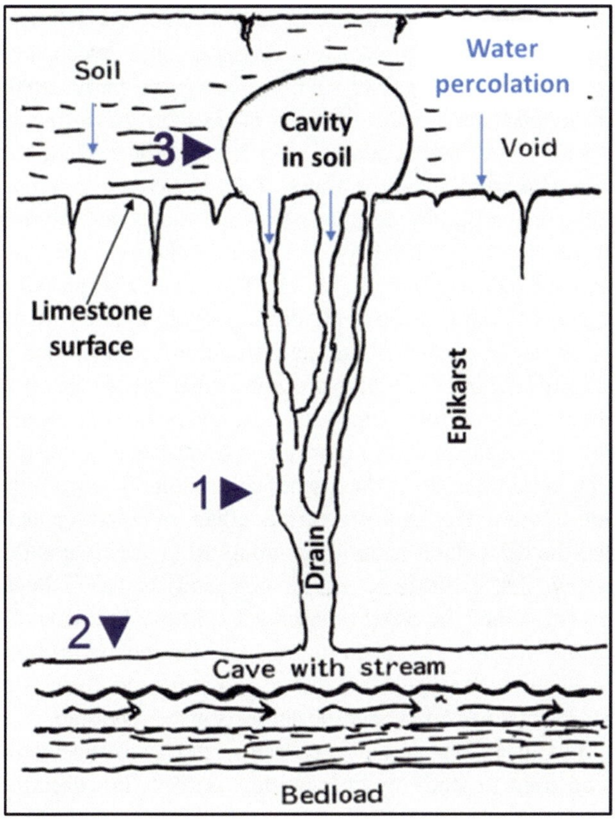

Fig. 1.4 Schematic diagram showing essential features of a cover-collapse sinkhole (modified from White 1988) (1—downward drainage "shaft" in limestone; 2—horizontal conduit system providing lateral groundwater flow and sediment transport; 3—upward stoping erosion in overburden sediment leading to sinkhole collapse)

water quality changes. In addition, remedial measures could reduce greatly, but not eliminate entirely, the probability of sinkhole occurrence or severity. Sinkhole collapse monitoring provides essential data for early warning of sinkhole occurrences. The sudden occurrence of a cover-collapse sinkhole tends to create bowl-shaped, funnel-shaped, or cylindrical depressions on the ground surface.

Sinkholes do not occur randomly. They occur only where hydrogeologic conditions have created solution-enlarged pathways in soluble rock. Solution conduits provide channels for water to move and erode cover sediment into deeper dissolved voids. Limestone within the Appalachian region generally dissolves at an approximate rate of 1–1.5 in. per 1000 years. As such and within project life cycles, limestone dissolution is in most cases not the concern except for paleo-karst cave collapses that occurred in geological history. The existing epikarst conduits play the most important role in providing pathways for overlying sediment to undergo subsurface erosion to create a sinkhole.

Years of experience suggest that the timeframe for sinkhole development is variable, sporadic, and unpredictable. Large downward water gradients or large fluctuations along the soil-rock interface are significant contributors to sinkhole development. As the soil cavity forms above the solution pipe from subsurface erosion, it may gradually enlarge upward as its roof continues to erode.

- If the sediments covering the limestone are relatively non-cohesive, the soil cavity may erode upward rapidly without growing wider. As the roof of the cavity crumbles and sediments are deposited on the floor of the cavity, the cavity may simply migrate upward without increasing in size, like a bubble rising through the liquid. For example, along the west coast of Florida where 5–10 in. of clean sand overlies the limestone, more than 100 sinkholes collapsed following a 10-inch rain event.
- More cohesive strata within the overburden may temporarily impede the upward erosion, forming a temporarily stable soil arch and causing the cavity to grow laterally with slabbing and raveling failures. Cavity shape tends to correlate with soil type and often resembles an inverted teardrop with more cohesive soils developing wider cavities. As the upward progression of erosion reaches a more competent layer, the roof of the cavity may flatten out, with lateral growth continuing until the walls taper down toward the bedrock. Eventually, the upward erosion of the soil void may leave only a thin roof of sediments that are not strong enough to support their self-weight. The result is a sudden collapse. The upward collapse may develop over the years. As a result, in terrain underlain by clay over limestone, air-filled soil voids may be present for extended time periods, which provides an opportunity to detect them using geophysical techniques before they collapse.
- Cover collapse can be a repetitive process. Collapse locations are localized over epikarst drains. If a drain is plugged with sediment, erosion ceases temporarily, and the sinkhole may fill. This can be a long-term condition. For collapse to continue, sediment must be removed from the deeper voids. The timing of repeated collapse is irregular and unknown.
- Subsidence rather than collapse sometimes results in the formation of sinkholes. Movement of unconsolidated materials into the bedrock, where the overlying materials are not strong enough to maintain a cavity roof, results in a subsidence sinkhole at the surface. Damages to structures by such sinkholes are not catastrophic but cumulative.
- Extreme weather conditions and human activities may intensify the sinkhole development processes and thus shorten the timeframe of sinkhole occurrence. Human activities include intrusive investigations and engineering constructions such as drilling activities, modification of surface water drainage, groundwater pumping, grouting, and excavation may accelerate the sinkhole

development processes. Extreme weather conditions involve high levels of precipitation and rapid changing in groundwater levels.

1.2.2 Research Status of Karst Collapse Mechanisms

The seriousness of the karst collapse did not receive attention until the 1970s. In 1973 the International Association of Engineering Geology held the first international symposium on "Karst collapse and subsidence, engineering geological problems related to soluble rock" in Hannover, then West Germany. Since 1984, the multidisciplinary international symposium on "Sinkholes and their engineering and environmental impacts" has been held approximately every two years. The research on karst collapse has gradually transitioned from descriptions of karst collapse characteristics and development regularity to investigation and exploration of geological conditions, risk assessment, and risk management.

The study of karst collapse in China can be divided into four periods. Before the 1980s, sporadic research was conducted on karst collapse induced by mining operations in karst areas. In the 1980s, karst collapse investigations were carried out in the Yangtze River basin, typical mines, well fields, and railway lines including the Guikun Railway. During this period, small scales of physical models were constructed to simulate the karst collapse processes. In the 1990s, karst collapse investigations were focused on six cities—Wuhan, Yulin, Tangshan, Tongling, Guilin, and Shenzhen. During this period, the Institute of Karst Geology in Guilin, China built a large physical simulation laboratory to simulate the sinkhole-forming process and to study the main factors contributing to the collapse occurrence in these cities. Such factors as mine drainage, pumping, rainfall, and the rises and falls of river level were investigated in the simulations. Since 2000s, the China Geological Survey has initiated a nationwide inventory of karst collapses. As of 2020, the karst collapse mapping on a scale of 1:5000 was completed over 30,000 km^2 of the typical karst collapse areas of south China (Lei and Dai 2018). Since the 2000s, the Natural Science Foundation of China has increased funding in karst collapse research and has launched 17 projects. Fifteen of these projects were on karst collapse mechanisms induced by pumping, pile foundation, shield, tunnel construction, mine drainage, train vibration, extreme climate, and natural conditions, whereas two projects were on the prediction and prevention of karst collapses.

The research on the mechanism of karst collapse began in the 1930s. Scholars from the former Soviet Union first proposed the theory of subsurface erosion, which is the phenomenon of soil particles moving under the action of seepage. Subsurface erosion includes both mechanical erosion and chemical dissolution. Since the early 1980s, researchers have made in-depth studies on the formation mechanism of karst collapses. Xu (1981) proposed the theory of vacuum suction through the experiences with karst collapses caused by dewatering in coal mines and ponding in reservoirs. Su (1982) put forward the theory of gas erosion and gas explosion. In the 1990s, Chen (1994) proposed the pressure difference as the driving force in karst collapse based on experience with the construction and operation of railways. Chinese scholars proposed other factors that may affect sinkhole formation such as gravity, blasting, vibration, unloading, and root erosion, depending on the site-specific conditions. In 1990s, American scholar Tharp (1999) put forward the hydrofracturing effect of karst collapse and the calculation method according to the hydrodynamic theory. The hydrofracturing of rock (soil) bodies is intensively studied in water conservancy dams. Studying the occurrence and expansion of cracks in the rock (soil) body revealed the mechanical response and structural changes of the rock (soil) body under the action of water pressure.

In-situ testing, triaxial apparatus, and centrifuges have been used to study hydrofracturing in soils. For example, two sets of in-situ hydrofracturing tests were conducted on the Teton Dam in the USA after the dam failure in 1976. The excavation of the test sections showed that almost all the test sections formed cracks and were vertical in the splitting surface, which was almost perpendicular to the dam axis. Hydrofracturing played a significant role in the dam failure process. The indoor tests were initially conducted using a model test tank. A triaxial apparatus was used to study the hydrofracturing damage mode, direction of splitting surface, and cracks propagation. Centrifuge tests were applied to studying the hydrofracturing performance of cohesive soils in the last decade. The hydrofracturing damage mechanism of soil includes shear failure and tension failure. The radial and tangential stresses around the perimeter of a hole lead to shear damage. The theory of tension damage is mainly because hydrofracturing occurs when the minimum effective stress becomes negative and exceeds the tensile strength of the soil. Some scholars also believe that hydrofracturing is a combination of the two mechanisms. Therefore, the tensile strength of soil is an essential mechanical parameter to control the hydrofracturing of soil. Wang et al. (2018) pointed out that there are original fractures in underground soil. Under the action of seepage water pressure, these fractures undergo expansion, and the final penetration process depends on the wedging action of water pressure. However, there is no sufficient evidence to prove the rationality. The Mohr–Coulomb strength theory applicable to the judgment of shear damage mechanism, although it was based on the elastic–plastic mechanics theory, is not readily applicable to explaining the crack expansion, while the judgment criterion of hydrofracturing corresponding to the tension damage mechanism is divided into the total stress criterion and effective stress criterion. All the tests above show that the initial cracking pressure of soil is related to the stress distribution of surrounding rock, shear strength of undrained soil, the ratio of center pore diameter to sample diameter, over consolidation ratio of soil, initial stress ratio, and compression speed.

The laboratory tests conducted by Liu et al. (2018) concluded that there was a positive correlation between the critical water pressure of hydrofracturing and the material strength and axial compressive stress. The axial compressive stress had a greater impact on the critical water pressure of hydrofracturing than the material strength. The critical water pressure of hydrofracturing has a negative correlation with the initial crack opening and initial crack length. With the increase of water pressure inside the seam, the fracture tip of the initial fracture of the rock mass expands, forming the damage degradation zone and macroscopic fractures. These processes lead to the instability and failure of the rock mass. The theoretical model can be divided into two categories. One assumes the rock mass as a fractured medium and considers that water exists and moves through the fractures of the rock body. The other assumes the rock body as an equivalent continuous medium and introduces the concept of damage to describe the development of fractures in rock mass in response to the change of rock stress. The water pressure is regarded as volume tension and establishes the relationship between the damage variable and the permeability coefficient. Damage mainly refers to the occurrence and expansion of fractures in the rock mass, which geometrically means the reduction of bearing area. Xie and Su (2004) pointed out the following assumptions in hydrofracturing study in a fractured rock mass with seepage-stress coupling:

- Rock near the fracture is impermeable or has no permeability.
- Water penetrates in a uniform medium, which is a hypothetical flow pattern.
- Water flow in fractured rock mass is a mixture of fractured flow and porous-medium seepage. When the fractures in rock mass are large, the conditions are more conducive to fracture flow or pipe flow.

Table 1.1 summarizes that the current karst collapse formation mechanisms are macroscopic hypotheses based on the geological conditions and influencing factors of the collapse. As the main driving force of the collapse, the effect of groundwater activity is manifested in two aspects: dissolution and transportation of water and sudden change of water pressure in karst conduits caused by water level change. Therefore, the research on the mechanism of karst collapse must analyze flow dynamics in the karst conduits. He and Xu (1993) believe that the real mechanism of collapse can be explored by studying the hydrodynamic conditions of flow and water–gas pressure characteristics in karst conduits. This is also the reason why the dominant karst collapse mechanisms involve subduction, vacuum negative pressure, pressure difference, and hydrofracturing. These processes are strongly associated with changing underground hydrodynamic conditions. In fact, the damage mechanism of changing groundwater dynamic conditions can all be attributed to the seepage deformation and failure of rock and soil, i.e., transport of soil particles or the whole soil/rock mass under the action of groundwater seepage force (hydrodynamic pressure) causes the deformation and damage of the formation. In the process of karst collapse formation, the action mode and direction of groundwater seepage on karst cavity roof rock/soil body can be different because of the change of groundwater dynamic conditions.

The mechanism of limit equilibrium theory of soil mechanics considers the problem of roof stability of karst cave, which is the last stage in the process of karst collapse development. The change of surface load only shortens the time of the collapse occurrence on the ground. Since the effect of groundwater is not considered, it is still open to discussion whether this process can be considered as the main mechanism of collapse.

The karst collapse mechanisms discussed above involve different disciplines such as hydraulics, soil mechanics, and seismology. These mechanisms are not only related to the change of hydrostatic pressure but also related to the effect of hydrodynamic pressure. Hydrodynamic pressure can be a new challenge and opportunity for the study of karst collapse mechanisms in the future.

1.2.3 Research Trend on Karst Collapses

With the application of fine optical fiber, accelerometer, sonar, and other measurement technologies, the monitoring frequency of the groundwater and gas pressure (or groundwater level) has increased from low frequency (such as 24 h) to high frequency (such as 5 min). The high-frequency monitoring is conducive to the determination of the macroscopic karst collapse discrimination index and the microscopic mechanism analysis of karst collapse under hydrodynamic pressure.

1.2.3.1 Study on Karst Collapse Criteria

Since the current research on karst collapse mechanism is still at the stage of qualitative macroscopic analysis, there is a long way to go to quantify these collapse-causing factors, such as the thresholds of hydraulic gradient, negative

Table 1.1 Types of collapse mechanism in literature

Mechanism	Description	Driving force
Seepage erosion	Large seepage pressure or hydraulic gradient tends to result in piping and flowing of soil. The seepage action of groundwater flow causes the destruction of the soil layer, produces the soil hole, and finally collapses	Hydraulics-Darcy's Law-Seepage deformation destruction
Vacuum suction	This mechanism is more applicable to karst cavities under sealed conditions. Negative pressure suction erosion occurs when the cross section of the water pressure changes from low to high or the groundwater level drops suddenly. The soil around the cavity peels off continuously and eventually collapses	Hydraulics-Darcy's Law-Seepage deformation destruction
Pressure difference	When the change of groundwater level causes enough pressure differences in a closed karst cavity, cavitation and subduction will occur on the over layer. The soil is damaged by multiple effects, resulting in ground collapse	Hydraulics-Darcy's Law-Seepage deformation destruction
Hydrofracturing	The rapid rise and fall of groundwater level and the load caused by air explosion increase the instantaneous pore water pressure in the soil, leading to the expansion and closure of soil fractures. The fractures continue to expand, or the soil pore wall flakes off to form the ground collapse	Hydraulics-Darcy's Law-Seepage deformation destruction
Gas out-bursting	After a water-flooded cave is blocked, the water level of the underground river rises rapidly in response to rains. The upward positive pressure jacking force is generated on the cover layer. When the jacking force is large, the blocking body will collapse. Or man-made blasting, earthquake, or extreme climate can cause the collapse in the cavern roof. Collapsed rock-soil mass blocks karst conduits, resulting in high gas pressure in the karst conduits and causing blasting and water hammer effect in the cover layer. The collapse occurs when the strength of the cover layer is weak, and cohesion is reduced	Geotechnics-limit equilibrium theory-fluid mechanics, seismology infiltration
Chemical erosion	The dissolution of artificially discharged acid-based solution or groundwater makes the karst cavity expand or strength decrease. Collapse occurs when the weight of the soil is greater than the bearing capacity of the soil cavity roof	Chemical reactions
Resonance	Collapse occurs when the fluctuation frequency of railway transportation or groundwater level is the same as the natural frequency in the cover layer	Physical forces
Liquefaction or vibration	The external force vibration or earthquake action on the soil leads to liquefaction and softening effect of saturated sand in the karst area. The shear strength drops to zero instantly, and the overlying soil flows into the karst cavity, resulting in collapse	Seismological activities

pressure value, pressure difference, or resonance frequency that may lead to the occurrence of karst collapse.

The seepage defamation of soil is extensively and systematically studied in the field of geotechnical engineering for water conservancy and hydropower. The critical seepage slope is used as the criterion to judge the seepage deformation and failure of soil. The method and index for judging the type of permeability deformation were developed according to the particle size distribution of the soil. The test protocol of permeability deformation of coarse gain soil was formulated. However, there is no relevant research on coupling the critical seepage slope with groundwater dynamic conditions. Since 2000, the Institute of Karst Geology of the Chinese Academy of Geological Sciences has developed a bench-scale test device for permeability deformation of undisturbed soil samples under the condition of fracture development (Lei et al. 2005). The relevant critical indicators are under ongoing studies. By assuming that the soil pore is equivalent to a circular pipe, Lei et al. (2004) proposed the expression of the critical condition of the potential erosion according to the pipe flow theory. The test device and the test method of the potential erosion were also designed. Based on the percolation head of seepage and the seepage flow, the critical slope of the seepage deformation and failure of the soil was determined. After the pore diameter of the soil was determined by the electronic scanning microscope, the critical fluctuation velocity of the groundwater caused by the potential erosion of the soil was calculated. The data were compared with the fluctuation velocity of the groundwater pressure measured in the field to determine whether the soil would be damaged under the exiting hydraulic conditions and then develop to the ground collapse. The main problem of this method is that the sample for the soil pore diameter is not undisturbed. It is still difficult to establish an effective and real relationship between the critical seepage slope index and the fluctuation rate of groundwater/gas pressure measured in the field.

The critical water pressure (initiation pressure) of hydrofracturing of rock and soil is the criterion to determine the splitting failure of rock (soil) under water pressure. However, the criterion to determine the minimum water pressure (occurrence of hydrofracturing) that causes hydrofracturing is not uniform. It is necessary to combine the concept of hydrofracturing critical water pressure with the failure index of soil permeability deformation so that an effective correlation with the dynamic conditions of karst groundwater might be established.

Since 2012, the Institute of Karst Geology has built karst collapse monitoring demonstration stations in Jinshazhou of Guangdong province, Meitanba of Hunan province, Shizishan of Anhui province, Laibin of Guangxi province, Zhongliangshan of Chongqing to monitor karst groundwater dynamic conditions (Jiang and Lei 2017; Jiang et al. 2018; Pan and Jiang 2018). The water and gas pressure in the karst conduit systems are being monitored by boreholes buried with vibrating string-type pore water pressure sensors. Measurements are taken at intervals of 5 to 20 min in a closed state. The monitoring programs provide continuous monitoring for many years to capture the abrupt changes of water (gas) pressure in the karst conduit system prior to and during the formation of karst collapse. For example, the monitoring station in Laibin obtained pressure data in the underground river system when a karst collapse occurred in response to an extreme rain event. The sudden change of groundwater gas pressure in different parts of the karst system is different during rainstorms. The increase of pressure and dissipation characteristics have distinctive characteristics. The monitoring station at Dachengqiao obtained for the first time the entire process of karst water level changes in response to coal mine dewatering and closure. The sudden changes in the karst water level and collapse process occurred in the dewatered area at two stages: dewatering during mining and water level rebound after mine closure. Abrupt changes are significant in the process of karst collapse formation. The monitoring effort helps achieve the purpose of grasping the dynamic conditions of groundwater that may trigger karst collapse. Analysis of the data collected at these monitoring stations also shows that the main cause of the collapse is the sudden change in water and gas pressure. The magnitude of the sudden change value is closely related to the monitoring frequency and whether the monitoring points are restored to the original pressure state.

Therefore, the karst collapse study requires expansion of the monitoring program on karst groundwater/gas pressure and determination of parameters of permeability deformation damage index. Results from further studies would shed light on the karst collapse mechanism.

1.2.3.2 Research on Damage Mechanisms Associated with Dynamic Water Pressures

The research on the infiltration deformation damage of rock (soil) body mainly considered the effect of hydrostatic pressure, and basically does not consider the effect of hydrodynamic pressure. In recent years, the effect of hydrodynamic pressure within the geotechnical body gained attention. The research on hydrofracturing of high-speed rail tracks shows that the following:

- The maximum value of water pressure inside a crack appears at the crack tip point.
- With the increase of loading frequency, the pore water pressure inside the crack shows an increasing trend.
- The pore water pressure has a quadratic relationship with the loading frequency.

The optical fiber water pressure sensing high-frequency sampling technology (100 times/second) was used in 2018 and 2019 to monitor the water and gas pressure in a karst cavity drainage test model. A pulsation phenomenon was discovered on the water and gas pressure in the sealed karst cavity. The pulsation can be considered as hydrodynamic pressure characteristics with a certain frequency. Figure 1.5 shows an example of high-frequency data. The variation range of hydrodynamic pressure (100 cm) is much larger than that of hydrostatic pressure (60 cm from water level to outlet height). At the same time, it was found on three occasions that the sudden drop of water/gas pressure was accompanied by an obvious resonance effect. Therefore, under the action of dynamic load and pressure pulsation, the fracture failure mechanism of rock and soil still needs further study.

Basic Theoretical Research on Pressure Pulsation

Pressure pulsation is the phenomenon that pressure changes cyclically with time. It means that the pressure acting on the object is not uniform, and there is a relatively concentrated or large pressure at a certain position. The single continuous time of this pressure is short, and it occurs periodically. The pressure is not static, but dynamic with a certain frequency. In addition to the typical airflow pressure pulsation, which is an inherent characteristic of piston compressors, pressure pulsation is a major problem faced by hydropower buildings (pressure pipes, turbines, or pumps). The pressure pulsation characteristic has become an important indicator of the performance of large power-generation units. For example, the tailwater pressure of hydropower plants and the water level fluctuation of regulating wells have pulsation characteristics. The main current research on pressure pulsation includes the following aspects: description of theoretical equations, analytical, methods of pressure pulsation, resonance damage, prevention, and control measures of pressure pulsation.

The mathematical equation of pressure pulsation mainly adopts the theory of fluid dynamics, and its propagation mainly includes the seepage model, transient flow model, and oscillatory type model. The three models help explain pressure pulsation propagation in the geotechnical body at the different stages:

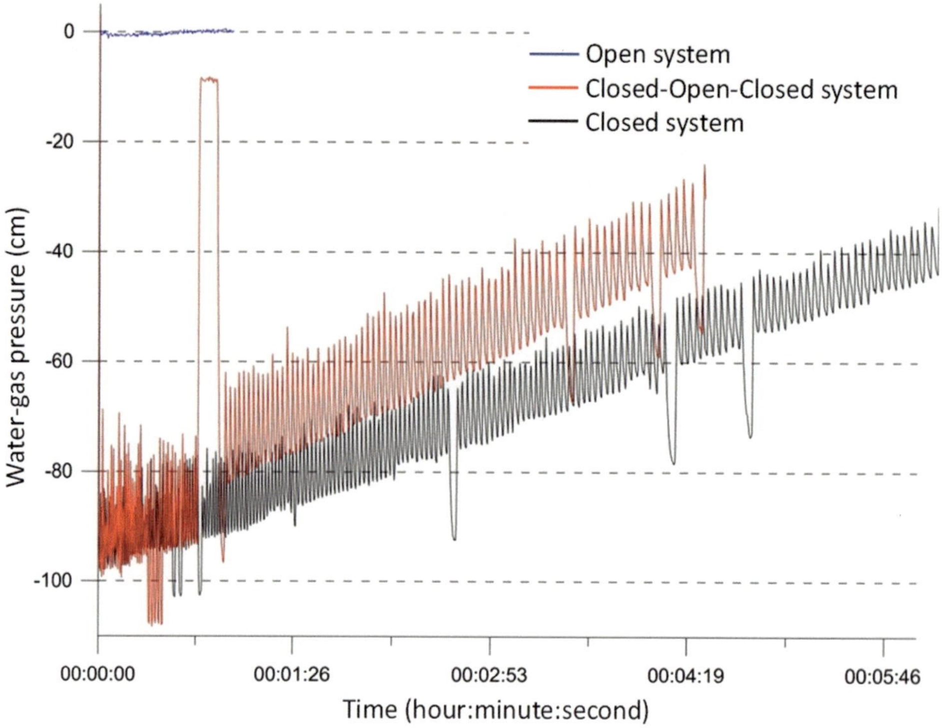

Fig. 1.5 Pressure pulsation observed in a karst cavity in response to water level drops (frequency between 0.98 and 2.88 Hz)

- When the pressure pulsation head is less than the resistance (depending on the permeability coefficient), the pulsation pressure is transmitted into the geotechnical body at the percolation velocity.
- When the pulsation water pressure is greater than the resistance, the pressure pulsation is propagated by pressure wave, not by the velocity of the water body. The velocity of the pressure wave is related to Poisson's ratio, elastic module, pipe thickness, and inner diameter. But the velocity of pressure has nothing to do with water velocity. The maximum, minimum, cavitation volume, and flow rate of transient water pressure can be calculated by the continuous equation and motion equation of the transient flow model.
- The oscillating model describes the oscillating motion (resonance) of water caused by fluctuating pressure differences.

The signal analysis methods of pressure pulsation are mainly based on the Fourier transform of the frequency domain segmentation time-domain inversion method and cross wave transformation method to analyze the source of disturbances that cause pressure pulsation, estimate the main distribution frequency of pressure pulsation, reveal the impact of different frequency components on the extreme value of water pressure of pulsation. The more mature transient flow theory is the water hammer or water strike effect in pressurized pipelines. The continuity equation and the motion equation used in the calculation of water strikes in circular pipes have been successfully applied in scientific research and pipeline engineering. The calculation starts from the calculation of the hydraulic envelope, carries out the calculation of the highest and lowest values of pressure, and provides the calculation results of different time frequencies, such as nodal pressure, flow rate, and flow velocity change process. The existing and widely used commercial software Bentley Hammer uses a transient hydraulic analysis algorithm for equation solution, which can calculate the instantaneous head and air content change process in the whole section of the pipeline during the continuous change of time serious in milliseconds, i.e., the process of pressure fluctuation generation, development, and continuation provide full process data analysis support for pressure change. In addition, computational fluid dynamics transient dynamic response simulation software (e.g., Fluent, ANSYS-CFX, AMEIM) can calculate the pressure pulsation value of the flow field and the trend of change, and frequency domain analysis of the pressure pulsation. The impact analysis of factors affecting the pulsation frequency in the conveying pipeline system, to find out a source of influence, to establish the formula for low-frequency pressure pulsation in the conveying pipeline.

According to the existing hydrodynamic pressure pulsation research results, as a naturally formed karst conduit system with extremely irregular morphology, water and gas pressure pulsation is inevitable in the process of groundwater movement. Under the action of water and gas pressure pulsation, the rock (soil) body directly overlying the karst cavity will probably produce resonance damage or cavitation damage like hydraulic structures, resulting in the reduced tensile strength, and finally damage and collapse. Therefore, the above pressure pulsation theory provides a good research basis for the hydraulic calculation of water and gas pressure pulsation in karst conduits.

Cavitation Damage

The high-frequency pulsation of water and gas pressure will cause cavitation damage to the carrying media. In fluid mechanics, it is also known as air blasting, water hammer, or water hitting. It is widely recognized in water pipelines, pumps, turbines, and other fluid machinery and industrial equipment. Cavitation is due to the destruction of the fluid continuity in the flow channel and the formation of localized low pressure, which makes the liquids vaporize and cause explosive growth of micro-bubbles. The micro-bubbles rapidly grow into large bubbles. The large bubbles collapse when they move with water to points of high pressure. The impact of pressures up to hundreds of times the atmospheric pressure on the pipe walls causes the erosion of coating materials. The generation and collapse of micro-bubbles are random and transmitted through fluid. Research on micro-bubbles is more in-depth because the conditions that influence their formation and collapse are relatively simple in fluid mechanics. Decent understandings are accomplished in the determination of cavitation, water impact wave velocity measurement, water impact mathematical models, and theoretical calculations.

German scientist Momber (2004) conducted an experimental study on water flow cavitation action to crush concrete and sort aggregates (municipal construction waste reuse). A water flow cavitation chamber was used to test five groups of concrete with different compressive strengths. The water flow pressures were 1.1 MPa and 0.17 MPa before and after the weir, respectively. The test results showed that the lower the compressive strengths of concrete are, the greater the cavitation rate will be. The cavitation of motor strength amounted to 160 mg/s, and the damage occurred mainly at the contacts of aggregate and motor.

In the monitoring and diagnosis of cavitation, the main methods used include the following:

- Energy method: When cavitation occurs, the characteristic curve drops sharply. Usually, the point where the head drops 3% in the stage of sharp change of head curve is taken as the critical point of centrifugal pump cavitation occurrence.
- Noise method: Pump cavitation will produce noise. Some researchers use sound pressure sensors for the probe placed at a certain distance from the pump body. When the pump cavitation develops to a strong stage, the vapor bubble breaks through the pump body, transmitting to the pump outside. The sensor picks up the signals, which are used to conduct noise spectrum analysis to determine whether cavitation has occurred.
- Vibration method: Use an accelerometer as a probe and fix it on the pump casing or pump shaft. Collect the vibration signal inside the pump. Consider the vibration caused by the motor, centrifugal pump, fan, etc. as the background vibration signal. The vibration signal caused by the bubble breakage is used as the cavitation fault signal to determine whether cavitation occurs in the pump.
- Pressure pulsation method: When cavitation occurs, the pressure position of the flow feud flow field in the centrifugal pump is significantly different from the normal operation. The dynamic test and spectrum analysis of the pump inlet and outlet pressure can be used to determine whether cavitation occurs.

Injection of air into water conveyance structure is a common practice to reduce cavitation. Wang (2001) and Mitosek (2000) carried out the experimental study of ventilation to reduce cavitation damage. Zhao (2016) systematically summarized the basic ideas of reducing cavitation damage in over-water buildings through ventilation aeration design, including that aeration facilities should be set up to reduce cavitation damage in large flow and high flow velocity conditions of drainage buildings. The openings and vents should be set between the maintenance door and the working door of the dam's hidden hole. The vents should also be set after the inlet gate of the pressure pipeline.

In addition, we cannot ignore that water hammering in long-distance water transmission projects can cause damage to pumps and transmission pipes. International financial institutions (such as the world bank and the Asian bank) require that design studies on water hammer analysis, hazard prediction, and water hammer protection be submitted throughout the assessment of loaned water conveyance projects. To effectively prevent the influence of vacuum negative pressure and water supply hammer when the pump is shut down during maintenance, the water injection valve or air injection valve is added at the sudden change of pipeline elevation. It is pointed out in the research experiment that the larger the diameter of the air injection valve is, the smaller the vacuum value in the air cavity is. If the inlet diameter is large enough, the vacuum value in the air chamber can be 0, or at one atmospheric pressure. However, the research results have no specific diameter size or relative proportion value of the air injection valve.

In geology, scientists began to study the phenomenon of cavitation in water by simple tests in 1662 (Caupin and Herbert, 2006). Gas cavities also exist in saturated dense sands in undrained triaxial compression tests. Gawin and Sanavia (2010) carried out numerical simulations of gas cavities saturated pore media using finite element equations. In recent years, high-pressure gas masses have been encountered in geological drilling in the Thames Tunnel in the United Kingdom, and the oxygen content of the gas masses is low. Standing et al. (2013) conducted a study on this issue. The groundwater level in the London basin dropped significantly to 70 m in the eighteenth century because of industrial development and gradually recovered in the mid-1960s. The gas generation and accumulation process after the aquifer water level dropped and then recharges were simulated. The simulation results show that:

- The cavitation phenomenon will occur during the aquifer water level drop stage, and the released air will rise to the bottom of the week permeable layer and accumulate. When the aquifer is recharged and the groundwater level rises, the air is pressurized to form a high-pressure air mass.
- The amount of gas formed by the cavitation phenomenon is proportional to the particle size of the sample.

The theories on cavitation damage and test and mitigation measures discussed above can be used as useful references to study the mechanism and prevention of karst collapses.

Resonance Damage

Resonance is the phenomenon of mutual vibration when the frequency of the excitation force is close to or equal to the natural frequency of the structure. Under the action of continuous resonance frequency, the vibration of the structure is unstable when the resistance of the structure is not enough to consume the energy of mutual excitation, and the vibration amplitude will grow larger until damage. Resonance damage is a problem that must be considered in civil engineering, especially with bridges. Standards for the resonance of high-speed railways have been promulgated around the world. In addition, the problems of flow pipeline vibration have also been receiving attention and research from the academic and engineering communities. Resonance studies of pressurized pipelines are mainly carried out through indoor and outdoor tests and finite element numerical simulations. Understanding the natural frequency, dumping ratio, and model vibration mode of the pipeline provides the basis for the vibration characteristics analysis, vibration fault diagnosis, and prediction of the pipeline system. Li et al. (2008) believe that the resonance damage of the pipeline results from the natural frequency of the water body being close to the frequency of the excitation force. Therefore, he believed that the causes of resonance of the conveying pipeline include two aspects:

- The resonance of the water body in the pipeline is caused by the excitation force frequency of the water (gas) pressure fluctuation in the pipeline system, which causes the vibration of the pipeline.
- The resonance of the pipeline is caused by the frequency of the multiplier of the excitation force which is basically the same as the inherent frequency of the pipe. As the pipeline is a homogeneous body, the inherent frequency of the pipeline system is related to the stiffness and quality of the pipeline, and the stiffness is related to the pipe diameter, wall thickness, number of elbows, and the condition of the support and hanger. Therefore, the measures to solve the vibration phenomenon mainly include increasing vibration-damping devices or dampers in the weak links of the pipeline, reducing elbow support, and hanger treatment.

In karst areas, the natural karst cavity system (karst cave, karst fissure, karst fractures) is irregular in shape. The Quaternary soil (filling material) in the cavity is complex and includes clay, gravel clay, silty clay, sand, or gravel. The following questions need to be answered through research:

- How should the intrinsic frequency of the geotechnical body at the top of the cavity of a karst piping system and the intrinsic frequency of the groundwater flow be measured?
- What is the relationship between the intrinsic frequency of the groundwater flow and the frequency of water and gas pressure pulsation?
- Does the resonance phenomenon lead to the multiple increase of the void water pressure in the rock (soil) body?
- Does the vertical vibration of the excitation force lead to the deformation and failure of the rock and soil body at the top of the cavity?

Because the current theory of the karst collapse mechanism is still in the hypothetical stage and lacks mature guidance on the collapse mechanism, it has become a technical bottle neck in the prediction and prevention of karst collapse disasters. With the improvement of monitoring technology for groundwater dynamic conditions of karst collapses, as well as the accumulation of monitoring data of indoor and outdoor groundwater/gas pressure, we believe that the sudden change of karst water and gas pressure is the main cause of karst collapses. Although the mechanism is the seepage deformation and failure of rock (soil), the discriminate index still needs further study. In addition, comprehensive analysis of model tests and application of fluid mechanics and other relevant disciplines are essential to determine the hydrodynamic pressure-induced cavitation damage and resonance failure mechanism. The relevant pressure pulsation theory, testing method, and prevention measure can be introduced to the mechanism monitoring and prevention of karst collapse in further research.

1.3 Vulnerability to Groundwater Contamination

Heterogeneities created by the multiple porosity and permeability of karst aquifers, and especially by multiple discrete conduit and fracture flow paths, are the major factors contributing to the overall higher vulnerability of karst aquifers to groundwater contamination, and to the difficulties encountered in detecting, assessing, and remediating contaminant occurrences. One widely reported, tragic incident involved microbial contamination of public water supply wells in Walkerton, Ontario, Canada, in the spring of 2000, in which over 2300 people were sickened, and seven died, because of groundwater contamination by *Escherichia coli* (*E. coli*) and *Campylobacter jejuni* bacteria (Worthington et al. 2012). Subsequent investigations revealed that the source of the contamination was livestock waste entering the karstic limestone aquifer through rapid infiltration of surface runoff. Field testing using dye-tracing methods demonstrated the occurrence of flow velocities greater than 300 m/d and contaminant travel times of 5 to 26 h in an area that conventional groundwater model applications had been delineated as having a 30-day time-of-travel. The groundwater model simulations did not include karst features, thus did not represent the flow system. Obviously, conventional groundwater modeling misrepresented the time required for contaminants to reach points of contact because karst features were not included in the model (Worthington et al. 2012).

Because of the complexity that arises from the inherent characteristics of karst aquifers, a conceptual site model (CSM) needs to be constructed from site-specific conditions and data available but may include sufficient information to describe the following:

- Traditional geologic/hydrogeologic data including vadose zone and aquifer thickness; estimates of porosity, hydraulic conductivity, transmissivity, and storage coefficients; and recharge and discharge areas.
- The type of karst aquifer (e.g., diffuse flow, free flow, or confined flow).
- Recharge and discharge zones illustrating, at least qualitatively, the relationship between subsurface flow, and recharge and discharge points.
- The flow path associated with each monitoring well. If applicable, the CSM should clearly depict the potential for conduit flow paths to short-circuit contaminants away from a monitoring location.
- Portions of the watershed that contain potential receptors.
- The potential for contaminated sediments to accumulate in flow conduits when applicable.

Figure 1.6 presents schematic transport processes of dissolved contaminants in a hypothetical karst setting. Contaminant transport through conduits occurs rapidly, over long distances, and contaminants within them may be dispersed into multiple plumes of varying concentrations that take unpredictable or unknown pathways to the aquifer's discharge points at wells, springs, or surface waters. Contaminants may be rapidly diluted within or flushed from conduit systems, but residual contaminants may be stored and released gradually or effectively immobilized within lower permeability, slower velocity zones of the karst aquifer matrix. Contaminant flow paths, flow velocities, residence times, travel times, and concentrations may change significantly under different flow conditions. Contaminant transport, storage, fate, and remediation are especially complicated where the karst aquifer is extremely heterogeneous and/or the contamination involves mixtures of contaminants having different physio-chemical characteristics.

Contaminant distribution in karst aquifers suggests that residual contaminant storage may occur in the regolith, in the epikarst, in diffuse rock fractures, and in conduits. In many karst sites, the epikarst has a high potential to trap residual contaminants due to its highly weathered, irregular surface. Residual contamination may also occur in conduit sediments. Flow channels include in-filled pockets and zones where porous media flow on a micro-scale dominates transport thus fostering a slow dissolution and desorption of contaminants so that secondary sources persist despite high flow rates in channels.

The contamination source includes non-point contamination sources such as waste disposal, accidental spills, or leaks of contaminants, mine drainage, industrial wastewater discharge, and point contamination sources such as fertilizer and herbicide applications in agriculture and roadway runoffs. Contaminant-laden media, such as soil, sediment, or water, introduced directly into a karst conduit system by storm-induced runoff entering open sinkholes and swallow holes may be transported by turbulent flow far into the karst aquifer system where they may act as subsurface sources of leachable chemical and microbial pollutants. High concentrations of suspended sediment or turbidity, especially during storm events, is a hallmark of sinkhole-dominated karst recharge and is a surrogate indicator of karst water vulnerability to contamination by surface runoff. In many karst aquifers, even those unaffected by anthropogenic contaminants, groundwater quality is naturally affected by suspended sediments or high turbidity. It is often a factor limiting groundwater resource development, affecting the water quality of water supply wells and springs.

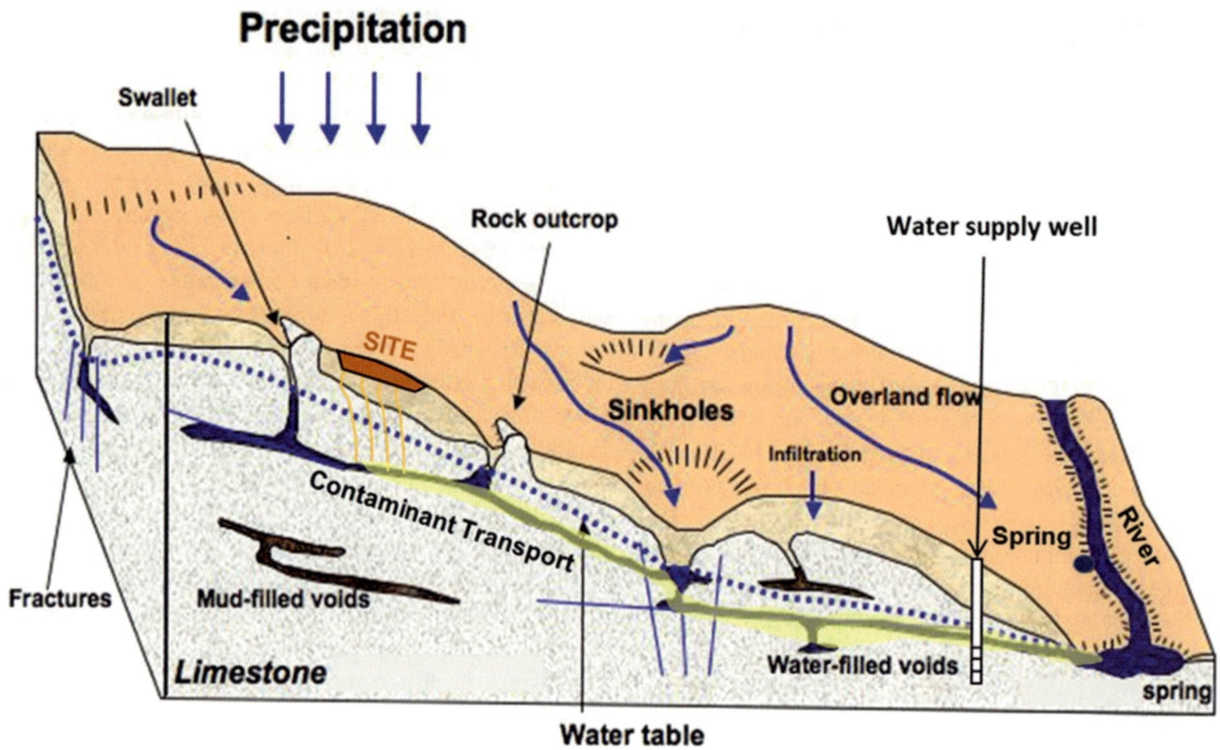

	Contaminant Hazard Factor	Migration Pathway Factor	Receptor Factor
High	Significant Contaminant Levels Concentrations > 100 X Screening Levels	Evident Migration – Analytical data or observable evidence indicates that contamination in the groundwater is moving or has moved away from the source area	Identified Receptor – There is a threatened water supply well downgradient of the source and the groundwater is a current source of drinking water or source of water for other beneficial uses such as irrigation/agriculture/biota
Medium	Moderate Contaminant Levels Concentrations > 10 X Screening Levels	Potential Migration – Contamination in the groundwater has moved only slightly beyond the source (i.e. tens of feet), could move but is not moving appreciably, or information is not sufficient to make a determination of Evident or Confined	Potential Receptor – There is no threatened water supply well downgradient of the source and the groundwater is currently or potentially usable for drinking water, irrigation, agriculture, or biota
Low	Minimal Contaminant Levels Concentrations > 1 X Screening Levels	Confined Migration – Information indicates that the potential for contaminant migration from the source via the groundwater is limited	Limited Receptor – There is no potentially threatened water supply well downgradient of the source and the groundwater is not considered a potential source of drinking water and is of limited beneficial use

Fig. 1.6 Generalized groundwater contamination processes in a karst setting

References

Caupin F, Herbert E (2006) Cavitation in water: a review. C R Phys 7(9):1000–1017

Chen GL (1994) Causes and prevention of Karst collapses. China Railway Press

Gawin D, Sanavia L (2010) Simulation of cavitation in water saturated porous media considering effects of dissolved air. Transp Porous Medial 81(1):141–160

He YB, Xu C (1993) On the hydrodynamic factors of karst collapses. J Hydrogeol Eng Geol 5:39–42

Jiang XZ, Lei MT (2017) Formation mechanism of large sinkhole collapses in Laibin, Guangxi. Environ Earth Sci 76(24):1–13

Jiang XZ, Lei MT, Gao YL (2018) New karst sinkhole formation mechanism discovered in a mine dewatering area in Hunan, China. Mine Water Environ 37(3):625–635

Kim YW, Lee YS (2008) Damage prevention design of the branch pipe under pressure pulsation transmitted from main steam header. J Mech Sci Technol:647–652

Lei MT, Dai JL (2018) Environmental geological survey report on Karst collapse in Western Hunan, Eastern Hubei, Northern Anhui (scale in 1:50,000). Karst Geology Institute, Chinese Academy of Geological Sciences

Lei MT, Li Y, Jiang XZ, Gan FP, Meng Y (2004) Preliminary study on monitoring and forecasting technology and method of karst collapse disasters, a case study in Zhemu Village, Guilin City. Chin J Geol Hazards Prevention S1:148–152

Lei M, Gao Y, Jiang X, Hu Y (2005) Experimental study of physical models for sinkhole collapses in Wuhan, China. In: Proceedings of the tenth multidisciplinary conference on sinkholes and the engineering and environmental impacts of Karsts, ASCE Geotechnical Special Publication, pp 91–102

Li GX, Li WP, Li H (2008) Natural frequency analysis of water flowing in liquid-filled straight circular pipes. Res Dev Hydrodyn Ser A 02:134–140

Little JR (1984) Relationship of modern sinkhole development to large-scale photolinear features. In: Proceedings 1st multidisciplinary conference on sinkholes and the engineering and environmental impacts of Karst, Balkema, Rotterdam, pp 189–196

Liu DT, Shen ZZ, Xu LQ, Qiu LT, Jiang T (2018) Influencing factors and mechanism of critical water pressure in hydrofracturing of rock mass. J Hydraulic Water Transp Eng 4:30–37

Lloyd JW, Harker D, Baxendale RA (1981) Recharge mechanisms and groundwater flow in the chalk and drift deposits of southern East Anglia. Q J Eng Geol 14:87

Mitosek M (2000) Study of transient vapor cavitation in series pipe systems. J Hydraul Eng 126(12):904–911

Momber AW (2004) Aggregate liberation from concrete by flow cavitation. Int J Miner Process 74:177–187

Palmer AN (2004) Growth and modification of epikarst. Karst Waters Institute, Special Publication 9

Pan ZY, Jiang XZ (2018) Mechanism of sinkhole formation during groundwater-level recovery in Karst mining area, Dachengqiao, Hunan province, China. Environ Earth Sci 77(24):799–812

Standing J, Ghail R, Coyne D (2013) Gas generation and accumulation by aquifer drawdown and recharge in the London Basin. Q J Eng GeolHydrogeol 46:293–302

Su JS (1982) Gas and water pressure problems of karst water pipeline flow. In: Proceedings of the second Karst academic conference of the geological society of China, Science Press, 241pp

Tharp TM (1999) Mechanics of upward propagation of cover-collapse sinkhole. Eng Geol 52:23–33

Wang GY (2001) Effect of ventilation on cavitation erosion and high-speed photographic observations. J Hydroelectric Power Gen 1:48–57

Wang JX, Chen JS, Tao H, Shi X (2018) Advancement in research on hydrofracturing tests of earth core dams. J Appl Basic Eng Sci 26(01):132–144

White WB (1988) Geomorphology and hydrology of Karst Terrains. Oxford University Press, New York

Williams PW (1985) Subcutaneous hydrology and the development of Doline and Cockpit Karst. Zeitschrift Fur Geomorpohologie 29:463–482

Worthington SRH, Smart CC, Ruland W (2012) Effective porosity of a carbonate aquifer with bacterial contamination: Walkerton, Ontario, Canada. J Hydrol 464–465:517–527

Xu WG (1981) Discussion on vacuum suction effect in ground collapses in karst mining areas. Geol Rev 27(2):174–180

Xie XH, Su BY (2004) A review of research on hydrofracturing of fractured rock mass. Rock Mech 2:330–336

Zhao YX (2016) Ventilation design of water-passing buildings. J Yunnan Hydropower 32(6):96–100

Zhou W, Beck BF, Arthur JP, Stephenson BJ (2002) A groundwater tracing investigation as an aid of locating groundwater monitoring stations on the Mitchell Plain of southern Indiana. Environ Geol 41:842–851

2 Monitoring and Early Warning of Karst Collapses

Monitoring and early warning of karst collapses have not been easy tasks due to the hidden and sudden nature of most karst collapses. A suitable stability monitoring technique must be capable of covering a relatively large area and should provide continuous information in real time. More attention has been paid to groundwater monitoring in karst aquifers for the development and utilization of groundwater resources. In addition, more monitoring work has been carried out on ground subsidence associated with underground salt solution mining such as at the I&W Facility in Carlsbad, New Mexico, or the Bayou Corne sinkhole in Louisiana (Jones and Blom 2015). In China, the karst collapse monitoring began in the 2000s. Monitoring of karst collapse has been conducted in combination with ground survey and geotechnical engineering investigation to provide technical support for the prevention and control of karst collapse. Monitoring can provide data to evaluate the collapse risks and help determine the effectiveness of selected long-term mitigation techniques. The following objectives can be achieved through karst collapse monitoring:

- Characterize the recharge and discharge conditions of the karst water system and the dynamic change characteristics of karst groundwater.
- Understand the development pattern and development trend of karst collapses, and provide an evaluation basis for the investigation, design, and construction of collapse prevention and control measures.
- Study the main influencing factors and formation mechanism of karst collapses and provide early warning for possible risks of karst collapse disasters.

The karst collapse research team with the Institute of Karst Geology of the Chinese Academy of Geological Sciences established ten long-term monitoring stations for karst collapses. Also compiled are the "Karst Ground Collapse Monitoring Specifications T/CAGHP075-2020" and "Karst Ground Collapse Prevention Engineering Exploration Specifications T/CAGHP076-2020". The monitoring methods mainly involve karst groundwater pressure, photoelectric distributed soil deformation, and ground surface deformation. The threshold values of monitoring indicators and the monitoring and early warning of urban karst collapses are being studied.

2.1 Monitoring Stage and Classification

In areas that are susceptible to sinkhole formation, karst collapse monitoring is specially designed according to the site conditions and the engineering construction project. In areas where sinkholes have already occurred, the monitoring efforts may include emergency response monitoring, remedial investigation monitoring, and sinkhole remedial construction monitoring. Table 2.1 summarizes the various monitoring stages.

Table 2.2 presents the monitoring level of karst collapses according to the severity induced by karst collapses and the size of the losses that have been (or may be) caused after the disaster. The importance of karst collapse is divided into very important, important, and common, and they are defined in Table 2.3. The magnitude of collapse-induced loss is presented in Table 2.4.

2.2 Overview of Karst Collapse Monitoring Methods

The monitoring of karst ground collapse includes monitoring of hydrodynamic conditions affecting karst ground collapse, internal deformation monitoring of subsurface soil, and ground surface deformation monitoring. The corresponding monitoring methods are summarized in Table 2.5.

Different monitoring methods can be chosen based on the level of monitoring effort, which relies on systematic

© The Author(s), under exclusive license to Springer Nature Switzerland AG 2024
X. Jiang et al., *Monitoring and Early Warning Technologies on Karst Lands*, Advances in Karst Science, https://doi.org/10.1007/978-3-031-59045-0_2

Table 2.1 Stages of karst collapse monitoring

Area	Project purpose	Investigation stage	Monitoring stage
Areas with collapse risks	Construction project to be implemented	Site-specific characterization	Site-specific monitoring
Areas with the presence of collapses	Mitigation engineering measures to remediate hazards	Emergency investigation	Emergency monitoring
		Remedial investigation	Remedial investigation monitoring
		Collapse remediation	Collapse remedial construction monitoring

Table 2.2 Classification of karst collapse monitoring levels

Monitoring levels		Importance of receptor exposed to hazard		
		Very important	Important	Common
Severity induced by collapse	Catastrophic	Class I	Class I	Class 2
	Critical	Class I	Class 2	Class 3
	Marginal	Class 2	Class 3	Class 3

Table 2.3 Classification of the importance of receptors exposed to the hazard

Importance level	Example receptors exposed to hazard
Very important	Urban and village planning areas, radioactive facilities, military and air defense facilities, nuclear power, high-speed railways, highways, railways, subways, airports, large-scale water conservancy projects, power projects, ports, mines, centralized water supply sources, waste treatment plants, water treatment plants, industrial buildings (span > 30 m), civil buildings (height > 50 m), long-distance oil and gas pipelines and gas storage reservoirs, schools, hospitals, theaters, and stadiums
Important	Newly built villages and towns, tertiary roads, medium-sized water conservancy projects, power projects, ports and wharfs, mines, centralized water supply sources, industrial buildings (span between 24 and 30 m), residential buildings (height between 24 and 50 m), landfills and waste treatment facility, water treatment plants, and municipal high-pressure gas pipelines
Common	Village residential areas, four-level roads, small water conservancy projects, power projects, ports, mines, centralized water supply sources, industrial buildings (span ≤ 24 m), residential buildings (height ≤ 24 m), landfills and waste treatment facility, water treatment plants, and community gas pipelines

analysis and understanding of the geological background conditions and dynamic conditions of the karst collapse development in the monitoring area. A monitoring work plan is recommended to discuss the monitoring objectives, methods, and data analysis and may include the following:

Table 2.4 Classification of the magnitude of possible loss after a disaster

Severity level	Economic loss and number of people exposed to threat	
	Direct (potential) economic loss (ten thousand yuan)	Number of people
Catastrophic	≥ 500	≥ 100
Critical	100–500	10–100
Marginal	≤ 100	≤ 10

- Overview of karst collapse conditions and karst hydrogeological and geotechnical conditions at the site and surrounding area.
- Monitoring level.
- Monitoring purpose, basis, and content.
- Arrangement and protection of monitoring points.
- Monitoring method, accuracy, frequency, and monitoring period.
- Monitoring data analysis and processing and reporting procedures on abnormal findings.
- Monitoring personnel and equipment allocation.
- Standards of operating procedures of instruments.
- Expected results.

2.3 Monitoring of Hydrodynamic Conditions

The hydrodynamics monitoring focuses on groundwater and gas activities. Groundwater seepage is the main driving force for the formation and evolution of karst collapses. Therefore, through the monitoring of karst groundwater gas pressure, the groundwater level in the overburden, and precipitation, changes in the dynamic conditions of the subsurface can be obtained in real time, and the dynamic conditions of the sinkhole development can be monitored. The monitoring instruments and equipment are either manual or automated.

The groundwater and gas pressure monitoring technology in karst conduit systems was first proposed by the collapse research group of the Institute of Karst Geology, Chinese Academy of Geological Sciences. The idea was derived from bench-scale physical model tests of karst collapse. Established in 1991, the physical model laboratory is believed to be the largest physical model laboratory of karst collapses in the world. By 1997, the simulation of typical karst collapses in six cities, Wuhan, Tangshan, Xiangtan, Guilin, Yulin, and Tongling, had been completed successively. The monitoring means include piezometric tubes near the bedrock surface, mercury column U-shaped tubes in karst cavities, and deformation monitoring inside the soil and

Table 2.5 Monitoring contents and techniques

Monitoring content	Monitoring indicators	Monitoring method	Monitoring levels		
			Class 1	Class 2	Class 3
Hydrodynamic condition	Groundwater levels in overburden and karst bedrock	Monitoring wells, pressure transducer, water level gauge	√	√	√
	Gas pressures in overburden and karst bedrock	Monitoring wells, pressure transducer	√	√	√
	Water quality monitoring	Turbidity or total suspended solids at wells or springs	√	√	
	Precipitation	Rain gauge	√	√	
Internal erosion of soil	Soil movement	Brillouin optical time-domain reflectometer (BOTDR)	√		
		Time-domain reflectometry (TDR)	√		
		Borehole tiltmeters (BTMs)	√		
		Borehole extensometers	√	√	
	Changes in karst or soil cavity size, soil density, soil conductivity, and raveling zones	Geophysics including ground penetrating radar (GPR), microgravity, electrical resistivity imaging, microseismic array	√		
Ground deformation	Horizontal and vertical displacements of the ground surface	Tape extensometers	√	√	
		Optical leveling techniques	√	√	√
		Real-time kinematics (RTK)/Total robotic station (RTS)	√	√	
		Global positioning system (GPS) surveys	√		
		Interferometric synthetic aperture radar (InSAR) or Light detection and ranging (LiDAR)	√	√	
		Horizontal in-place inclinometer (HIPI)	√	√	
	Building cracks	Crack gauge monitoring	√		
	Surface changes	3D laser scanning monitoring	√	√	√
		Drone monitoring	√	√	√
		Surveillance camera system	√	√	√
	Precursors of karst ground collapse including dry well (spring), ground fissure, raised floor, cracked wall, and others	Geologic mapping/reconnaissance, visual observation, tape measurement, percussion	√	√	√

on the surface. It was found in the test that there is a water level difference between the piezometric tube and the mercury column U-shaped tube. After analysis, the piezometric tube monitoring is an open system (connected to atmospheric pressure), and the mercury column U-shaped tube is a closed system, which better reflects the effect of water–gas pressure in the karst cavity on the bedrock surface soil. Since 1998, through the transformation of the laboratory model, a piezometer with high-frequency sampling and automatic recording has been used to record the relationship more effectively between the change of water–gas pressure in the karst cavity of the collapse model and the formation of the collapse pit.

2.3.1 Construction of Monitoring Well for Karst Groundwater Gas Pressure Monitoring

The Karst Collapse Group in the Institute of Karst Geology established the first karst collapse monitoring station in the Zhemu collapse area of Guilin in 2000. The sensors were embedded in the borehole by backfilling in the first application. This method has three shortcomings: (1) The sensor must be pulled out when backfilling the water-permeable medium layer by layer. (2) During this process, the sensor is easily damaged. (3) The sensor cannot be recovered after the monitoring work is completed, resulting in sensor losses and high monitoring costs. To facilitate the installation, recovery, and reuse of the sensor, the on-site monitoring well installation process has been improved, and a protective casing was designed and used to effectively isolate the water–gas pressure inside the monitoring well from the outside and avoid the impact of the external environment on the monitoring data. The pressure transducers can be retrieved if needed.

Figure 2.1a, b show the construction specifications for the monitoring well in different geological conditions. Figure 2.1c shows a typical completed installation with the data logger, power supply, and data transmission module. A data logger was used to receive and store data on water pressure through the probe and transmitted data to a terminal computer via a data transmission module. Storage

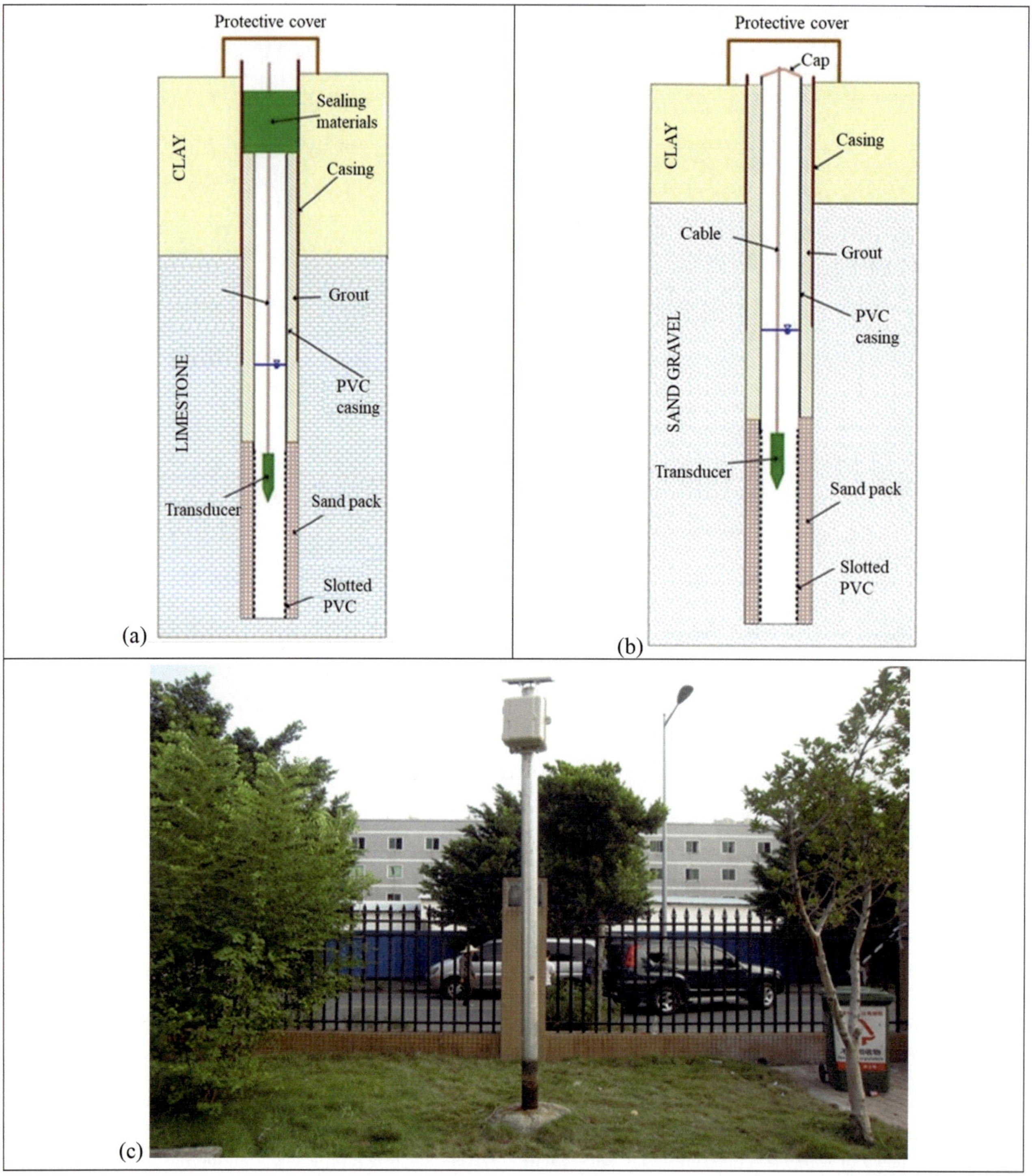

Fig. 2.1 Illustrations of transducer installation in different conditions. **a** Karst conduit system; **b** Porous-medium aquifer; **c** Completion of installation with data collection and transfer unit and solar panel for the power

battery supplied electric power to the data logger and data transmission module and connected to a solar panel for recharge.

The monitoring system mainly includes three parts: sensor installation, automatic collection equipment installation, and orifice sealing requirements. The key part is orifice sealing, which is different from conventional groundwater level monitoring. The sensor is installed on site (refer to the installation manual of the piezometer used), and the orifice can be sealed only after the manual data collection and debugging are normal. The sensor installation must meet the following requirements:

- Before installation, the sensor should be soaked in water for more than half an hour, so that the air in the sensor cavity can be discharged.
- The depth of the borehole and the depth of the water level should be measured, and the placement depth of the sensor should be designed.
- Mark the water level buried depth and placement depth mark on the sensor cable.
- Place the sensor near the water surface and determine its zero reading (water surface).
- After the sensor is placed to the design depth, it is fixed and read.
- The orifice should be sealed with expansion foam, glass glue, etc.

It is advisable to adopt dry drilling for the covering section of the monitoring hole, and clear water drilling or low water pressure drilling for the bedrock section; the diameter of the final hole should not be less than 91 mm. The drilling should be kept vertical and should be drilled with the casing during the hole-forming process. After the hole is finalized, a PVC protective tube should be put in the casing, and then the casing should be pulled out; while the casing is pulled out, the protective pipe should not be affected by influences. Finally, the length of the protective steel casing is retained to be determined according to the specific situation.

The diameter of the protective casing should not be less than 70 mm, and it should be placed at the bottom of the hole. The protective casing consists of three parts: the lower part is a 3–5 m impermeable sand settling pipe with a plug, the middle part is a slotted pipe (screening interval), and the upper part is an impermeable protective casing. The depth of the fixed section is 1–2 m greater than the depth of the bedrock surface.

The lower part of the fixed section of the protective pipe is bound with a water plug, and the cement mortar is slowly and evenly poured to the ground along the gap between the casing and the protective pipe. The purpose is to isolate the internal and external environment of the protection pipe, truly and fully reflect the change information of the water–gas pressure in the karst conduits, and pouring cement mortar can also enhance the strength of the disturbed soil layer on the bedrock surface.

The depth of monitoring wells should be determined according to the type of karst aquifer to be exposed and its buried depth and thickness. Drilling should pass through karst fissure development zones or karst pipelines (broken core or low core extraction rate, water gushing or leaking during drilling, drill drop, etc.) to achieve the purpose of monitoring the dynamic changes of karst groundwater and gas; If there is any connection between the stratum and the karst aquifer, the depth of the Quaternary water level monitoring hole should be drilled to a depth of 1 m below the floor of the lowest aquifer. For the monitoring area of different engineering activities, the specific depth of the monitoring well should be determined according to the buried depth and thickness of the aquifer:

- The exploitation of groundwater in the water source area induces karst collapse, and the depth of the monitoring borehole should not be greater than the floor of the aquifer where the groundwater is exploited.
- For karst subsidence induced by dewatering of mine or tunnel engineering, the depth of the monitoring hole should be greater than 20–50 m below the dewatering aquifer; since the annual fluctuation of groundwater level in most mines is dewatering and drainage area reaches 50 m, the depth of monitoring borehole shall be based on cost and the purpose is to reach the maximum depth as possible.
- Karst collapse is induced by foundation pit engineering construction, and the depth of monitoring boreholes should be greater than 10 m below the floor of the construction layer, such as foundation pit dewatering and tunnel excavation, the karst aquifer affected by engineering activities should be clearly defined, and the depth of monitoring holes should be designed. The 10 m here is only considered according to the maximum drawdown caused by general construction.
- The monitoring boreholes in other areas should be 10 m lower than the static water level, and the Quaternary monitoring wells should penetrate the aquifer floor by 0.5–1.0 m.

2.3.2 Layout Principles of Monitoring Points

2.3.2.1 Monitoring Area

The "Technical Specifications for Building Foundations in Karst Areas" BG/T51238-2018 stipulates that the engineering influence area refers to the area around the construction site that is disturbed and damaged during the construction of underground projects and may cause disasters. It is concerned with safety construction issues. Guide the construction monitoring work, avoid engineering hidden dangers in advance, feedback on the insufficient construction design plan, and provide a basis for plan improvement. Second, clearly define the scope of relevant responsibilities, urge relevant construction units to perform their duties, and ensure the safety of surrounding areas; the impact of underground projects is divided into primary, secondary, and possible impact areas. In the above regulations, the main impact area and responsibility area are defined in the plane range of 3 times the depth of the foundation pit, the groundwater monitoring work is arranged within the red line of the foundation pit, the buildings and ground deformation are arranged within the scope of the main influence area, and the secondary impact area and There are no clear requirements for the scope of the possible impact area and the definition of responsibilities. However, in the karst area, the disturbance of karst water caused by pumping and drainage of underground engineering construction has far exceeded the influence range of 3 times the depth of the foundation pit. As listed in Table 2.6, the karst groundwater is disturbed by the underground engineering in the karst area, and the distance to induce the karst collapse is as far as 1.5 km. Most of the karst collapse and land subsidence mainly occur within 300 m.

Due to the limitations of the current regulations on the scope of groundwater influence, the construction of karst underground projects has caused many geological environmental problems in recent years, such as the construction of the Jinshazhou section of the Wuhan-Guangzhou high-speed railway, Longgang Technical Secondary School, and Baihuiwei. Therefore, the determination of the monitoring range of karst groundwater gas pressure can effectively divide the scope of relevant responsibilities and provide a scientific basis for the safe drawdown control of karst groundwater.

According to the karst collapse monitoring the case of Guangzhou Jinshazhou Tunnel Construction and Longgang Subway, the monitoring range of karst groundwater gas pressure can be controlled within 500 m outside the red line of the underground engineering site, and the specific responsibility area can be defined according to the geological complexity of the site and engineering activities The depth of underground construction and the prevention and control measures to be taken are determined.

2.3.2.2 Monitoring Point Density

The scope of the monitoring area should cover the entire karst hydrogeological unit or karst water system where the site is located, and the monitoring area can be divided into key monitoring areas and general monitoring areas:

- Key monitoring areas should include areas with strong karst development, including strong groundwater runoff zones, structural zones, and pure carbonate rock distribution zones; areas affected by existing karst ground collapse; and important engineering site distribution areas.
- General monitoring areas include areas other than key monitoring areas.

According to the project points that induce (trigger) karst collapse and the direction of karst groundwater runoff, monitoring points are arranged to form a monitoring network that is perpendicular to the flow direction of groundwater and can control the scope of project influence.

When there is a Quaternary pore water aquifer, the monitoring points include monitoring holes for karst aquifers and monitoring holes for Quaternary pore water aquifers. The distance between monitoring points is determined with reference to Table 2.7. In addition, real-time kinematic (RTK) positioning, or total station should be used to measure the coordinates and elevation of monitoring points.

The karst groundwater system has heterogeneity in spatial development. To control the main flow direction of the karst groundwater system and the influence range of water–gas pressure caused by sudden changes in the groundwater level, one or two observation lines should be laid out in parallel and perpendicular directions to the groundwater flow. There are no less than 2 monitoring points for each survey line.

The use of drilling to form standard monitoring holes is conducive to obtaining accurate geological data and

Table 2.6 Range of groundwater influence by underground engineering projects

Engineering case	Location	Collapse time	Excavation depth (m)	Maximum length of groundwater influence range (m)	Engineering activity
Longping station	Longgang District, Shenzhen	March 2021	20	800	Dewatering in excavation pits and tunnels
Shumacheng station	Longgang District, Shenzhen	May 2020	21	700	Dewatering in excavation pit
Jingshazhou	Jianshazhou, Guangzhou	August 2007	40	1200	Dewatering in tunnel
Longyuan road	Longgang District, Shenzhen	December 2019	8	500	Pile installation
Luangjiang Shangwuo mine	Huizhou, Guangdong Province	October 2018	60	1500	Dewatering in quarry
Wanhao excavation	Guigang, Guangxi	April 2013	22	400	Dewatering in excavation pit

Table 2.7 Monitoring point spacing for dynamic condition monitoring at different monitoring levels

Monitoring level	Class 1		Class 2		Class 3	
	Key monitoring area	General monitoring area	Key monitoring area	General monitoring area	Key monitoring area	General monitoring area
Monitoring point spacing (m)	50–100	500–1000	100–200	500–1000	200–500	1,000–10,000

monitoring data, and better interpreting the mutation process and causes of karst groundwater pressure. However, for collapse areas with clear influencing factors and collapse disasters are not serious, existing, and available motorized wells can be selected as monitoring wells to save costs and expenses.

Since the monitoring technology of groundwater gas pressure in karst collapse can only analyze the soil deformation and damage within the influence range of the karst pipeline crack where the monitoring hole is located, the monitoring range of the karst collapse monitoring hole is limited. The technical specification for groundwater monitoring engineering (GB/T51040-2014) stipulates that the layout density of basic monitoring stations in karst collapse areas is between 10 and 20 per km^2, and the dynamic observation regulations for urban groundwater must be greater than 12 per km^2 in the most complex areas. When the map scale is 1:5000, the unified observation point density of the groundwater level is 8–12 points per km^2. According to the data from Ningxiang and Jinshazhou karst collapse monitoring stations, the impact range of underground pumping and drainage activities can reach 1.54 km, which is related to the supply of groundwater and the displacement of the project. According to statistics, the collapse caused by the construction of the Jinshazhou high-speed railway tunnel is mainly distributed within 200 m near the drainage point, and the longest distance is 500 m. Therefore, it is recommended that the spacing of monitoring holes in the direction of groundwater flow be at most 300 m.

2.3.3 Monitoring Frequency and Maintenance Requirements

In fact, people have realized that the monitoring frequency of conventional groundwater level observation is closely related to the error of monitoring data. In 2001, Charles conducted a daily, monthly, and quarterly comparative analysis of the groundwater level collection frequency of observation wells in western Massachusetts, USA. The results show that monthly groundwater level changes reflect seasonal changes and long-term trends but miss short-term effects such as pumping or recharge; less frequent groundwater level monitoring often leads to underestimation of water level fluctuations. The relationship between groundwater level monitoring frequency and monitoring error tends to increase with the increase of the monitoring time step and is related to topography and hydrogeological conditions. For the low mountain and hilly types, it is recommended that the groundwater level monitoring frequency be once every 3 days. The technical specification for groundwater monitoring engineering requires monitoring data 6 times a day (GB/T51040-2014), that is, collecting data every 4 h. In view of the engineering impact, the monitoring frequency of the groundwater level has been increased. For example, the monitoring of groundwater in karst tunnels is more than 3 times a day, and it is intensified when the change is large. The key area of Jinshazhou collapse in Guangzhou is monitored every hour. The occurrence of two karst collapses was predicted by the critical daily comprehensive groundwater level change. The Edwards Aquifer Authority in Texas, USA, has monitored the groundwater level of the Edwards karst aquifer for 84 years. The current groundwater level monitoring interval has been increased from the original 5 days to the present 15 min.

Likewise, the higher the frequency of monitoring data collection, the higher the probability of capturing changes in karst groundwater pressure before and after karst collapse. However, high-frequency data collection has a great relationship with the type of sensor used and the collection memory. To determine the optimal sampling frequency for karst collapse monitoring, Fig. 2.2 plots the data collected at the Jili Monitoring Station. The interval sampling method was used to extract the sampling intervals of 10 min, 20 min, 30 min, 40 min, 1 h, 1.5 h, 2 h, 3 h, 6 h, 12 h, and 24 h of data series to compare the peak amplitude of groundwater gas pressure and the peak velocity change and use the two peaks as a reflection to trigger karst collapse important indicators. The results in Table 2.8 show that when the sampling interval is greater than 20 min, it is obviously impossible to capture the complete peak value of groundwater gas pressure change; when the sampling interval is greater than 30 min, the peak value of groundwater gas pressure change rate is also significantly reduced. When the sampling interval is greater than 1 h, the abnormal changes cannot be collected. Therefore, optimal sampling intervals of 5–20 min are proposed. The number of sudden changes in water–gas pressure captured is much higher than in other sampling intervals.

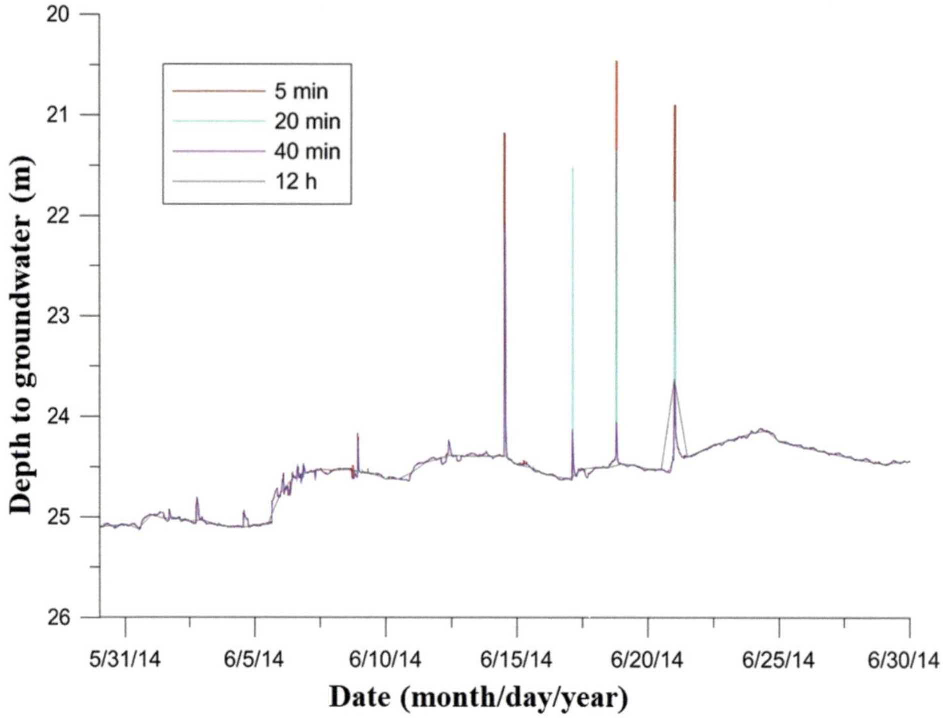

Fig. 2.2 Effect of sampling frequency on variation of groundwater gas pressure at Jili monitoring station

The monitoring time interval of karst groundwater gas pressure should not exceed 20 min. When the daily variation of karst water–gas pressure is greater than 2 m, the monitoring time interval should be shortened. When directly collecting data on site, the consistency of parameter settings should be checked, and the data should be checked in detail to eliminate errors. When the client automatically collects and transmits data, every data point should be screened to check for abnormal data.

The monitoring points shall be regularly maintained. The maintenance tasks include periodic checking and replacing the power supply system of the data acquisition system, checking the sealing of monitoring equipment and orifices every six months, and conducting inspections of karst ground collapse in the monitoring area every 3 to 6 months.

Table 2.8 Statistical characteristics of different sampling frequencies

Sampling interval	Maximum change rate (cm/min)	Maximum change amplitude (m)	Number of anomalies (amplitude > 2.5 m)
5 min	64.8	4.7	4
10 min	38.8	4.7	4
20 min	19.65	4.7	4
30 min	13.16	4.7	2
40 min	7.78	4.7	1
1 h	6.6	4.7	1
1.5 h	0.89	1.5	0
2 h	0.693	1.5	0
3 h	0.488	1.5	0
6 h	0.255	1.5	0
12 h	0.124	1.5	0
24 h	0.062	1.47	0

2.3.4 Groundwater Turbidity Monitoring

Monitoring of groundwater turbidity is mainly arranged at the outlet of underground rivers (springs) and monitoring wells screened in karst conduits. The monitoring data shows that the turbidity of the underground river has a good correlation with the rainstorm. The groundwater turbidity spikes are indicative of subsurface erosion that can cause karst collapses. Figure 2.3 shows the bench-scale seepage deformation test on core samples and intact soil samples. Figure 2.4 presents snapshots of an erosion test. Based on the observed phenomena in the model test, when the hydrodynamic conditions near the bedrock surface change, groundwater turbidity can be divided into two types:

- Intermittent turbidity: In the moment of water pressure change and negative pressure, a small part of the soil is destroyed, and then gradually becomes clear (Fig. 2.4b, c). The duration of this process is relatively short.
- Persistent turbidity: When a large water–gas pressure change occurs, most of the soil is eroded away, and the water is completely turbid. The soil erosion lasts until the water level drops to the bedrock surface, or the soil is destroyed during seepage deformation (Fig. 2.4d).

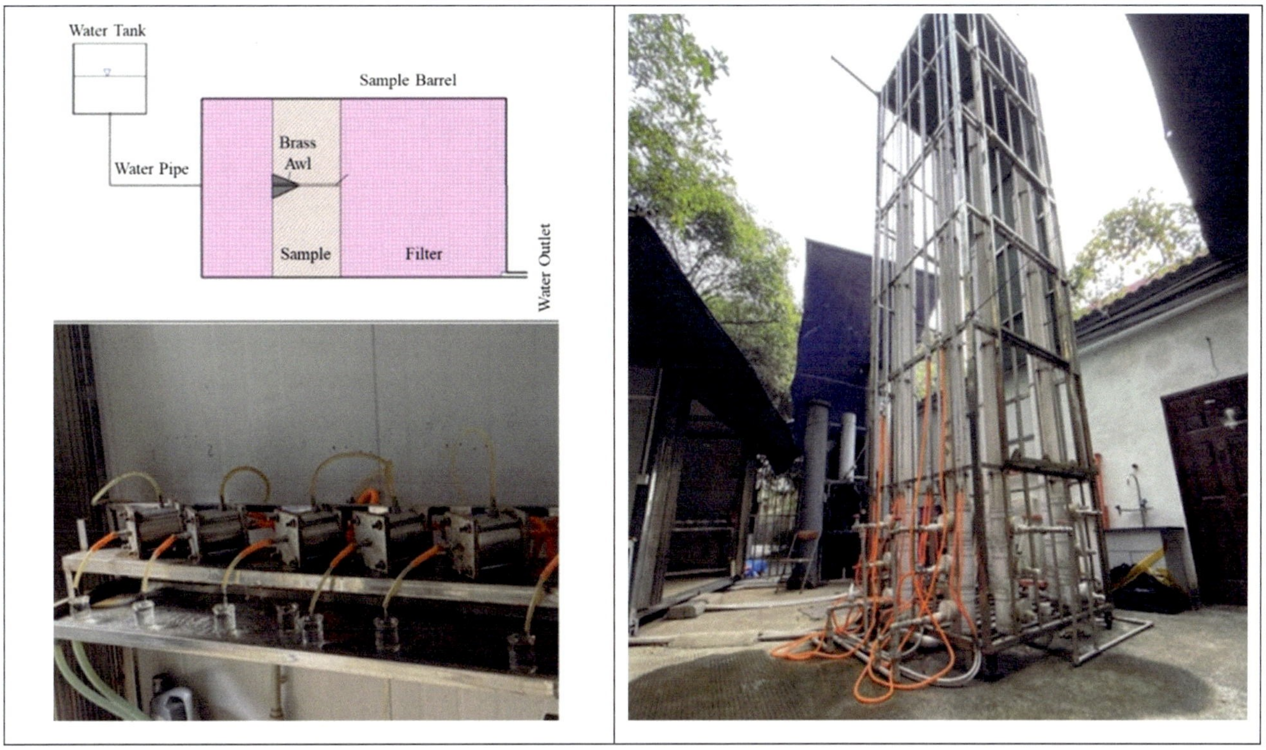

Fig. 2.3 Principle of seepage deformation test and apparatus (upper left: schematic of seepage deformation test, lower left: setup for large non-disturbed samples; right: hydraulic head setup for seepage deformation test)

The two types of intermittent and persistent turbidity represent the process of critical soil destruction and complete destruction, respectively. Therefore, during the non-rainfall period, the change of groundwater turbidity in the collapse area can reflect the formation process of the soil cave (collapse pit) in the study area. A multi-parameter water quality sensor or turbidity sensor is often used to measure the turbidity of groundwater in a monitoring well or spring. Analysis of suspended solids in groundwater samples is another option to obtain turbidity data. The data collection frequency for in-situ monitoring should be consistent with that of the hydrodynamics monitoring.

2.4 Monitoring of Physical Property Changes in Soil

The soil internal erosion monitoring focuses on soil transport processes in the subsurface. Since the karst collapses are formed from the formation of soil cavities overlying karst conduits and develop upwards, the spatial development position of soil caves can be discovered in time by monitoring the internal deformation of the soil through a selection of the appropriate monitoring method. Due to the hidden characteristics of karst collapses, it is difficult to precisely locate the location of the soil cavity. First, the scale of the soil cave is generally small, and the distance between the measuring lines must be small enough, otherwise, it is easy to miss. Secondly, the requirements are high to monitor the site, and the internal erosion may occur near the surface or at great depths. The soil erosion monitoring methods include geophysical methods, time-domain reflectometry (TDR), Brillouin optical time-domain reflectometer (BOTDR) methods, borehole inclinometer, and borehole extensometer. The geophysical methods, TDR, and BOTDR are more applicable to monitoring soil erosions at shallow depths, whereas the borehole instrumentation can be used in depths of several hundred meters.

2.4.1 Geophysical Methods

Geophysical methods are often used for site characterization and delineation of sinkhole-prone areas. However, they have been applied for monitoring purposes in recent years, especially the time-lapse geophysical surveys. Application of the same geophysical method to the same transect periodically or keeping the geophysical sensors in place for periodic data collection helps understand the changes in the subsurface conditions over time, thus evaluating ground stability. In general, no single geophysical method is sufficient for reliable subsurface erosion detection. The use of

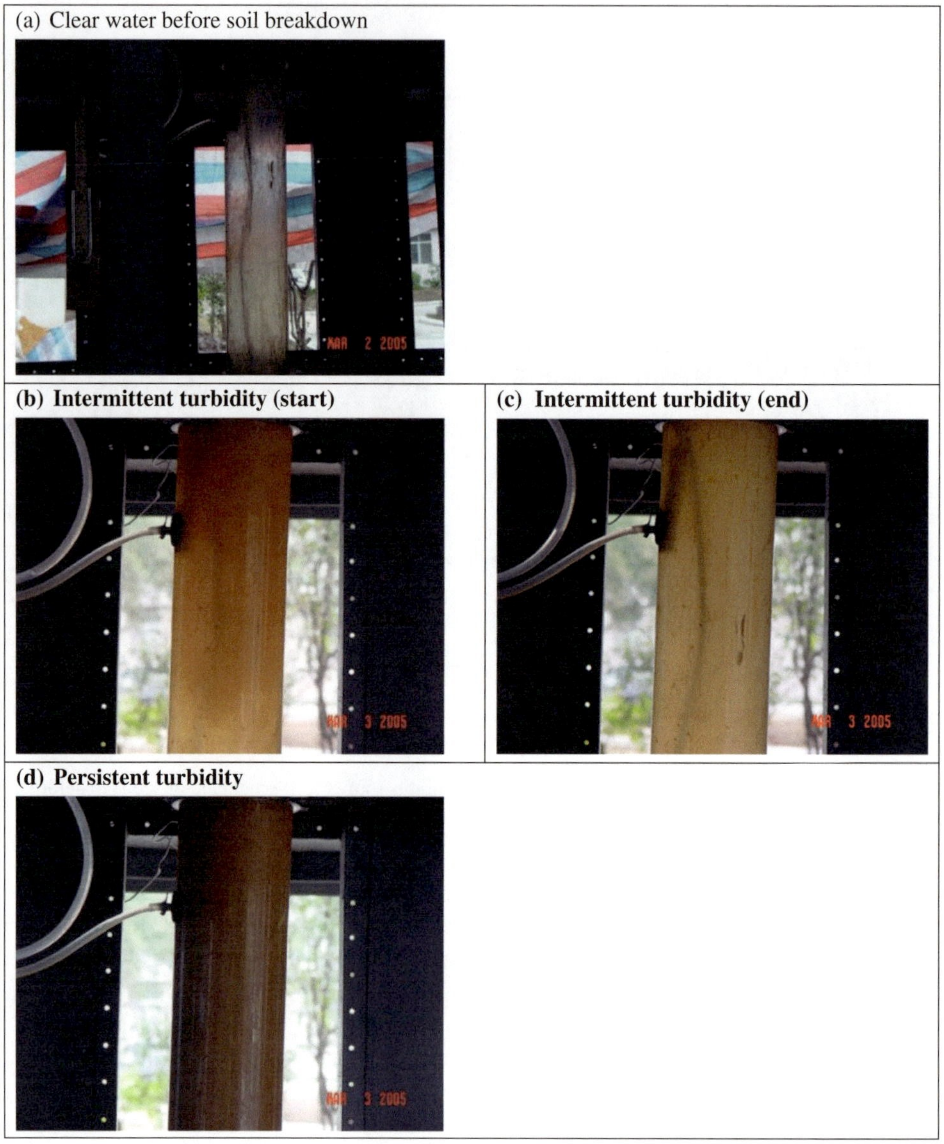

Fig. 2.4 Turbidity observations during erosion tests of soil samples

two or more techniques is preferable followed by drilling evaluation to confirm any geophysical anomalies. Based on our experience, the following four geophysical techniques have proven relatively reliable for karst collapse detection: microgravity, electrical resistivity imaging, GPR, and microseismic. Each of these techniques is discussed briefly in the following sections.

2.4.1.1 Ground Penetrating Radar

The monitoring depth of GPR may reach 5 m if the subsurface soil is predominantly sand. The survey line should be arranged in an area where the terrain is relatively flat, there are no obstacles on the ground, and the ground radar antenna is easy to move. There is no strong electromagnetic wave interference such as metal components or radio frequency sources; the monitoring lines should be arranged in parallel, and the distance between the measuring lines should be between 0.5 and 3 m according to the scale of the hidden soil cave.

Successful application of GPR to monitoring requires the selection of the most appropriate equipment parameter. The selection of the operating frequency of the radar host antenna should be comprehensively determined based on factors such as monitoring depth, resolution, medium characteristics, and whether the antenna size meets the site conditions. It is best to choose a shielded antenna with a frequency of 100–500 MHz and select 6 to 10 times the frequency of the antenna as the sampling rate. When antennas with multiple frequencies can meet the requirements of the detection depth, a high-frequency antenna should be selected.

The endpoints and inflection points of the monitoring line should be buried with piles and rocks for protection, and GPS, RTK, or robotic total stations should be used for position data. During the GPR survey, metal objects, high-voltage lines, and wires near the measuring line should be removed or avoided. The equipment supporting the antenna should be made of insulating materials, and the antenna operator should not wear objects containing metal components and should keep a relatively fixed distance from the working antenna. During the detection process, the plane of the working antenna should be kept basically parallel to the ground, and the distance should be relatively consistent. When continuous measurement is used, the moving speed of the antenna should be uniform and match the scanning rate of the instrument; when spot measurement is used, the point distance should not be greater than 1 m. Where anomalies occur indicative of karst or soil cavities, two sets of orthogonal directions should be used to scan separately to determine the cavity boundaries.

In areas prone to karst collapse, more than one measurement should be carried out every two months, and at least no less than two times a year. When abnormal changes are detected in the monitoring data, the measurement frequency should increase.

2.4.1.2 Electrical Resistivity Imaging

A resistivity survey measures variations in the voltage potential of an electrical field induced by a current source. Variations in the material resistivity through which the current is passed result in anomalies of the potential distribution. This variation in measured resistivity is indicative of changes in the composition or structure of the subsurface material. For example, an air-filled cavity or void can be a zone of near-infinite resistivity, whereas a water- or clay-filled void will be a zone of low resistivity with respect to the surrounding rock. Therefore, anomalous resistivity highs or lows are indicative of subsurface material changes and potentially indicative of open and filled subsurface cavities, respectively. The resistivity imaging is reasonably quick, and data processing is straightforward. Although the interpretation of the data is not unique and requires confirming information based on additional geological or geophysical assessments, the time-lapse imaging can show changes in the subsurface conditions over time.

Depending on the pair of current and potential electrodes used for each measurement, the electrodes are configured in various arrays such as Schlumberger, Wenner, pole-dipole, and dipole–dipole. For karst collapse monitoring, the dipole–dipole array has been proved to be the most effective (Zhou et al. 2002). In addition, the electrodes can be placed on the ground surface or in boreholes to generate different resistivity profiles. In surface application, the depth of investigation is directly proportional to the length of the array of electrodes, but the data resolution becomes poorer with depth. Cross-borehole imaging, on the other hand, can be accomplished when the resistivity electrode cables are placed into multiple boreholes. The borehole imaging provides far better data resolution at depth than the surface resistivity survey, while the boreholes provide details on the lithology and structure of the subsurface.

2.4.1.3 Microseismic Monitoring

Most solids emit low-level seismic signals when they are stressed and deformed, or collapse. We refer to this phenomenon as acoustic emission/microseismic activity, although other terms such as seismoacoustic activity, sub-audible noise, elastic radiation, or micro-earthquake have been used. The technique was originally developed to predict and reduce the incidence of rock bursts in hard rock and coal mines. In recent years the application of microseismic techniques to monitoring collapse processes has increased (Trifu and Shumila 2010). One of the major advantages of using microseismic monitoring over other geotechnical monitoring techniques is its ability to delineate the area of instability.

In essence, the measurement of microseismic activity involves the deployment of highly sensitive geophones that collect the acoustical signals generated by microseismic events. The geophones can be uniaxial or triaxial and can be installed in boreholes at various depths, depending on the location of the potential source area. It is extremely important to install geophones in the most suspected area of sinkhole development. The small observed events may be due to a weak source located close to the sensors or due to a strong source located a considerable distance away. In addition, many background activities can be included in the data, which do not different collapse activities from the background ones. Figure 2.5 shows an example of the instrument setup. In practice, an array of microseismic stations should be established to ensure the areal coverage. The spacing between the stations should be determined from an attenuation study at the specific site (Hardy et al. 1986).

According to the microseismic events caused by the collapse, the mobile seismic digital observation platform can be used to monitor in real time, accurately interpret the epicentral position and focal depth, and establish relationships between the earthquake and the collapse. The mobile seismic station includes seismic sensors and data loggers. In a typical application, four seismometers are used concurrently to form a seismic digital observation array. A seismic mobile station can be arranged in each of the four directions east, west, north, and south. It is recommended that the instrument is installed on a concrete platform to ensure leveling of the seismometer. The instrument shall be calibrated before data collection. The recommended technical parameters of the seismic sensor (three-way accelerometer) are as follows:

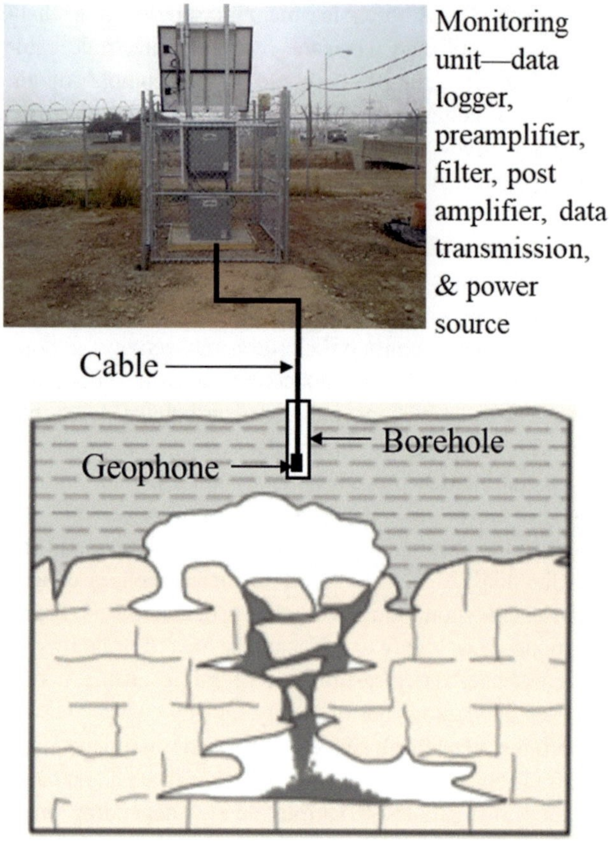

Fig. 2.5 Schematic setup of microseismic monitoring for karst collapses

- Frequency response DC ~100 Hz.
- Standard speed response output 2–100 Hz.
- Output sensitivity 2 × 1000 V/m/s.
- Full-scale output ± 10 V differential.
- The lowest spurious resonance frequency > 450 Hz.
- Linearity > 90 dB.
- Quadrature rejection > 65 dB.
- Temperature sensitivity < 0.6 V/10 °C.
- Nulling range of heavy hammer ± 3°.

The data collection frequency varies but is recommended every second. The observation stations should be maintained regularly including removal of debris, replacement of battery, and calibration.

2.4.1.4 Microgravity Monitoring

Theoretically, there is a density difference of more than 0.5 g/cm^3 between the karst or soil cavity and the surrounding environment. An anomalously low value of gravity might indicate the presence of an open cavity. The scale and characteristics of the concealed karst (or rock salt goaf) are identified and monitored with a high-precision, high-accuracy microgravity method. Perturbations in the gravity field caused by features such as a soil-filled cavity, an increase in soil cover, a decrease in soil density or a variation in the water table are small typically in one-hundred millionth in reference to the total gravitational field of 980 Gals. A Gal is equal to 0.01 m/s^2 and a μGal is 10^{-6} Gal. The μGal is the measurement unit used for microgravity surveys. According to DZ/T0171-2017 "Code for Large-Scale Gravity Survey", the accuracy requirements are shown in Table 2.9.

The CG-5 Autograv gravimeter by Scintrex Ltd. is a commonly used gravimeter. The sensing element in this gravimeter is based on a fused quartz spring system. The gravitational force on the proof-mass is balanced by the spring and an electrostatic restoring force. The position of the mass, which is sensed be a capacitive displacement transducer, is altered by a change in gravity. An automatic feedback circuit applies DC voltage to the capacitor plates producing an electrostatic force on the mass which brings it back to a null position. The feedback voltage, which is a measure of the relative value of gravity at the reading site, is converted to a digital signal and then transmitted to the instrument's data acquisition system for processing, display, and storage. The CG-5 Autograv has a reading resolution of 1 μGal and a standard deviation of less than 5 μGal.

Gravity measuring point observation can be arranged in a transect or grid, with a point distance of 2.5–20 m. Data filtering is a common practice to correct the impact of shallow interference factors (topographic relief, buildings and structures, ponds, ditches, foundation pits, construction waste, etc.) and measurement errors on Bouguer gravity anomalies. Profile fitting and inversion are used to quantitatively calculate the structural shape, spatial position, buried depth and undulation characteristics of the density interface, and the occurrence elements of fault structures, etc. This method first analyzes the main factors restricting the change of the gravity field according to the change characteristics of the gravity field of the section and designs the initial geological model based on the existing geological, geophysical, and drilling data in the work area, and then according to the density characteristics of the geological bodies in the area, establish a geological-physical–mathematical model for quantitative calculation, and modify the model repeatedly according to the difference between the theoretical anomaly curve and the measured anomaly curve, so that the theoretical curve and the measured curve can achieve the best fit, so as to obtain the structural characteristics and stratigraphic characteristics on the two-dimensional section.

2.4.2 Distributed Photoelectric Sensor Monitoring

The purpose of distributed photoelectric sensor monitoring is to monitor the development process of hidden soil caves. Therefore, the layout of optical fiber (or optical cable)

Table 2.9 Accuracy requirements of microgravity measurements

Overall accuracy of Bouguer anomaly	Mean square error of microgravity measurement	Mean square error Bouguer anomaly correction	Mean square error of topographic correction	Mean square error of latitude correction	Mean square error of elevation	Mean square error of coordinates
10^{-5} m/s^2					m	
±0.025	±0.015	±0.010	±0.017	±0.005	±0.050	±0.1

and coaxial cable measuring lines should be arranged 5 m below the foundation of linear projects under construction and important structures. Early warning can be provided before the foundation is affected.

2.4.2.1 TDR Monitoring

TDR utilizes a pulsed electromagnetic signal along a coupled coaxial cable to detect reflected changes resulting from deformation. Both travel time and signal strength are measured. Travel time is used to determine position, with signal strength being an indication of the severity of the deformation. Signal strength is an approximate measure of deformation; therefore, the TDR instrumentation should be viewed as a means to detect but not fully quantify deformation. The TDR system is designed such that should sufficient ground deformation occur to create a crack in the cemented soils that crack would transmit itself through the controlled low strength material (CLSM) that the cable is encased in and be detected by a signal sent through the TDR cable. While the TDR system has proven effective in detecting cracks that form in the CLSM backfill material, it has proven to be more sensitive than anticipated. Cracks detected have been the result of shrinkage cracking within the CLSM material and not the cemented soils.

TDR has been applied to real-time monitoring of ground deformation and changes under a highway (O'Connor et al. 2001), and early warning of sinkhole collapses associated with void propagation in solution mining (Renner et al. 2014). Jiang et al. (2011) conducted a series of geotechnical and material tests in both laboratory and field settings and proved that coaxial cable could be used to monitor and forecast karst collapse under suitable conditions. The coaxial cable grouted with cement mortar was paved under places where soil voids tend to occur. The coaxial cable with cement would be fractured by the impact of overlying soil at which the soil void underneath the coaxial cable developed to a certain scale (Fig. 2.6). The location of the soil void could be determined through the TDR curve image.

TDR is a remote electronic measurement technology. It was first used in the electric power and communication industries to determine the faults and breaks of communication cables and transmission lines. The principle is that the electric pulse excited by the TDR tester propagates along the coaxial cable in the form of electromagnetic waves. When the coaxial cable undergoes local shearing and stretching deformation, it will cause changes in the local characteristic impedance of the coaxial cable, and the electromagnetic wave will pass through these reflections and transmission occurs in the area and is reflected in the reflected signal. According to the return time and reflection coefficient of the reflected signal, the position and amount of deformation of the coaxial cable can be determined.

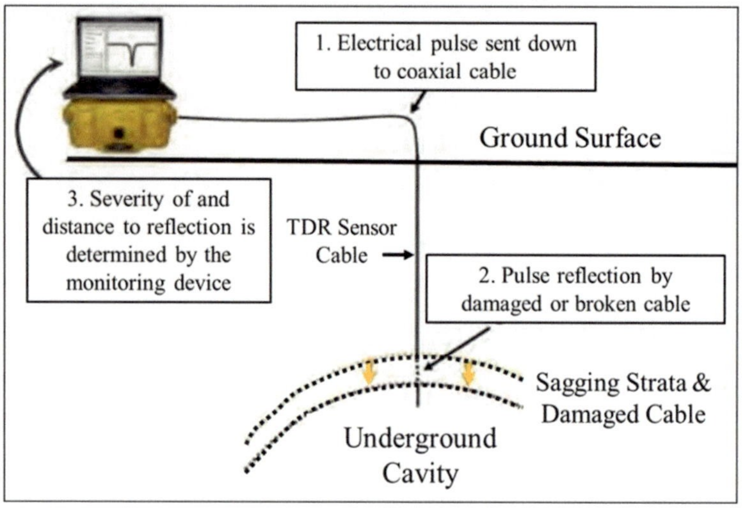

Fig. 2.6 Principle of TDR monitoring technology for void propagation monitoring

- TDR equipment and coaxial cables are mainly used. The choice of coaxial cable type should have little difference with the mechanical properties of the buried soil, and the reflection loss should be less than $0.1\rho/100$ m; the characteristic impedance of the coaxial cable should not exceed $50 \pm 3\ \Omega$; the elongation of the coaxial cable under the clamp mode should not exceed 50%.
- For coaxial cable laying, according to actual conditions, choose horizontal trenching or non-excavation directional drilling to lay coaxial cables.
- The monitoring frequency should be no less than 3 times a year, and one measurement should be added in time when there are abnormal changes in other monitoring data.
- Attention should be paid to checking the moisture-proof and waterproof conditions of the coaxial cable connectors, and timely updating damaged connectors. Surface anomaly surveys should be carried out along the survey line every 4 months.

2.4.2.2 BOTDR Monitoring

Distributed optical fiber sensing and monitoring technology uses optical cables as both sensor sensors and information transmission channels. Based on the Brillouin scattering phenomenon, the scattered wave frequency is only affected by temperature and axial force, and distributed sensing is realized after temperature compensation. Optical fiber sensing includes BOTDR and OTDR (optical time-domain reflectometry). They were first used to detect the quality of optical fibers, such as measuring the breakpoint position of the optical fiber, the axial strain of the optical fiber, and the loss of the optical fiber. If the optical fiber is buried in the rock and soil mass, the deformation and destruction of the rock and soil mass will cause the corresponding strain or even breakpoint of the optical fiber. Therefore, by measuring the strain at different positions of the optical fiber or determining the position of the breakpoint, the corresponding rock can be calculated. The amount of deformation of the soil and the location and scale of the damage can achieve the purpose of continuous monitoring of the deformation and damage of the rock and soil. Jiang et al. (2006) applied BOTDR to measure the sinkhole formation process in a large sand box and demonstrated that BOTDR would be a viable technology to monitor or predict sinkhole collapse. Guan et al. (2015) tested the responses of different optical fibers to the void size change in the application of BOTDR.

The fiber optic monitoring cables can be buried horizontally or vertically, as shown in Fig. 2.7. The optical fiber buried horizontally should be designed and arranged according to the characteristics of the site and the monitoring purpose. The distance between the measuring lines should be less than 3 m. The number of drilled holes shall be determined according to the scale of the cave, and the depth of the holes shall not be less than the buried depth of the roof of the cave (or soil cave) to be monitored.

Technical Requirements for Optical Fiber Horizontal Embedding

- Before installing the optical fiber, the length of the required optical fiber should be calculated in combination with the monitoring line; at the same time, the equipment should be connected for testing and the actual length should be recorded.
- It can be buried by digging grooves on the ground or directly bury the optical fiber during the construction of the foundation project.
- Technical requirements for ground trenching and embedding.
- Determine the position of the fiber optic connector for subsequent measurement and maintenance.

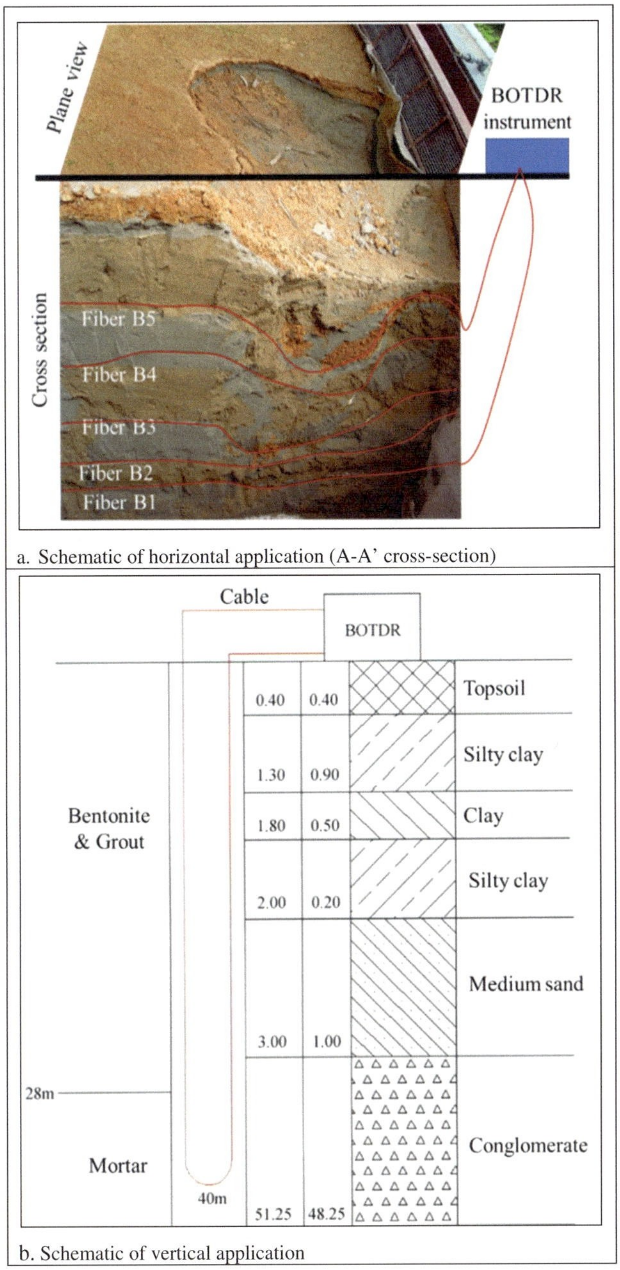

Fig. 2.7 Placement of optical cables in application of BOTDR to monitoring sinkhole development

- To locate the monitoring line, use putty powder to draw a line on the ground and insert a label for identification.
- Dig trenches along the line and place the excavated soil in layers.
- The pull measurement line records the thickness and properties of the soil layer from the surface to the bottom of the groove, the location of the soil hole, and other characteristics along the groove, and determines the calibration position of the optical fiber.
- Level the bottom of the groove, place the optical fiber, reserve 2 m of optical fiber at the calibration position, fix it with sand or crushed clay at an interval of 2 m, and measure the optical fiber.
- According to the position of the survey line, record the fiber at the corner, the calibration position, and the change position of the soil layer.
- Use RTK or total station to measure the coordinates of the endpoints and corner points of the monitoring line.
- Use the heating method to heat at the calibration position, measure the optical fiber, and read the calibration distance.
- Manually use pulverized clay or sand to backfill to a height of 20 cm and measure the optical fiber.
- Mechanically backfill to the surface, compact, and test fiber length and deformation.
- Technical requirements for embedding during construction.
- Determine the position of the fiber optic connector.
- Pull the test line, draw a line on the ground with putty powder, and determine the calibration position of the optical fiber.
- Lay optical fiber directly along the measurement line, reserve 2 m of optical fiber at the calibration position, and fix it with sand or crushed clay at an interval of 2 m.
- According to the position of the measuring line, record the code mark of the optical fiber at the corner, the calibration position, and the initial end of the buried section.
- Use RTK or total station to measure the coordinates of the endpoints and corner points of the monitoring line.
- Use the heating method to heat at the calibration position, measure the strain of the optical fiber, and read the calibration distance.
- The optical fiber is covered with mortar, and the thickness of the mortar should be greater than 2 cm; the ratio of cement to sand should be between 1:3 and 1:4; the flexural strength of the mortar should be lower than 2 MPa.
- Test fiber length and deformation.

Technical Requirements for Vertical Fiber Burial

Generally, drilling is used for installation, the diameter of the final hole should not be less than 90 mm, and the verticality of the drilling should be less than 2%. According to the hole depth and hole size, the amount of grouting material and the length of the required optical fiber should be preliminarily estimated. The equipment is then connected to test and record the actual length of the optical fiber. The position of the code mark at the bottom of the borehole should be recorded. Also recorded is the position of the code mark of the fiber corresponding to the characteristics of the cave section, rock-soil boundary, and orifice. The optical fiber embedding process follows the following steps:

- Place the fiber to the design depth.
- Grouting in sections from bottom to top.
- Ensure that the optical fiber is not pressed, stepped on, or bent during the grouting process.
- After the grouting is completed, record the actual code mark of the optical fiber at the aperture, connect the instrument and equipment for measurement, and record.
- Install the orifice protection box.
- The monitoring frequency should be measured once a day in the morning and afternoon after the grouting, and last for 3 days. During the normal monitoring period, it should be no less than 3 times a year, and when other monitoring data changes abnormally, one measurement should be added in time.

For the maintenance of monitoring points, attention should be paid to the moisture-proof and waterproof conditions of optical fiber connectors, and the investigation of surface anomalies along the measuring line should be carried out every 4 months.

2.4.3 Deep Soil Deformation Monitoring

The measurement of deep displacement of underground soil has always been the focus of geological disaster monitoring, and inclinometers are widely used in landslide and foundation pit monitoring. Since the inclinometer uses a conduit with a fixed groove and a certain strength, it is difficult for the conduit to perceive the raveling and loss of the karst collapse soil. The traditional collapse mark monitoring can now realize automatic monitoring through the static level. Due to the uncertainty of karst collapse locations, the monitoring points of deep displacement must reach a certain density to predict the occurrence of collapse in advance, and the cost is relatively high, so it is difficult to be widely used in karst collapse monitoring at present.

Borehole extensometers have been used to measure deep soil deformation between two or more points (Fergason et al. 2015). The often-used types are anchored cable counterweighted extensometers and slip-joint casing extensometers. Figure 2.8a, b show their working principles. A cable counterweighted extensometer tool consists of a balance beam carrying a cable or a pipe, which is fastened at one end to an anchor weight located at the bottom of the compacting system, and at the other end to a counterweight keeping the cable at a constant tension. A sliding-joint casing extensometer is installed in a borehole drilled to a depth where the rock is stable. The borehole is then lined with a steel casing with slip-joints to prevent crumpling as collapse occurs. An inner pipe rests on a concrete plug at the bottom of the borehole and extends to the top. This inner pipe then transfers the stable elevation to the surface. Measurement of the distance from the inner pipe to the surrounding land surface provides the amount of compaction that has occurred over a given time interval. A computer-controlled system records the compaction data against time. The instrumental precision heavily depends on the actual extensometer implementation, but a nominal deformation resolution of 0.01–0.1 mm can be achieved over a depth from 200 to 1000 m.

The ground deformation monitoring focuses on collecting data on the ground surface. It is mainly based on observations that ground fissures, building cracks, and building tilts may be precursors of karst collapses. In addition to sophisticated monitoring methods, visual inspections by people can also be applicable.

2.5 Ground Deformation Monitoring

Since the karst ground collapse disaster is the result of the upward development of soil holes on the bedrock surface, it is difficult to achieve the purpose of early warning by ground deformation monitoring. The scope and development trend of karst collapse pits.

When the ground is deformed, it means that the karst collapse has developed to the ground, so ground deformation monitoring is suitable for short-term monitoring in the emergency rescue stage to guide rescue work. Or houses and roads in high-risk areas of karst collapse should be monitored within a certain period within a range of 100 m from underground construction projects. Both labor-intensive manual monitoring and sophisticated remote sensing technologies are parts of ground movement monitoring. Only manual monitoring is discussed in this section.

2.5.1 Leveling Survey

A leveling survey is a commonly used method for ground deformation monitoring. The monitoring network is laid out according to the class 3 leveling standard. Ground deformation monitoring should be mainly aimed at areas where collapse pits, ground fissures, and collapse zones have developed. The monitoring includes monitoring of surface vertical displacement and collapse ground fissures, building cracks, or inclination. The ground deformation monitoring area should expand 100–200 m outside any existing karst collapses, ground fissures, and collapse zones.

A monitoring network consisting of points, transects, or grids is established to control the entire collapse area. More than 3 monitoring transects are arranged with the long axis

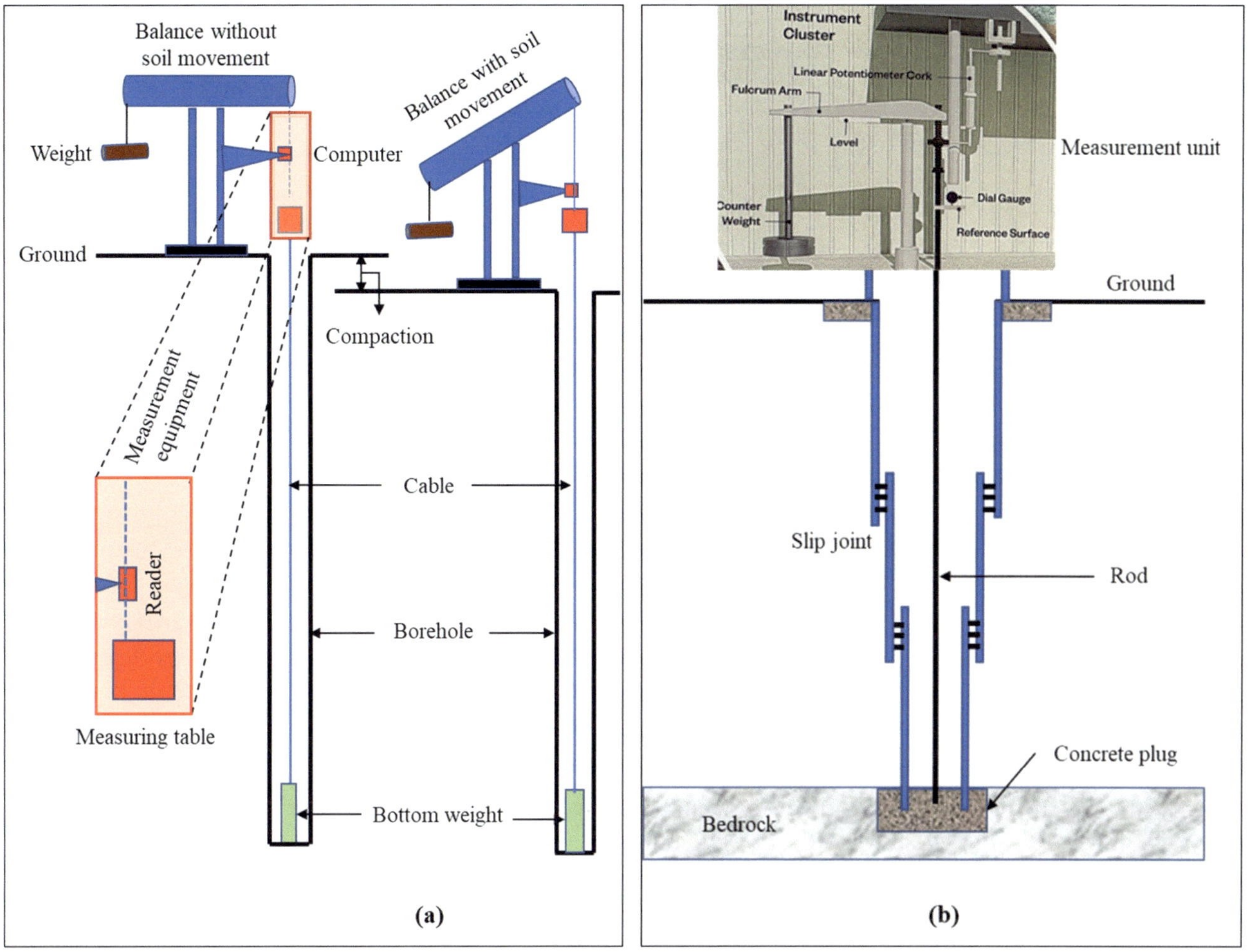

Fig. 2.8 Working principle of two types of borehole extensometer. **a** A single-layer cable counterweighted extensometer; **b** A slip-joint casing extensometer

of the collapse zone or the direction of the main geological structure line as the central axis. The survey line should be laid out along linear structures and avoid crossing obstacles such as rivers, lakes, and water bodies that are more than 500 m in length. The measuring points should ensure that the buried signs can reflect the ground deformation information at the location and are convenient for protection and maintenance.

The main and representative building cracks that have threatened the safety of buildings should be selected for monitoring. A measuring point should be arranged for each crack, and a crack meter should be installed for automatic monitoring. If necessary, crack warning alarms can be installed.

When conducting ground measurements, attention should be paid to maintaining the continuity of personnel and equipment. Relevant measurement specifications should be strictly followed and carried out in accordance with the requirements of the monitoring work plan. The recommended monitoring frequency should be no less than once a day in the first 10 days after the occurrence of karst collapses. The monitoring frequency may increase to 3 days between 11 and 20 days and further increase to 7 days afterward. Other frequencies such as once every month can also be used, depending on the site conditions.

2.5.2 Mass Inspection

Mass inspection is mainly to observe the changes in surface features by a group monitoring and prevention system and undertaken by local non-professional inspectors. Those people have knowledge of the locality. When the mass inspection finds major abnormal phenomena, it should be reported in time to carry out professional monitoring. Through inspections and visits to the points and residential areas of interest, precursors of karst ground collapse including dry wells (springs), turbid well water, sand spraying on the ground, concave water catchment on the ground, elevated floors, and cracked walls are documented. Table 2.10 presents an example form for consistency during mass inspections.

Table 2.10 Karst collapse hazard inspection form for local people

Geographic location	Anomaly on surface[a]	Extent of anomaly (length x width)	Engineering Activity[b]	Anomaly Timeframe	Timeframe of Engineering Activity	Inspection Frequency

[a]Anomalies on the surface include drying up of a well or spring, turbid water, water, or sand bubbling up, water sinking ground, floor suspension, and wall cracks

[b]Engineering activities include (1) blasting, (2) dewatering in mine, (3) excavation pits deeper than 1 m, (4) water supply well (daily production, well depth, overburden thickness, depth to water, pumping time), (5) irrigation point (depth to water, pump capacity, pumping time), (6) dewatering in excavation pit, (7) dewatering in foundation, and (8) dewatering in tunnel

Mass inspection routes include the front and back of buildings and roads, and the objects of inspection are cracks in walls, floors, windowsills, collapse pits, etc., noting whether old cracks are expanding and abnormalities such as groundwater catchment. Table 2.11 presents the various anomalies that may be observed and the measures that may be taken for the root-cause investigation. Nearby engineering activities should be inspected. During flood seasons when dewatering may be intensive, mass inspections in high-risk areas should be conducted no less than once a day, and no less than once a month during other periods.

Table 2.11 Principles of judging karst collapse hazards and responsive measures

Engineering activity	Judgement	Responsive measures
Surface anomaly observed within 20 m of an irrigation point	Strong probability	Increase inspection frequency (once per day)
Surface anomaly observed within 20 m of blasting	Strong probability	Stop blasting and increase inspection frequency (once per day)
Surface anomaly observed within 20 m of an excavation or foundation pit	Strong probability	Reporting to responsible authority and starting professional monitoring
Surface anomaly observed within 100 m of a new water supply well	Strong probability	Reporting to the responsible authority, evaluating of radius of influence by water supply well, and starting professional monitoring
Surface anomaly observed within 200 m mine dewatering	Strong probability	Reporting to the responsible authority, evaluating of radius of influence by water supply well, and starting professional monitoring

2.6 Integration and Comprehensive Analysis of Karst Collapse Monitoring Data

Karst collapse evaluation relies on multiple lines of evidence and comprehensive analysis of the various monitoring data. The internal connection and regularity between different types of data as well as the relationship between monitoring data and natural conditions, geological conditions, and various factors are the basis for the collapse risk analysis and forecast of its deformation and damage.

2.6.1 Integration of Karst Collapse Monitoring Data

Traditional data are recorded in forms or reports or data loggers for regular downloads, whereas telemetry monitoring has become more popular. Data can be transmitted to offices through wireless transmissions for instantaneous analysis. Table 2.12 presents some recommended procedures for data integration and analysis.

2.6.2 Comprehensive Analysis and Evaluation of Monitoring Data

2.6.2.1 Effectiveness Evaluation of Karst Groundwater Pressure Monitoring Points

The evaluation is mainly carried out from four aspects: the karst water system and location of the borehole, the depth of the aquifer, the condition of the exposed karst cave fissures, the measurable depth of the borehole and the thickness of the Quaternary overburden, and the relationship between the position of the protection pipe and the filter pipe. If the core is complete and the water is drilled into a borehole without leakage, the measurable depth of the borehole is less than the depth of the bedrock surface or the top of the filter pipe, these two types of boreholes should not be used as monitoring holes.

Table 2.12 Integration of multiple lines of data for karst collapse evaluation

Data type	Collation procedures
Karst water–gas pressure monitoring data	• According to the sensor calibration parameters and installation parameters, convert the collected original monitoring data into water head pressure, groundwater depth, and elevation in time to form time-series data, establish a database, copy and number the data for archiving • Draw the time-varying curves of karst groundwater gas pressure, water level elevation, and water–gas pressure change rate • Draw groundwater level contour maps, groundwater level change contour maps, groundwater level and bedrock surface relationship maps, etc. • Statistically analyze the monitoring data at monthly, quarterly, semi-annual, and annual intervals to obtain the maximum, minimum, and average values of karst water–gas pressure, water level elevation, and water–gas pressure change rate • Analyze possible causes of data anomalies
Rainfall monitoring data	• The data should be analyzed in time, and the hourly precipitation, 12-h precipitation, daily precipitation, and monthly rainfall should be calculated, respectively • Establish a rainfall database • Draw the relationship curve between karst water–gas pressure change and rainfall
GPR monitoring data	• GPR data processing methods include deleting useless traces, horizontal scale normalization, gain adjustment, terrain correction, frequency filtering, *f–k* dip filtering, deconvolution, offset homing, spatial filtering, and point averaging • The selection of processing methods and processing steps should be carried out according to the quality and interpretation requirements of field recorded data. When the reflection signal is weak and the signal-to-noise ratio of the data is low, it is not suitable to perform deconvolution and migration homing processing. Before *f–k* dip filtering and migration homing processing, useless traces should be deleted, and horizontal scale normalization and terrain correction should be performed • Draw geological radar profile images, grayscale or chromatographic images can be drawn during continuous measurement, and waveform images can be drawn during point measurement. The radar image should be marked with the survey line number, stake number, and depth time • Combined with the karst geological conditions, interpret the detection results, and draw the interpretation profile of radar geological results, including layer boundaries, soil caves or abnormal centers, ranges, and extension directions of soil disturbance zones • Compare the ground radar detection results at different times and analyze the change of the abnormal center of the soil hole or soil disturbance zone • Create a database and copy the numbers for archiving
TDR monitoring data	• Draw the coaxial cable time-domain reflectance curve directly based on the monitoring data • Compare and analyze the time-domain reflection curve of the coaxial cable at different times on the same monitoring line and record the distance of the new breakpoint of the coaxial cable, the actual position of the measuring line, and the measurement time • When the TDR measurements have abnormal breakpoints, it can be judged that underground karst soil caves are being formed • Create a database and copy the number for archiving
BOTDR monitoring data	• Export the data and convert the monitoring data obtained by the optical fiber strain analyzer into the vertical strain data of the soil • Eliminate the impact of instrument blind areas and optical fiber post-maintenance on monitoring data registration • Record the distance and actual position of the abnormal strain value, and the measurement time • Draw the strain curve diagram of the monitoring optical fiber and the time-series diagram of the monitoring optical fiber strain • Deformation damage judgment, when the breakpoint of the optical fiber is abnormal, it is directly judged that the karst soil cave is formed; the strain of the two comparisons before and after the results increases by 50%, it is inferred that the soil at this point is abnormal, and the formation of the soil cave needs to be further verified • Create a database and copy the numbers for archiving
Ground deformation monitoring data	• Carry out screening and analysis of the data to ensure the reliability of the data • Calculate the vertical displacement and horizontal displacement of the monitoring point, draw the displacement vector diagram of the monitoring point, and draw the contour map of the ground collapse in the monitoring area • Analyzing the time variation law of ground fissures and building cracks • Create a database and copy the number for archiving
Mass inspection data	• Organize manual inspection records and establish a database • Screen the manual inspection results and fill in the manual inspection results analysis table for karst ground collapse • Create a database and copy the number for archiving

The monitoring data of karst groundwater gas pressure has remained stable and without fluctuations under heavy rain conditions, indicating that its effectiveness is poor and cannot be used as monitoring holes; monitoring holes in the same karst water system should have synchronous rises under the influence of rainfall and human engineering activities and the decline process, there will be a certain lag between the upstream and downstream.

2.6.2.2 Monitoring Abnormal Judgment

There are two main types of anomalies in karst groundwater pressure monitoring points: abnormalities caused by sporadic or resonance of acquisition equipment, and abnormalities caused by changes in groundwater; under high-frequency monitoring conditions, the two are easy to judge, but in the case of low-frequency monitoring such as one monitoring value per day, the two anomalies are indistinguishable.

The abnormalities detected by ground radar scanning and monitoring are caused by underground pipelines, soil holes, saturated soil disturbance zones, surface wires or roofs, etc. For pipelines, the direction and buried depth should be identified and eliminated. Soil caves and saturated soil disturbance zones have parabolic anomalies. The range and depth can be confirmed through the meter-shaped measurement.

Photoelectricity is embedded in distributed monitoring points, the measurement section is easily affected by temperature, and the drift increases, but after 2–3 m of embedding, it is basically not affected by temperature, and the abnormality is mainly caused by the deformation or breakpoint of the photoelectric sensor. In addition, the measurement abnormality caused by insufficient waterproof measures in the detection section will cause the monitoring point to be scrapped.

The effectiveness of ground deformation monitoring points should be combined with the monitoring purpose and object, which is different from the monitoring of consolidation settlement. For the deformation within the scope of the collapse pit (depressed area), the collapse trend will continue to increase, especially during the rainy season or when the groundwater level drops sharply. The monitoring points are arranged in the collapse pit area and determined according to the monitoring data.

2.6.2.3 Comprehensive Analysis and Evaluation

The objects of comprehensive evaluation include analysis of observed values of multiple measuring points in the same project, analysis of multiple monitoring items at the same site, analysis of monitoring values of various parts of the same ground deformation, and analysis of instrument fixed-point observation and inspection data.

1. According to the monitoring results of karst ground hydrodynamic conditions, analyze the risk of karst ground collapse.

   (1) Through the seepage deformation experiment of undisturbed soil samples, the critical water–gas pressure change rate and critical hydraulic gradient of soil seepage damage and formation of soil caves were obtained. Analyze the change rate of karst water–gas pressure in the monitoring area and the temporal and spatial distribution of the hydraulic slope of the Quaternary bottom soil layer. When the monitoring data is greater than the critical value, it enters the dangerous period of karst ground collapse and compiles the distribution map of the dangerous area. At project sites with limited spatial groundwater data, the hydraulic slope may not be reliable.

   (2) Analyze the dynamic changes in the relationship between the karst water level and the bedrock surface in the monitoring area. When the karst water level fluctuates up and down on the bedrock surface, it is in a dangerous period of karst ground collapse and compile a distribution map of the dangerous area.

   (3) Analyze the dynamic changes in the relationship between the karst water level and the Quaternary water level in the monitoring area and study the hydraulic connection between the karst water and the Quaternary water. When the two are closely connected and there is a large water level difference, it shows that the Quaternary aquifer and the karst aquifer "skylights" appear in the relative water-resisting layers between them, and the seepage of the soil body is enhanced, entering a dangerous period of karst ground collapse, and a distribution map of the dangerous areas is compiled.

   (4) Analyze the rainfall monitoring data and count the dynamic changes in daily rainfall. In areas where the karst groundwater level is lower than the bedrock surface when the daily precipitation is greater than 1/3 of the annual average precipitation in the area, it enters the dangerous period of karst ground collapse, Compile the hazard zone distribution map.

   (5) Regularly prepare distribution maps of karst ground collapse risk areas in the monitoring area under hydrodynamic action.

2. According to the monitoring results of internal deformation of the soil, delineate the hidden danger area of karst ground collapse:

   (1) According to the geological radar interpretation map of each monitoring line, delineate the spatial position of hidden soil caves and disturbance zones.

   (2) According to the time-domain reflection analysis chart of the coaxial cable of each monitoring line, delineate the spatial position of the breakpoint, and verify it through geological radar detection to determine whether it is a soil hole or a disturbance zone.

(3) According to the optical fiber strain monitoring curves of each monitoring line, delineate the spatial location of breakpoints and large strains, and verify through geological radar detection to determine whether it is a soil cave or a disturbance zone.
(4) Regularly compile distribution maps of soil caves and soil disturbance zones in the monitoring area.

3. According to the ground deformation monitoring results, analyze the development characteristics of karst ground collapse:

(1) Analyze the temporal and spatial variation characteristics of the vertical displacement and horizontal displacement of each monitoring point in the monitoring area and study the temporal and spatial variation law of ground deformation.
(2) According to the magnitude of ground deformation, delineate the affected area of karst ground collapse.
(3) After each round of monitoring work is completed, the distribution map of the affected area of karst ground collapse in the monitoring area shall be compiled.
(4) According to the results of manual inspections, preliminarily delineate the macroscopic anomalies on the ground; regularly draw the distribution map of the macroscopic anomalies in the monitoring area and analyze the temporal and spatial changes in the abnormal areas.

4. Comprehensive analysis of monitoring data

(1) Regularly integrate data maps such as the distribution map of karst ground collapse risk areas under hydrodynamic action, the distribution map of soil caves and soil disturbance zones, the distribution map of karst ground collapse affected areas, and the distribution map of ground macroscopic anomalies to form the risk of karst collapse in the monitoring area and hazard distribution map.
(2) Comprehensively analyze the relationship between the changing law of karst collapse dynamic conditions, hidden dangers of karst collapse, and macroscopic abnormal changes.
(3) Analyze the development trend of karst ground collapse.
(4) According to the dynamic change characteristics of the inducing or triggering factors of karst ground collapse and the law of hidden karst catastrophe, countermeasures and suggestions are put forward.

2.6.2.4 Preparation of Monitoring Reports

According to the importance of the monitored object (project) and the complexity of the collapse, the monitoring report generally includes daily monitoring reports, weekly reports, monthly reports, quarterly reports, annual reports, and final monitoring results reports. The daily or weekly monitoring report shall include information such as main monitoring data, simple analysis, and abnormal situations that may need to be reported. The main monitoring data, duration curves, and related curves should be reflected, and comprehensive analysis and evaluation should be carried out on the change law within this period and the risk of karst ground collapse. The main maps include karst collapse risk and hidden danger distribution maps in the monitoring area. The main content should include physical geography and geological situation, monitoring objects, causes of karst ground collapse, deformation or dynamic characteristics of activities, development trends, conclusions, and suggestions (disaster prevention and treatment measures, etc.). If there is a control project, the effect evaluation of the control project should be added. The main maps include a geological map, monitoring point network layout map, monitoring data analysis map, karst collapse risk and hidden danger distribution map in the monitoring area, annual statistical table of monitoring data, and others. In addition to the relevant content of the annual report, the final report of the monitoring project needs to give clear and conclusive content and make clear suggestions for the follow-up measures of the monitored objects.

2.7 Purpose and Classification of Early Warning of Karst Collapse

The purpose of early warning of karst collapse is to alleviate and prevent the occurrence of disasters as much as possible, and to ensure the safety of people and property. The conventional understanding of early warning of karst collapse is the specific location and time of occurrence of ground collapse pits, but the current technical means still have certain difficulties. The main reasons include:

- The location of hidden karst caves has a good up-and-down correspondence with ground collapse pits, but there is no good up-down correspondence between the location of hidden karst caves and collapse pits on the ground.
- The time from the development of soil caves on the bedrock surface to the appearance of collapse pits on the surface of karst collapse is related to the thickness, structure, and properties of the Quaternary overburden, and the intensity of dynamic inducing factors.

The monitoring stations and cases of karst collapse in the past ten years show that early warning of karst collapse can be realized in local areas through limited technical means; effective early warning of karst collapse risk can be carried out through dynamic investigation and control of karst collapse risk sources.

2.7.1 Early Warning of Disasters

Short-term early warning when the disaster is approaching, on the one hand, effectively improves the timeliness of karst collapse disaster early warning, and on the other hand, provides sufficient and reasonable time for personnel and property transfer or early disposal.

2.7.1.1 Early Warning Indicators

The occurrence of karst collapse disasters is concealed and sudden, and the regional statistical rules are more obvious. It is difficult to carry out early warning of disasters at specific locations. The current effective method includes applications of some geophysical methods. The GPR regularly scans the soil holes within 5 m below the ground and disposes of them before they cause damage to property and personnel. In addition, it is the point deformation monitoring of the surface and deep rock and soil. However, due to the uncertainty of the collapse position, the density and depth arrangement of point monitoring is more difficult.

According to the changes and trends of the karst geological environment and natural or man-made engineering activity conditions, the risk of karst collapse that may occur in the future is predicted and alarmed. Its significance is mainly to prevent problems before they happen through the control of risk sources, hidden danger points, and main influencing factors. The monitoring and control of influencing factors in risk early warning is relatively easy to implement, but it has higher requirements for ascertaining the karst geological environment.

Early warning indicators are an important part of monitoring, early warning, and forecasting, corresponding to the monitoring method. In addition, the index of borehole deposition degree is added to reflect the damage ability of groundwater to the overlying soil on the bedrock surface. The types of each early warning indicator are shown in Table 2.13.

Table 2.13 Early warning indicators and types

Early warning indicator	Risk early warning	Disaster early warning
Anomaly detected by GPR	–	√
Anomaly detected by BOTDR	–	√
Surface deformation	–	√
Deep displacement	–	√
Karst water–gas pressure monitoring	√	–
Groundwater turbidity	√	–
Precipitation indicator	√	–
Sedimentation in borehole	√	–
Inspection by local people	√	–

2.7.1.2 Methods of Early Warning Indicators

Methods of analyzing the early warning indicators include graphical methods, statistical methods, empirical analogy method, and modeling. The graphical method draws process line diagrams, correlation diagrams, distribution diagrams, and comprehensive process line diagrams, analyzes the changing size, regularity, and trend of abnormal observation values, and judges the factors and degree of influence on observation values. The statistical methods are to obtain the maximum and minimum values (including the time of occurrence), instantaneous amplitude, cycle or annual average, annual variation trend of the monitored value, etc., and statistically monitor the relationship between physical quantities. For the monitoring of GPR and internal deformation of the soil, the abnormality after review can be used as a hidden danger point of karst collapse, and a direct warning and disposal plan is given. According to the geological structure and nature of the existing collapse areas and the daily, monthly, quarterly, or annual variation of the groundwater level, the early warning indicators of the groundwater level drop in similar areas are estimated. Modeling involves establishing the model of rainfall and karst groundwater gas pressure change, establishing the model of drawdown, hydraulic gradient, flow velocity, and water level change.

2.7.2 Cloud Platform for Karst Ground Collapse Risk Prevention and Control

The construction of the karst ground collapse risk prevention and control system can realize the functions of automatic collection, automatic transmission, automatic storage, automatic early warning, and risk management of monitoring data. Through the network, technicians can timely understand the source of karst collapse risk, evaluate risks, and carry out early warnings. Control influencing factors and emergency response.

2.7.2.1 System Architecture

Based on the existing geological environment safety laws and regulations and information security system, integrating 5G, things, cloud, network, artificial intelligence, and other digital elements. A whole-process karst process of high-precision perception-real-time collection-intelligent research and judgment-dynamic early warning-emergency plan and emergency rescue is built. Figure 2.9 shows the components of a typical collapse risk management and control system.

2.7.2.2 Selection Principles of Early Warning Monitoring Points

Since the regional monitoring is based on points instead of areas, the drawdown of monitoring points within the scope of the groundwater sinking funnel is different. It is very

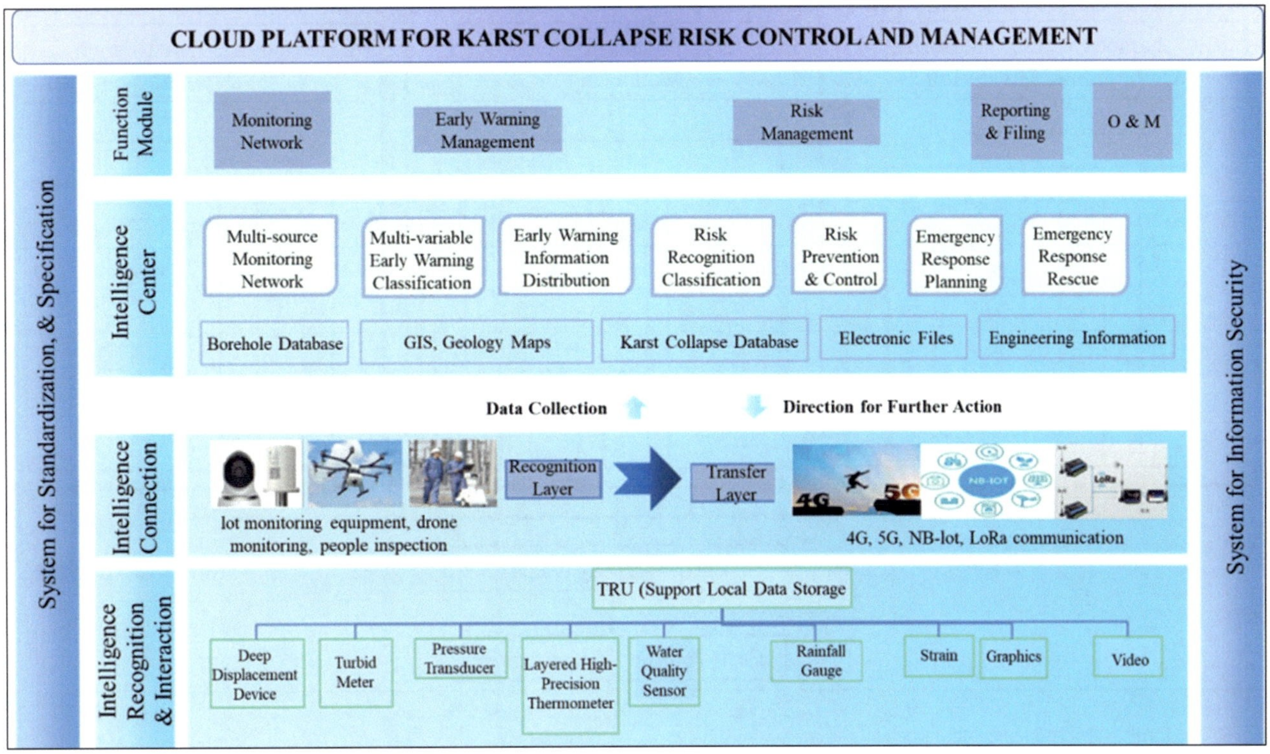

Fig. 2.9 Cloud platform framework for karst collapse risk control and management

important to select typical and important monitoring points and the calibration of early warning thresholds in a fast and timely manner. The selection principles of typical and important early warning and monitoring points are as follows:

- Select boreholes with high karst rate in line, and the selected monitoring points must have detailed engineering geological histograms and monitoring equipment installation information.
- Each karst water system has at least 2 monitoring points.
- For the point where the water level changes greatly in the karst water subsystem, the monitoring point is selected at an interval of 200 m from the center of influence to the long axis of the groundwater descending funnel. Generally, the number of monitoring points in the downstream area of the original groundwater is more than that in the upstream area of the groundwater.
- Select buildings and roads to focus on.
- Arrange along the direction of development of karst ditches and troughs, and areas near water-repellent boundaries (or relative water-repellent boundaries) (without groundwater recharge sources, and groundwater table drops rapidly).
- Due to the long impact period of underground engineering activities (excavation, pile foundation, grouting, etc.), the selection of early warning and monitoring points is also a dynamic process.

2.7.2.3 Early Warning Model and Classification

At present, groundwater monitoring and early warning in the field of geotechnical engineering are judged and early warned by a single indicator threshold of the cumulative change range of water level, without considering the short-term and long-term impact of rainfall on groundwater level, and the false alarm rate is high; the analysis results of monitoring data from existing karst collapse monitoring stations found that the density of groundwater monitoring points is sparse and the monitoring frequency is low, resulting in a lack of mutual verification and mutual feedback of the effectiveness of monitoring data.

The early warning model is divided into a single-variable model and a multi-control variable model. The multi-control variable model is linked with multiple indicators such as ground survey and deformation monitoring, dynamic field monitoring, and geological environment data, and then verified with multiple measurement items, that is, the threshold plus comprehensive research and judgment. The single-variable early warning model mainly sets indicators for the hydrodynamic field of karst collapse. Figure 2.10 presents the early warning model and classification scheme.

The early warning status is divided into three levels from small to large according to the severity, and the corresponding karst collapse risk early warning is also divided into three levels: early warning value, control value, and maximum allowed value.

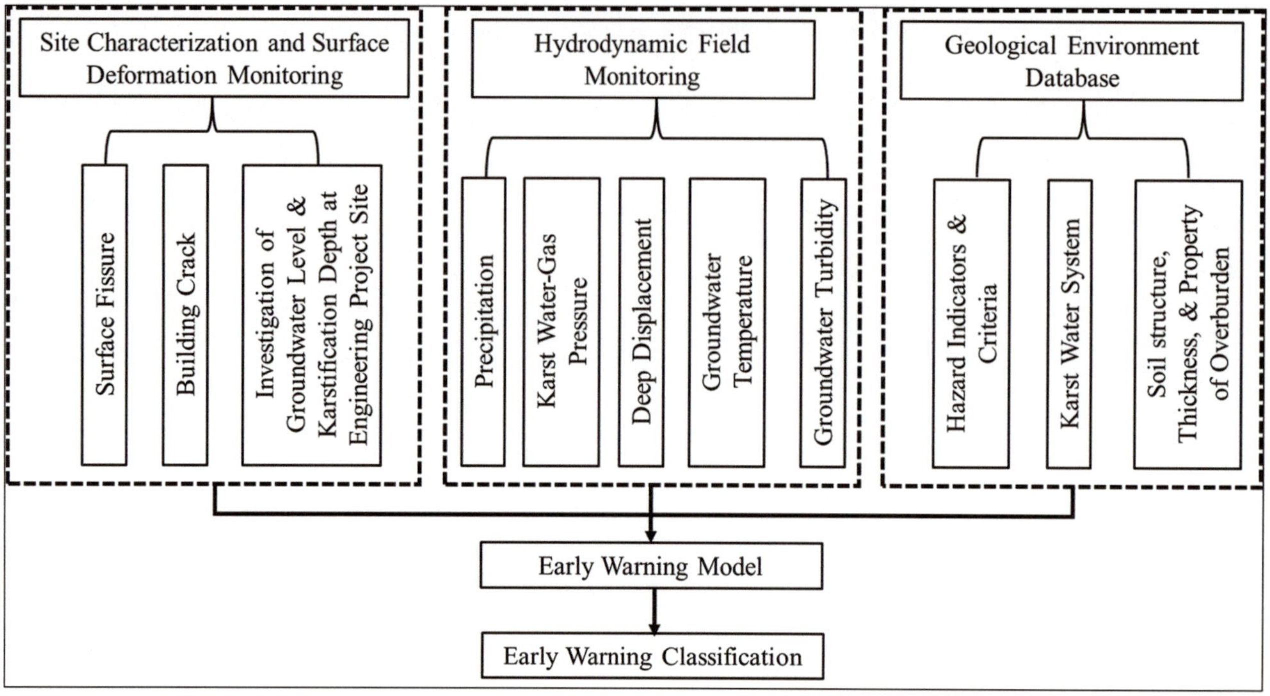

Fig. 2.10 Early warning model and classification

Table 2.14 Recommended early warning thresholds

Indicator	Early warning value	Control value	Maximum allowed value
Groundwater level (m)	Shallow bedrock surface	10 m above shallow bedrock surface	10 m below shallow bedrock surface
Instantaneous change rate of karst water–gas pressure (cm/min)	7	30	60
Cumulative groundwater level drop within 3 days (m)	3	5	8
Daily precipitation (mm)	50	100	200
Percentage of borehole sedimentation (%)	5	5–20	20

2.7.2.4 Early Warning Threshold

Risk Warning Threshold

The fluctuation of karst groundwater is the necessary dynamic condition for karst collapse. The development of concealed karst, the thickness, structure, and properties of Quaternary overburden are the fundamental conditions. Therefore, the risk warning threshold mainly refers to the monitoring indicators of karst groundwater and the factors affecting groundwater. Table 2.14 summarizes the recommended risk warning thresholds.

1. Elevation of karst groundwater level

The groundwater level of the karst fluctuates near the bedrock surface, and karst collapse is very easy to occur. In addition, the groundwater level is lower than the bedrock surface. During the rainy season, the sudden rise and fall of the karst groundwater pressure is also very easy to induce karst collapse. Therefore, the bedrock surface elevation can be used as a control value for monitoring and early warning. In addition, due to the large fluctuation of the bedrock surface in the karst area, the elevation of the shallowest bedrock surface within 100 m of the monitoring point, the average elevation of the bedrock surface + 10 m, and the average elevation of the bedrock surface − 10 m are considered as the warning value, control value and limit value, respectively.

2. Decline of karst groundwater level

For areas lacking high-frequency karst groundwater pressure monitoring, if there are groundwater level monitoring values, the karst groundwater level drop index can be used. The current groundwater level monitoring data interval is mostly 1–24 h, so the cumulative groundwater level decline in 2 days can be used to represent the groundwater level decline, rate, and trend.

Combining the results of the indoor soil seepage deformation test and the records of the field inspection of the collapse pits, when the cumulative groundwater level drop was greater than or equal to 3 m in 2 days, it was far greater than the critical head pressure value of soil seepage deformation in many areas, indicating that the karst soil caves (or surface collapse pits) are easy to form in the area; when the instantaneous karst groundwater pressure difference is greater than 5 m, which has exceeded the pressure value of the breaking water head of clay, karst soil caves (or surface collapse pits) can be formed quickly. Therefore, the cumulative drops of 3 m and 5 m are set as the warning value and the control value, respectively, and the maximum allowed value of 8 m is calculated based on 1.6 times of the control value.

3. Change rate of instantaneous karst groundwater gas pressure (cm water/min)

The instantaneous change rate of groundwater level is closely related to the monitoring frequency. The higher the monitoring frequency, the greater the rate of water level change, and the lower the monitoring frequency, the smaller the rate. The karst groundwater pressure monitoring data collection interval used in this paper is 2–5 min.

Combining the results of indoor soil penetration and deformation tests and the records of field surface inspections of collapse pits, when the instantaneous karst groundwater pressure rate is greater than or equal to 7 cm water/min, it indicates that karst soil caves (or surface collapse pits) may form in this area; when the instantaneous karst groundwater When the air pressure rate is greater than 30 cm water/min and 60 cm water/min, karst soil caves are easy to form and expand to form ground collapse sinkholes. Therefore, the above-mentioned values are respectively set as warning value, control value, and limit value.

4. Rainfall

The change of karst groundwater gas pressure is not only related to the pumping and drainage activities of the project but also closely related to the rainfall. According to the long-term monitoring data of Hunan Ningxiang, Guangxi Laibin, and Shenzhen Longgang Technical Secondary School, the daily rainfall of 50 mm has an impact on the groundwater level, but the possibility of karst collapse is small. When the daily rainfall is 100 mm, the change in water level is less than 3 m, and the daily rainfall (or two-day cumulative rainfall) exceeds 200 mm, and the risk of karst collapse in areas where the groundwater level is lower than the bedrock surface is extremely high. Therefore, 50, 100, and 200 mm can be used as the early warning value, control value, and maximum allowed value of rainfall.

For the newly established key areas, after a hydrological year of monitoring, the impact of daily rainfall (or 2-day cumulative rainfall) on groundwater levels can be graded and early warned.

5. Borehole sedimentation degree

Serious siltation in the boreholes indicates that human engineering activities on the site (such as pumping and drainage) have seriously disturbed the overlying soil and karst cave fillings, which can easily lead to collapse pits and cracks on the ground.

The sedimentation rate of boreholes can be calculated in the following three ways:

- The difference in hole depth before and after the pumping test is divided by the hole depth before the pumping test.
- Divide the difference between the depth of the final drilling hole and the measurable hole depth before the sensor is installed, divided by the depth of the final drilling hole.
- Divide the depth measurement difference between the two periods of the same monitoring hole by the hole depth measured in the first period, and the two periods can be measured monthly or quarterly according to the decline of the groundwater level and the water output of the foundation pit.

When the percentage of the above values is less than 5%, it is judged that the borehole sedimentation degree is low, which is set as the warning value; 5–20%, the borehole sedimentation degree is medium, and it is set as the control value; when it is greater than 20%, the borehole sedimentation The high degree is set as the limit value.

Disaster Warning Threshold

Disaster warning indicators mainly include GPR scanning anomalies, photoelectric distributed displacement monitoring, ground deformation monitoring, and point-like deep displacement monitoring indicators. The threshold settings are shown in Table 2.15.

Table 2.15 Recommended disaster warning thresholds

Indicator	Threshold	Early warning value	Control value	Maximum allowed value
Anomaly on GPR transect	Anomaly indicative of soil disturbance or void	Threshold value	–	–
Surface deformation at point	Collapse trend	Threshold value	–	–
Building crack (mm)	4	3	4	6
Deep rock/soil displacement at point	Measurement indicative of displacement	Threshold value	–	–
Displacement from BOTDR monitoring	Measurement indicative of displacement	Threshold value	–	–

Fig. 2.11 Relationship between early warning cloud platform and users

2.7.3 Early Warning Release and Response Management Measures

To deal with possible emergencies during on-site inspection and monitoring, effectively prevent, control, and resolve geological environmental safety risks induced by underground engineering in schist areas during construction, and ensure the safety of the surrounding environment, it is necessary for multiple parties to participate in karst collapse risks prevention and control. Figure 2.11 shows the relationship among construction parties, construction parties, monitoring parties, administrative management, and supervision units.

The alarm process and emergency management are as follows:

- Carry out patrol inspection and collect data, analyze, and deal with abnormalities in the inspection content, as presented in Table 2.16, and notify the owner and relevant units in time, and further review and verify.
- If the monitoring data reaches the warning value, all parties should be notified urgently, and the relevant personnel of all parties should be notified by text message or phone call at the first time, and emergency measures should be initiated (Fig. 2.12). The monitoring and early warning notification objects and countermeasures are shown in Table 2.17.

Table 2.16 Surface anomaly type and management

No.	Anomalous observations	In-situ emergency response measures	Risk evaluation
1	New ground fissures and building cracks	• Verify and determine if caused by geological factors • Establish rudimentary monitoring points and take measurements once per day • Eliminate water percolation	High risk when fissure or crack width greater than 5 mm
2	Ground collapse (water ponding or sinking)	• Verify the presence of concentric ground fissures • Delineate dangerous zone and fence off the area by adding 5 m • Install fences if appropriate to prevent unauthorized people from entering • Get rid of ponded water	High risk where buildings' roadways are within 10 m
3	Bubbling up water or sand on ground	• Delineate dangerous zones and fence off the area by adding 5 m • Install fences if appropriate to prevent unauthorized people from entering	High risk where buildings' roadways are within 10 m
4	Base floor suspended, new cracks on walls, or dramatic changes of existing cracks	Verify, survey, and delineate the extent of collapse	High risk especially when building does not have a framed structure
5	Engineering activities (blasting, water pipe leakage, dewatering in foundation pit, or piling)	Verify engineering activities	High risk when groundwater is significantly disturbed
6	Turbid water in wells	• Take notes • Take a sample if appropriate • Delineate the extent of observed turbid water if possible	High risk if the water is very turbid

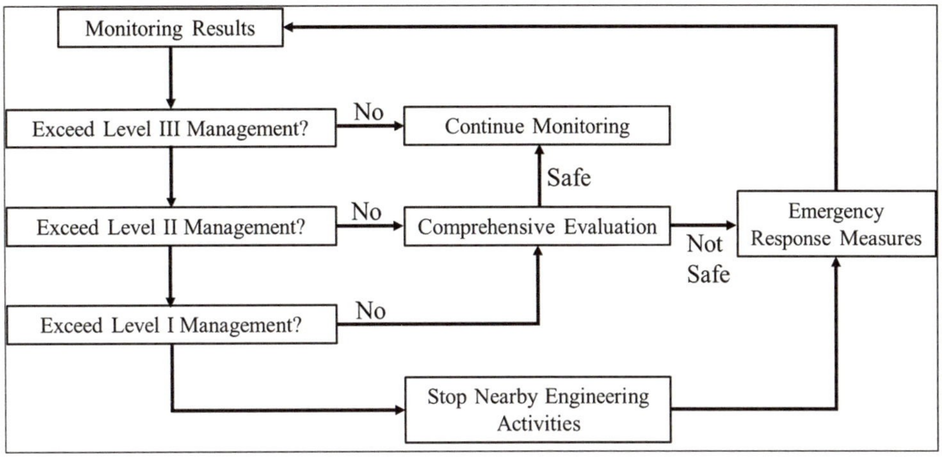

Fig. 2.12 Three-level early warning management

Table 2.17 Three-level early warning management approach and corresponding measures

Level of early warning	Monitoring values	Reporting mode	Responsible organization	Corresponding measures
Level III	$U_0 \geq$ early warning value	Verify, instrument checking, oral reporting, normal monitoring	Monitoring organization lead, site characterization organization lead, local municipality, or planning committee	• Pay close attention to groundwater hydrodynamics • Reinforce monitoring of ground collapse • Inspect nearby buildings and roadways
Level II	$U_0 \geq$ control value	Site inspection, Written report within 24 h increase the intensity of monitoring	In addition to those included in Level III reports to pertinent provincial departments	• Reinforce monitoring of groundwater hydrodynamics and ground collapse • Inspect nearby buildings and roadways • Limit surrounding engineering activities • Control groundwater level drop
Level I	$U_0 \geq$ maximum allowed value or an increase in ground collapse or building cracks is observed	Temporarily stop surrounding engineering activities and provide solutions to the problems	In addition to those included in Level III, report to pertinent professional organizations	• Stop engineering activities within 500 m • Fence off the area and prevent unauthorized people from entering • Backfill any collapses in the vicinity of a building to ensure the stability of the building

Note U_0—actual measurement

- The monitoring unit and the survey unit cooperate with the owner and related work units to carry out emergency rescue, focusing on strengthening the dynamic monitoring of groundwater, inspection of building cracks and ground collapse, timely fencing off dangerous areas, and evacuating personnel.
- Early warning cannot be eliminated until the danger is eliminated.

References

Fergason KC, Rucker ML, Panda BB (2015) Methods for monitoring land collapse and earth fissures in the Western USA. Proc IAHS 372:361–366

Guan ZD, Jiang XZ, Wu YB, Pang ZY (2015) Study on monitoring and early warning of karst collapse based on BOTDR technique. In: Doctor DH, Land L, Stephenson JB (eds) Sinkholes and the engineering and environmental impacts of Karst: proceedings of the fourteenth multidisciplinary conference, NCKRI symposium 5, Rochester, Minnesota, 5–9 October. National Cave and Karst Research Institute, Carlsbad, New Mexico, pp 407–414

Hardy HR Jr, Belesky RM, Mrugala M, Kimble EJ, Hager ME (1986) A study to monitor microseismic activity to detect sinkholes. Report submitted to U.S. Department of Transportation, DOT/FAA/PM-86/34, pp 199

Jiang XZ, Lei MT, Chen Y (2006) An experiment study of monitoring sinkhole collapse by using BOTDR optical fiber sensing technique. Hydrogeol Eng Geol 33(6):75–79

Jiang XZ, Lei MT, Dai JL (2011) A study of the monitoring deformation of sinkhole collapse using TDR time domain reflectometry. Hydrogeol Eng Geol 38(1):118–122

Jones CE, Blom RG (2015) Pre-event and post-formation ground movement associated with the Bayou Corne sinkhole. In: Doctor DH, Land L, Stephenson JB (eds) Sinkholes and the engineering and environmental impacts of Karst: proceedings of the fourteenth multidisciplinary conference, NCKRI Symposium 5, Rochester, Minnesota, 5–9 October. National Cave and Karst Research Institute, Carlsbad, New Mexico, pp 415–412

O'Connor K, Clark R, Whitlatch D, Dowding C (2001) Real-time monitoring of collapse along I-70 in Washington, Pennsylvania. Transp Res Record: J Transp Res Board 1772(1):32–39

Renner J, Lindgren E, Vessely M (2014) Using time-domain reflectometry to monitor subsurface void propagation. In: Proceedings of 65th highway geology symposium, Laramie, Wyoming, August

Trifu CI, Shumila V (2010) Microseismic monitoring of a controlled collapse in field II at Ocnele Mari, Romania. Pure Appl Geophys 167:27–42

Zhou W, Beck BF, Adams AL (2002) Effective electrode array in mapping karst hazards in electrical resistivity tomography. Environ Geol 42:922–928

3 Geophysical Techniques in Sinkhole Prediction and Monitoring

Identifying locations that are susceptible to sinkhole formation requires the ability to map the voids in the overburden soils and bedrock surface and any dissolution features within the bedrock. Geophysical methods offer a means to produce continuous profiles of the rock surface and possible locations for voids. Table 3.1 summarizes some geophysical methods that are applicable to sinkhole investigations. Most of these techniques are used for site characterizations, whereas a couple of them are gradually being used for monitoring and forecasting purposes. This chapter presents case studies to illustrate how ground penetrating radar (GPR) and earth resistivity imaging (ERI) have been used to predict and monitor sinkhole formation.

3.1 Application of Earth Resistivity Imaging to Sinkhole Prediction

3.1.1 Introduction to Earth Resistivity Imaging

Resistivity is a material property, and different materials, geological or otherwise, have different resistivity. The fundamental science behind the application of ERI is the empirical Archie's law, which is often expressed by:

$$\rho_0 = a\rho_w\varphi^{-m}$$

where ρ_0 is the bulk soil/rock resistivity; ρ_w is the water resistivity; a is an empirical constant ($0.6 < a < 1$); m is the cementation factor ($m = 1.3$ for unconsolidated and $m = 2.2$ for cemented or crystalline), and φ is the fractional porosity, a ratio between water volume and soil/rock volume.

According to Archie's law, electricity is conducted through interstitial water by ionic transport. Both electrical resistivity and fractional volume of water in the earth materials are the main factors affecting the bulk resistivity. Large fractional volumes of water with low resistivity tend to correspond to overall resistivity lows, whereas dried materials tend to correspond to resistivity highs. Although crystalline bedrock such as marble or phyllite may have a high resistivity value, the buck resistivity can be significantly decreased if the bedrock consists of a large fractional volume of water and the water resistivity is low. When the water resistivity remains constant across a site, it is the moisture content or fractional volume of water in the soil/bedrock that affects the resistivity values the most, not the soil/bedrock itself. Therefore, clay or water-filled raveling zones or voids tend to have much lower resistivity values, while air-filled cavities or competent rock tend to have higher resistivity values. Table 3.2 lists possible resistivity values for materials encountered at the site.

3.1.2 Site Background

The study site consists of a vehicle storage building, a salt storage building, an equipment storage building, and an office and shop building. The facility has been plagued by sinkhole/collapse problems with ten sinkholes recorded at the site. These sinkholes have been backfilled or remediated, and their characteristics are unknown. Stress fractures and cracks have developed in the exterior bricks on the northwest side of the office and shop building, in a wall separating the shop area from the office, around door and window frames in the office area adjacent to the shop, and on the floor in the vicinity of the freight elevator. To stabilize the shop and office building, a grouting program was implemented on the southern side of the building where the cracks were most noticeable. Approximately 840 cubic yards of grout were injected to a depth of 60 feet. Because of concerns over the ability to control the grouting pressure under the building footprint, the grouting operation was limited to 25 feet beyond the building.

Figure 3.1 shows the rock formations encountered by more than 50 test boreholes and 29 grout holes. The

© The Author(s), under exclusive license to Springer Nature Switzerland AG 2024
X. Jiang et al., *Monitoring and Early Warning Technologies on Karst Lands*, Advances in Karst Science, https://doi.org/10.1007/978-3-031-59045-0_3

Table 3.1 Commonly used geophysical methods for sinkhole investigation

Investigation purpose	Viable geophysical method
Overburden structure and thickness	GPR, seismic refraction, electrical resistivity imaging, seismic reflection
Depth to bedrock and bedrock undulation	Electrical resistivity imaging, Rayleigh wave seismic, seismic refraction, seismic reflection, ground penetrating radar, transient electromagnetic, audio frequency or controlled source audio frequency magnetotelluric
Soil cave, soft soil, collapse	Ground penetrating radar, seismic reflection, microgravity, Rayleigh surface wave, electrical resistivity imaging, radioactive method
Karst cave and fissure	Borehole geophysical logging, electrical resistivity imaging
Underground river and Karst reservoir leakage	Induced polarization, natural potential, borehole geophysical logging, electrical resistivity imaging, transient electromagnetic, audio frequency or controlled source audio frequency magnetotelluric, air-borne geophysics
Karst water flow velocity, flow direction, and discharge at springs	Induced polarization, natural potential, infrared thermography, air-borne geophysics
Paleochannel	Electrical resistivity imaging, seismic reflection, transient electromagnetic, audio frequency or controlled source audio frequency magnetotelluric
Weathered zone	Seismic refraction, electrical resistivity imaging, Rayleigh wave
Karst and non-karst contact and water rich zone	Electrical resistivity imaging, transient electromagnetic, audio frequency or controllable source audio frequency magnetotelluric, induced polarization, microdynamic
Fault, fractured zone, karst development zone, and groundwater preferential flow zone	Electrical resistivity imaging, seismic refraction, transient electromagnetic, controlled source audio frequency magnetotelluric method, radioactive geophysical, microdynamic, induced polarization, borehole hydrophysical logging
Rock mass loose zone, rock integrity, dynamic elastic mechanics parameters	Seismic refraction, acoustic logging
Morphology and filling characteristics of caves and crevices	Ultrasonic logging, sonar logging, borehole acoustic, or optical televiewer

Table 3.2 Possible resistivity values for materials at the site

Resistivity (ohm-m)	Materials
< 5	Possible interference from man-made objects or artifacts from too many iterations in data processing
5–50	Clay-dominated materials with high water content, saturated clay, clay-filled void, water-filled void
50–200	Moisture silt, fine sand, and weathered bedrock (phyllite or marble) below groundwater table
200–1000	Competent bedrock (phyllite or marble), dry silty materials, and possible small air-filled void above the water table
> 1000	Very competent rock (phyllite or marble), dry sandy materials, possible large air-filled void above the water table, or artifacts from too many iterations in the data processing

bedrock consists of inter-layered marble and phyllite. Either marble or phyllite or both could be encountered in one borehole. The marble/phyllite contacts indicate that they were inter-fingered. The data does not support a well-defined linear contact between these two formations. In general, between the two interpreted bedrock contacts lies a narrow marble strip that appears to underlie the middle part of the office and shop building. The orientation of the marble band is roughly northeast-southwest at the building but changes further away from the building. Superimposed on the bedrock map is a generalized contour map showing the inferred bedrock surface. The bedrock is overlain by unconsolidated clay, fine sand, and gravel with thickness varying from 20 to 70 feet. It is possible that rock floaters were encountered in some boreholes. The bedrock surface is irregular and characterized by pinnacles and cutters, which is typical in karst areas. It appears that bedrock lows are either in the marble or at the marble/phyllite contacts. The weathered phyllite or saprolite is friable and could often be penetrated by a split-spoon sampler. Disintegration of the saprolite occurred along foliation and joints. Although vugs were recorded in the phyllite, voids were mainly encountered in marbles or at the phyllite/marble contacts.

After the grout work, a gauging network was established to monitor the development of the cracks in the walls. The gauging data indicates that the cracks on the divider wall

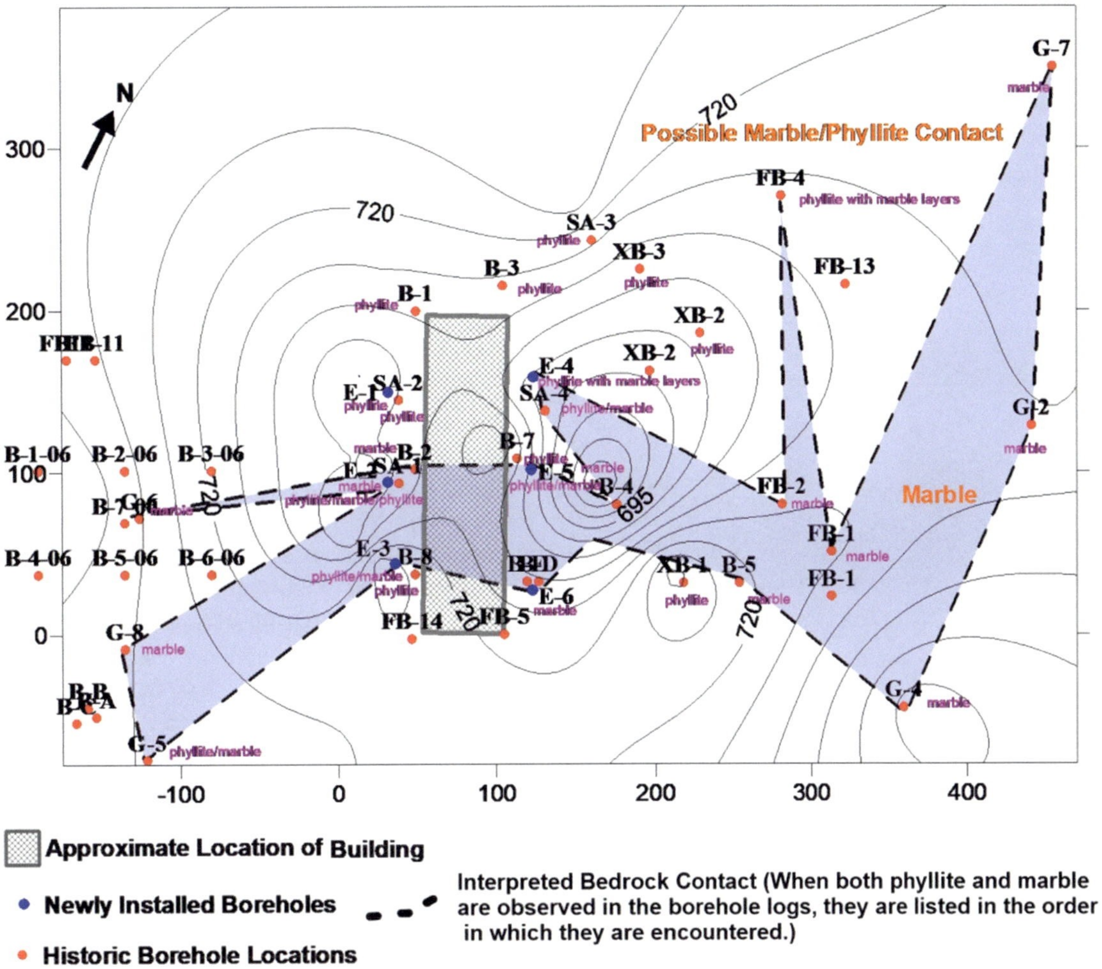

Fig. 3.1 Marble distribution interpreted from the borehole (cross-hole ERI was conducted in newly installed boreholes E-1 through E-6)

continue to expand. In addition, the divider wall began to show signs of tilting, and the exterior main door, adjacent to the divider, has been difficult to close due to the shifting of the frame. Because of continuous concerns over the integrity of the building and the safety of workers and visitors, sinkhole investigation and monitoring were conducted to help evaluate options for either mitigating the sinkhole problems or relocating the facility. The investigation and monitoring activities included geophysical surveys, lineament trace analysis, installation of monitoring wells, aquifer testing, groundwater level measurements, and groundwater turbidity monitoring. For the geophysical surveys, the ERI technique was applied on the ground surface and in boreholes for sinkhole evaluation.

3.1.3 Surface ERI

Figure 3.2 shows the data collection sequence for the dipole-dipole array in a two-dimensional resistivity imaging investigation. The spacing of electrodes was selected based on the desired depth of penetration, the required resolution, and the type of array. The electrode spacing and dipole separation were constant for each data layer and increased with each successive layer. Larger electrode spacing provides data from greater depths, but with lower resolution.

The measured apparent resistivity is a volume-averaged value affected by all the geologic layers through which the induced electric current flows. The data points are plotted in the middle of the four selected electrodes at a depth or distance proportional to their spread.

3.1.3.1 Data Collection

The 8-channel AGI SuperSting® was used for data collection in this investigation. An automatic selection of four electrodes was used for each measurement. Figure 3.3 shows the field setup of the instrument. The resistivity data were collected along six traverses (Fig. 3.4). Lines 1 and 3 are parallel to each other and oriented southeast-northeast. Lines 2 and 5 are parallel to each other and

Fig. 3.3 Surface ERI field setup

oriented approximately north-south. Lines 4 and 6 are parallel to each other and oriented northwest-southeast. The unit electrode spacing was 10 feet for all traverses. Fifty-six electrodes were used in all traverses except for Line 2, which used 45 electrodes because of space constraints. On paved areas, holes approximately 1 inch in diameter and 10–12 inches deep were drilled to facilitate the contacts between the electrodes and the ground.

3.1.3.2 Data Processing

Once the data were collected along each traverse, it was downloaded to a computer for editing and review. The recorded data were in the format of apparent resistivity. Apparent resistivity could be considered a weighted average of resistivity distribution between the four measuring electrodes. To create a resistivity image or volumetric resistivity distribution the apparent resistivity was inverted using the 2-D EarthImager® software. An example of the processed profiles is shown in Fig. 3.5.

Three graphs are presented representing a single ERI traverse. The top one shows the apparent resistivity distribution measured in the field; the middle graph shows the calculated apparent resistivity and the bottom graph shows the geologic model of true (model) resistivities that produced that apparent resistivity distribution. The small squares on the horizontal axis of each graph represent the electrodes. The difference between the top and the middle graphs is represented by the Root-Mean-Squared (RMS). The smaller the RMS is, the closer the model resistivity approaches the true resistivity. RMS values less than 15% are typically considered to be representative of a good model match. The RMS values of the six traverses range from 4 to 12, indicating that the model resistivity distributions are representative of the true resistivity distributions.

3.1.3.3 Data Interpretation

The true resistivity distribution along each traverse results from the measured apparent resistivity data which are average values representing a volume of geologic material. The volume-averaging inherent in resistivity methods tends to obscure small-scale irregularities in the geologic interfaces. The data are more generalized at greater depths. Like other geophysical techniques, the interpretation of resistivity anomalies is not unique, especially in areas with complex

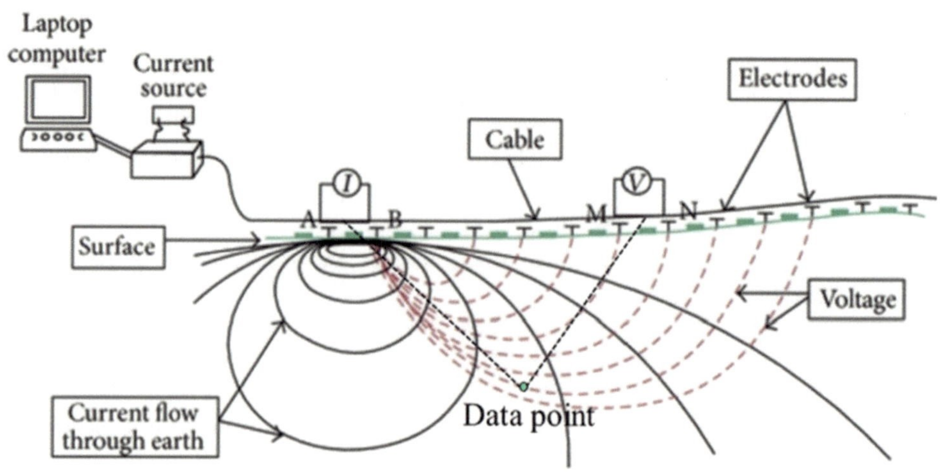

Fig. 3.2 Surface ERI data collection process in dipole-dipole array

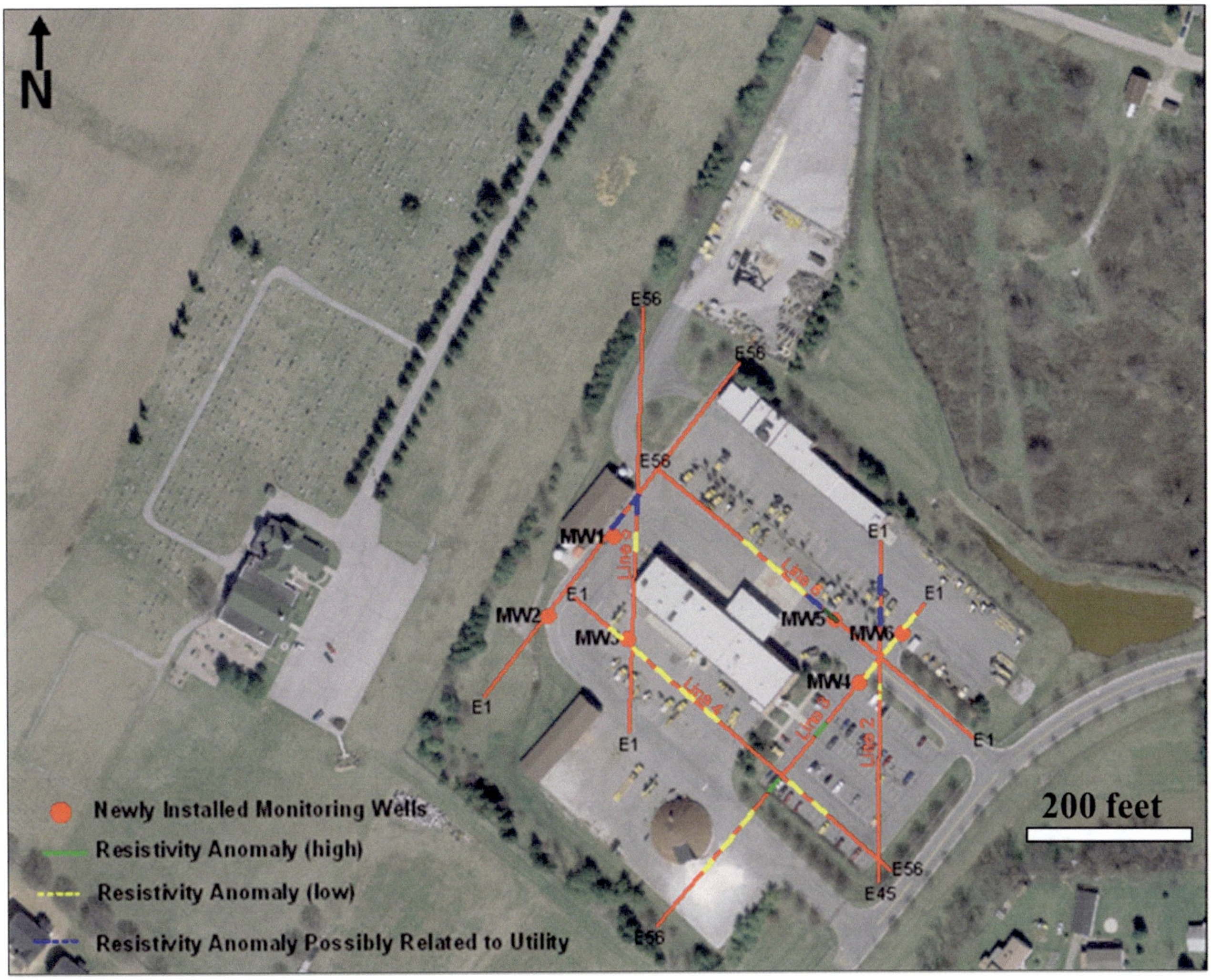

Fig. 3.4 ERI transects and anomaly interpretations

geology. Isolated, near-surface areas of high resistivity may be caused by less moisture silt, oxide nodules, or bedrock fragments. An apparent resistivity low may be caused by the presence of a clay- or water-filled fracture, cavity, or highly fractured but water-saturated rocks. A large void or multiple small voids may produce the same resistivity anomalies. Although the resistivity imaging is two-dimensional, significant variations in subsurface resistivity in a direction perpendicular to the survey can cause distortions in the lower sections of the model. This effect is most pronounced when the resistivity line is near a localized feature with different electrical properties than the material along the line. Some of the resistivity highs or lows shown on individual profiles may be caused by features that are laterally offset from the profile. As a result, when resistivity lines cross each other, the resistivity profiles should be similar but may not be the same at their intersections. The different results at the intersections may be indicative of the heterogeneous and anisotropic subsurface conditions.

Interpreting resistivity data requires understanding how the image is created. The algorithm used in the EarthImager® is a "smoothing algorithm" meaning that a target is delineated by a number of equi-resistivity lines. The lines describe the transition from one material to the other. However, if a target is too small, the equi-resistivity lines will not reach the target resistivity. Because the profiles are presented by resistivity contours, the size of an anomaly over a geologic feature may be larger than the actual size of the feature.

Figure 3.4 shows the interpreted resistivity anomalies. The anomalies in dashed yellow and dark green lines are likely related to geologic features, while the anomalies in dashed blue lines are likely caused by buried utilities. Interpreted resistivity anomalies are present around the building. The only area free of resistivity anomalies is to the northwest of the building. The geophysical interpretations were refined by the boring logs of the monitoring wells. On average, the bedrock appears to be buried much deeper northeast of the

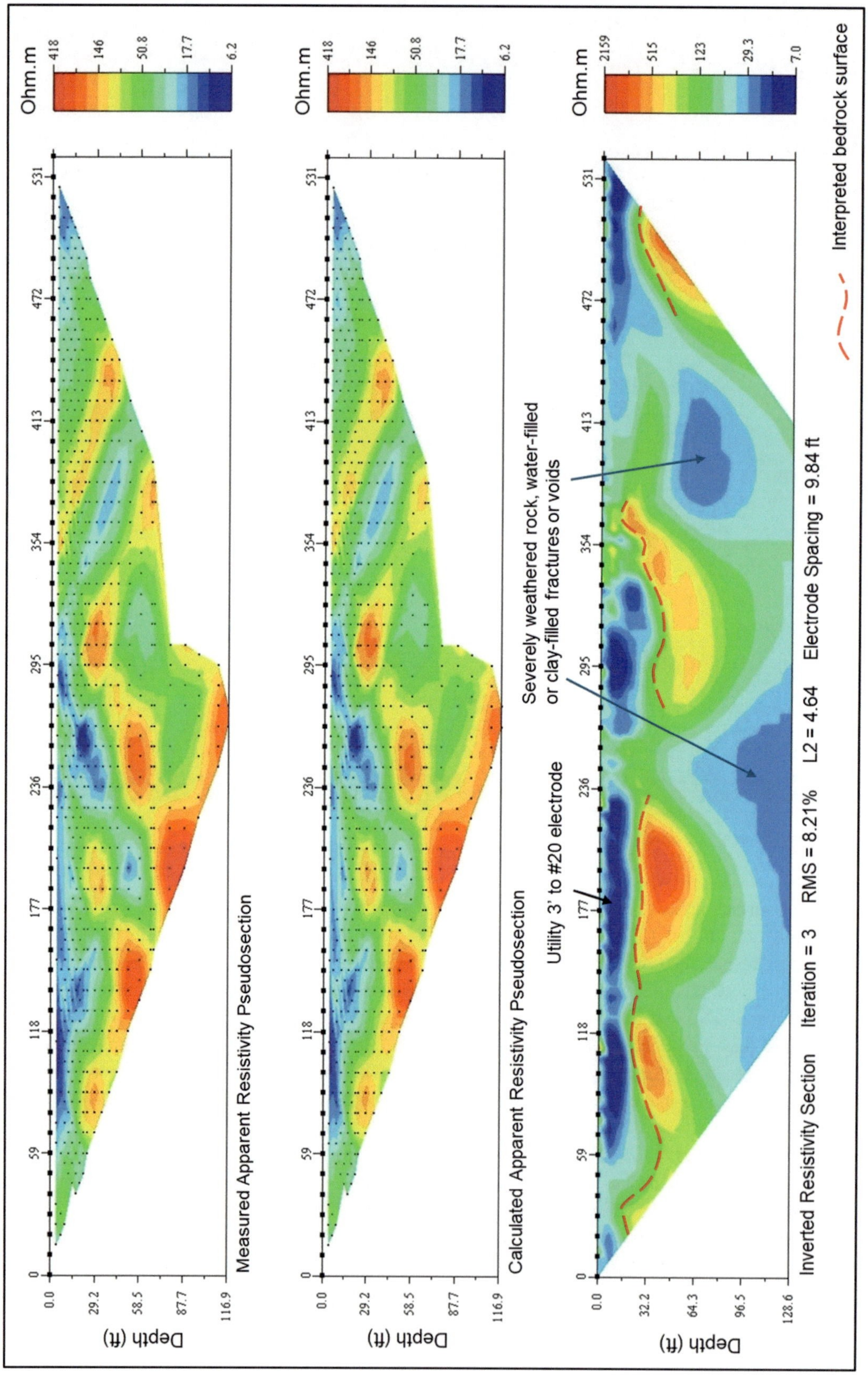

Fig. 3.5 ERI profile for Line 4

building. The resistivity survey indicates the existence of large dissolution features at depth (below 55 feet) such as clay- or water-filled fractures or voids, or severely weathered but saturated bedrock along Lines 3, 4, 5, and 6.

The anomaly on Line 3 was verified by monitoring wells MW4 and MW6. At MW4 a 2.5-foot void was encountered between 55.0 and 57.5 feet below ground surface (bgs), and phyllite was encountered at 58 feet bgs. At MW6 the bedrock was encountered at 51.5 feet bgs. Multiple water- or clay-filled voids with a cumulative size of 5 feet were encountered in the marble. The well logs at MW4 and MW6 indicate that the resistivity anomaly is caused by thick overburden and the presence of water- or clay-filled voids. The anomaly on Line 4 was verified by monitoring wells MW3. A 20-inch void was encountered between 40.0 and 41.75 feet bgs but no bedrock was intercepted at 46 feet.

3.1.4 Cross-Hole ERI

3.1.4.1 Borehole Drilling and Completion

Six boreholes were drilled near the shop and office building. The borehole locations E-1 through E-6 are shown in Fig. 3.6 as well as in Fig. 3.1. To acquire reliable data for the survey, the boreholes were located within 100 feet or less from one another. The boreholes located on the same side of the building were positioned approximately 50 feet apart from one another. Extending from one side of the building to the other, the length measured to be approximately 90 feet. The spacing between boreholes E-5 and E-6 required a slightly longer distance, approximately 80 feet, due to the roof extending over the fuel dispensing station. The borehole depth was designed to be drilled to 100 feet bgs. However, all the boreholes were drilled to 105 feet bgs to accommodate the possible accumulation of loose materials at the bottom.

Hollow stem auger methodology was utilized to advance the borings down to bedrock. Standard penetration tests were conducted to record blow counts. Split spoon samples were collected every 5 feet to characterize the overburdened soil. A drill rig equipped with 6.25-inch augers was used to advance the borings to refusal. Upon augur refusal, the drilling tool was exchanged for rock core drilling with the wire-line method. Potable water was used to lubricate the coring bit and flush cuttings to the surface. Continuous 5-foot rock cores were collected until a terminal depth was reached. The rock cores were logged, cataloged, and placed into dedicated boxes. The rock recovery and rock quality designation index was calculated for all the core runs. After the boreholes were completed and cleared of cutting, 0.020-inch slotted schedule 40 PVC pipes were installed in each borehole. The annular space was filled with clean #2 drillers' sand and extended from the bottom of the boring to near the ground surface. After the cross-hole geophysics was completed, all wells were abandoned and backfilled.

3.1.4.2 Instrument Setup

Figure 3.7a shows the general layout of electrodes for cross-hole mapping, where *A* denotes the positive current electrode, *B* is the negative current electrode, *M* is the positive potential electrode, and *N* is the negative potential electrode. When a constant injection current is applied between electrodes *A* and *B* across the two boreholes, the voltage measurements between *M* and *N* depend on the geophysical properties of materials between *A*, *B*, *M*, and *N*. By employing a string of multiple electrodes in each borehole and alternating the electrodes between current injection and voltage measurement, we produce a 2-D image of the geo-electrical properties between the boreholes. There are different ways to arrange the *A*, *B*, *M*, and *N* electrodes. Dipole-dipole configuration was used for data collection. Approximately 1200 data points were collected for each dataset. Fifteen sets of resistivity data were collected between the six boreholes.

The AGI SuperSting 8-channel resistivity instrument (Fig. 3.7b) was used for data collection. Twenty-eight electrodes with 3.28 feet (1 m) spacing were used in each borehole. Collection of reliable data requires that the electrodes be installed in full contact with the soils or rocks, so that transmitting electrodes can inject an electric current into the formation and potential electrodes can sense the electric field. Contact to the formation was provided by water in the hole. In the unsaturated zone, a small amount of water was injected into the borehole to facilitate the contact between the formation and electrodes. Slotted PVC casing (Fig. 3.7d) was used all the way. The passive cable and electrodes were then hung inside the slotted PVC casing.

3.1.4.3 Data Processing

Once the data was collected for each borehole pair, it was downloaded to a computer for editing and reviewing. The recorded data was in the format of apparent resistivity. Apparent resistivity could be considered a weighted average of resistivity distribution between the four measuring electrodes. To create a resistivity image or volumetric resistivity distribution the apparent resistivity was inverted using the 2-D EarthImager software. A 2-D profile was produced for each pair of boreholes. Because E-1, E-2, and E-3 lied in a straight line, the data between E-1 and E-2, E-2 and E-3, and E-1 and E-3 were combined to produce a single profile. Similarly, the data between E-4 and E-5, E-5 and E-6, and E-4 and E-6 was combined to produce a single profile. Two select processed profiles are presented in Fig. 3.8.

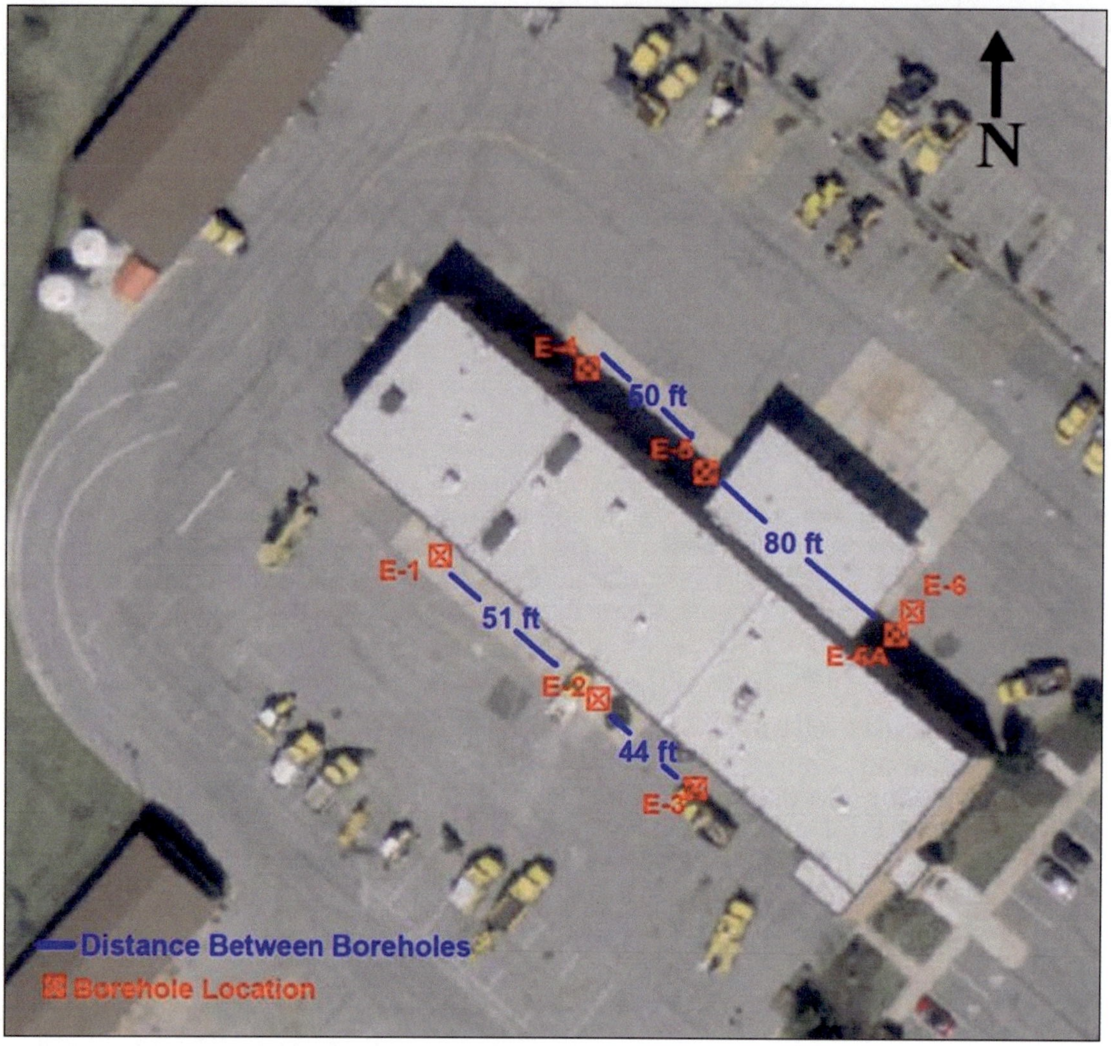

Fig. 3.6 Borehole locations for cross-hole resistivity imaging

3.1.4.4 Data Interpretation

When interpreting resistivity data, it is important to understand how the image is created. The algorithm used in the EarthImager is a "smoothing algorithm" meaning that a target is delineated by equi-resistivity lines. The lines describe the transition from one material to the other. However, if a target is too small, the equi-resistivity lines may not reach the resistivity value of the target. Because of the mixture of materials in the subsurface, there is no one-to-one correspondence between a resistivity value and a material. The following summarizes the significant findings as interpreted from the cross-hole resistivity survey:

- The inverted resistivity profiles show a wide variance of the subsurface conditions. Only two profiles: E-1—E-6 and E-2—E-6 indicate layered structures that dip west, while the layered structure is not obvious in all other profiles. This lack of layered structure may be caused by the fact that the rock formations at the site are metamorphic rather than sedimentary. The geologic processes have fractured, folded, and displaced the metamorphic formations.
- Isolated resistivity lows occur consistently in profiles E-1—E-4, E-1—E-5, E-2—E-4, E-2—E-5, and E-3—E-4. The resistivity lows have circular shapes and may indicate a possible existence of large voids underneath the building. The voids can be either water-filled or clay-filled. The voids appear to be below the top of bedrock, between 40 and 75 feet bgs on the northwest side of the investigation area. The voids appear to be in a northwest-southeast orientation on the northwest side of the investigation area. On the southeast side of the investigation area, the low-resistivity anomalies are not pronounced but it is possible that the voids are oriented in the west-east direction.
- Profile E-1—E-2—E-3 indicates a low-resistivity area around E-2, where a 10 feet void was encountered.

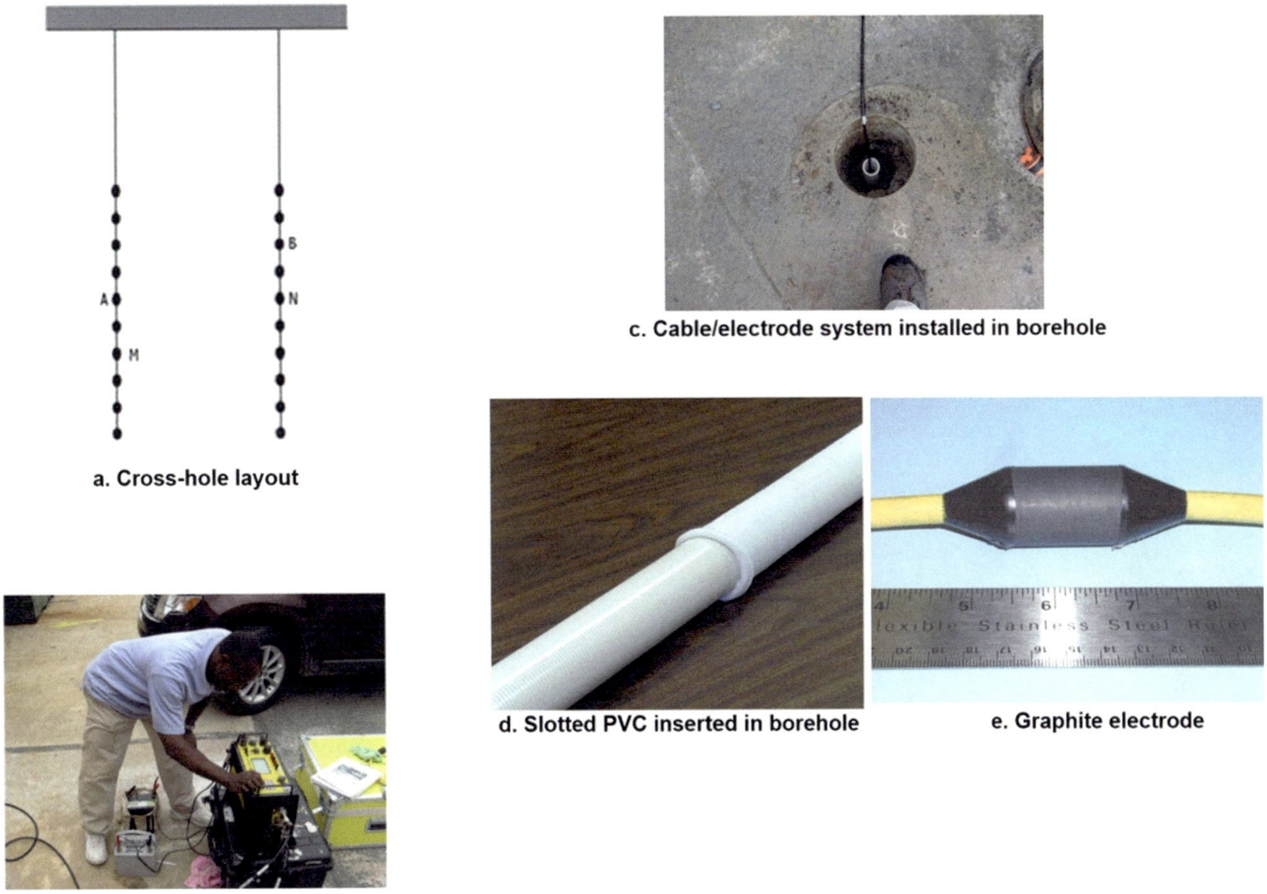

a. Cross-hole layout

b. SuperSting 8-channel instrument

c. Cable/electrode system installed in borehole

d. Slotted PVC inserted in borehole

e. Graphite electrode

Fig. 3.7 Cross-hole ERI setup

Between E-1 and E-2, the formations seem to be more resistive except for a low-resistivity area at approximately 60 feet bgs. A similar low-resistivity zone is also noticeable at the same depth between E-2 and E-3.

- Profile E-4—E-5—E-6 shows a lateral change of the subsurface conditions at E-5. Present between E-4 and E-5 is a low-resistivity zone between 30 and 55 feet bgs, which is probably overlain by competent rock, sandy material, or an air-filled void. The rock below 55 feet is relatively more resistive around E-5. The voids encountered at E-4 appear to have a limited lateral extent toward E-5. The resistivity distribution between E-5 and E-6 is approximately opposite to that between E-4 and E-5. The resistivity contrast along one profile is possibly caused by the change of the earth materials, which may result from a fault. The water levels measured at E-4, E-5, and E-6 suggest that E-5 is located at a local groundwater divide. This divide may be related to the geologic structure.
- The geologic feature that causes the resistivity to low at approximately 75 feet bgs between E-5 and E-6 may also cause the resistivity low at a similar depth on the profile between E-3 and E-5. However, this feature does not seem to be repeated on the profile between E-3 and E-6. The profile between E-3 and E-6 is relatively homogeneous except for a few small, isolated resistivity lows, mostly in the vicinity of E-6. One resistivity low is at approximately 40 feet bgs, where a void of 2.5 feet was encountered at E-6. The uniform but high resistivity values in the vicinity of E-3 suggest that the formations are more competent in the vicinity of E-3.

The cross-hole resistivity profiles show the resistivity lows and highs along cross-sections between borehole pairs at different depths. To better visualize the anomalous areas, the data was further processed to illustrate the resistivity distribution across the investigation area at discrete depths. Appendix D includes the plane view maps at 6 different depths: 10, 20, 30, 45, 60, and 72 feet.

In the cross-hole resistivity survey, the data was available only along the cross section between each borehole pair. As a result, the rest of the data in the plane view maps was extrapolated. Because the density of the actual data was small, and the majority of the data was extrapolated,

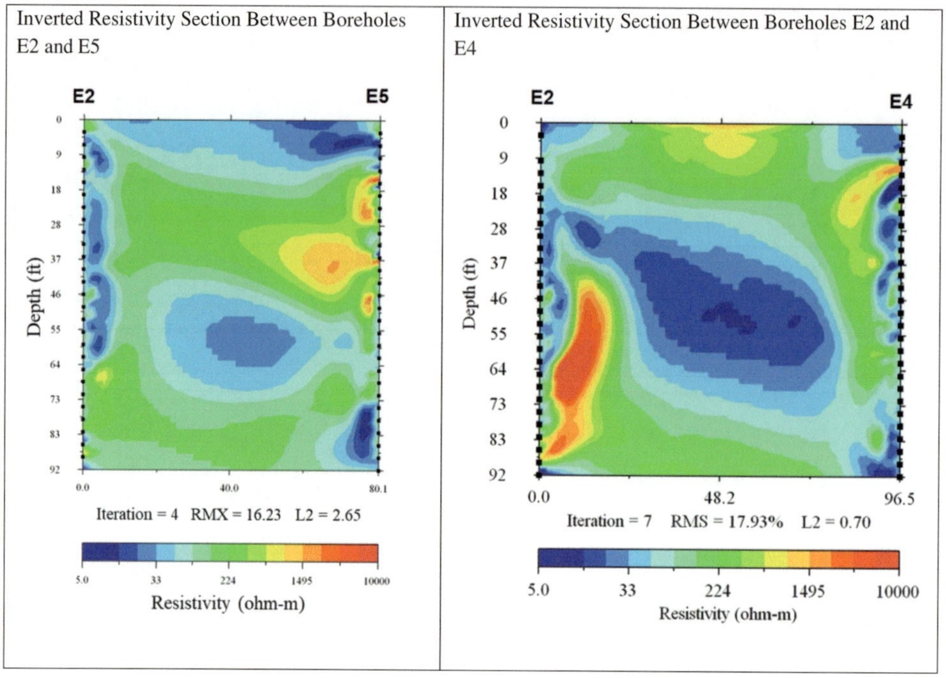

Fig. 3.8 Example of cross-hole resistivity images

these maps can only be considered schematic. However, the change with depth in the total area of the resistivity lows is still noticeable. The resistivity lows (< 50 Ω m) may correspond to clay- or water-filled voids in the bedrock or moisture or saturated clay in the overburden.

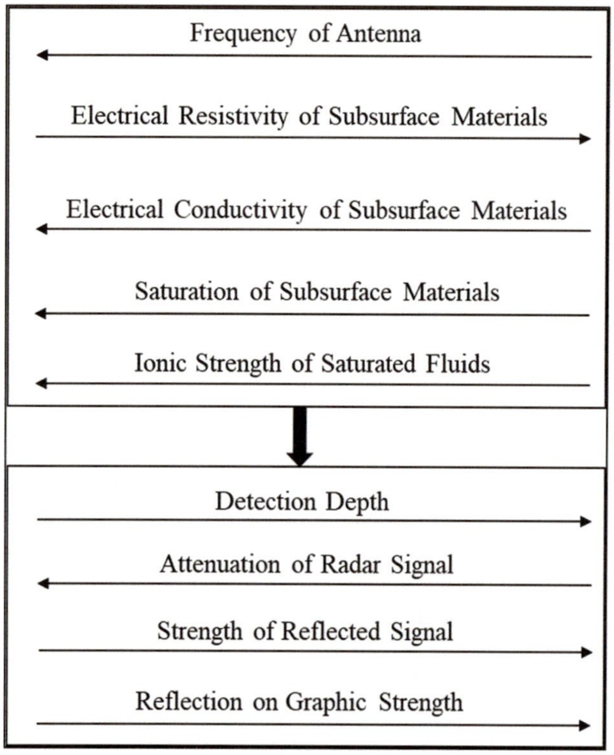

Fig. 3.9 GPR responses to formation resistivity, ionic strength of formation fluid, and signal frequency

- The plane maps at 60 feet and 72 feet bgs represent the resistivity distribution in the bedrock. The resistivity lows are more extensive at 60 feet bgs. At 72 feet bgs, the total area occupied by the resistivity lows decreases significantly.
- The plane maps at 30 and 45 feet bgs probably represent the transitional zone from overburden to bedrock. The distribution of resistivity lows does not overlap with each other. At 45 feet bgs, the largest resistivity low is located at E-4, whereas the largest resistivity low is shifted to E-6 at 30 feet bgs.
- The plane maps at 10 feet and 20 feet bgs represent the resistivity distribution in the overburden. At 20 feet bgs, the resistivity lows are limited to a narrow band in the middle of the investigation area, running diagonally from E-6 to E-1. However, at 10 feet bgs, the resistivity lows occupy a larger area and are oriented approximately in the west-east direction. The resistivity lows at 10 feet bgs may indicate the weak zones at the near surface, which can be directly related to the structural damage within the office and shop building.

3.1.5 Summary of Findings

The study site is an active facility and susceptible to sinkhole formation because the marble underlying the site is conductive to karst development. Of the 12 boreholes (six for monitoring well installation and six for cross-hole imaging) drilled during the investigation, marble was encountered in most of them. Historic and new boring data suggests a marble band underlying the office and shop building. The orientation of the marble band is roughly northeast-southwest at the building but changes further away from the building. The data collected do not support a well-defined linear phyllite/marble contact. Based on results from surface and cross-hole earth resistivity imaging, the area with sinkhole risk was delineated for continuous monitoring and remediation of the risk area.

3.1.5.1 Results from Cross-Hole ERI

Cross-hole resistivity imaging was conducted through six boreholes (E-1 through E-6) drilled outside the building for the imaging purpose. Sizable voids were encountered at E-2, E-4, and E-6. The voids are below the top of bedrock and below at least 40 feet. The voids at E-2 and E-6 seemed to be water-filled, while the voids at E-4 were partially clay-filled. The clay retrieved from E-4 was soft and saturated with water. The cross-hole resistivity survey indicates the possible existence of large dissolution features such as voids underneath the building. The voids appear to be in a northwest-southeast orientation on the northwest side of the investigation area. On the southeast side of the investigation area, low-resistivity anomalies are not as pronounced but it is possible that the voids are oriented in the east-west direction. On average, the voids appear to be between 40 and 75 feet bgs and below the top of competent bedrock on the northwest side of the investigation area. The interpreted possible voids are 3-dimensional and are at different depths on different resistivity profiles. The resistivity distributions in the plane view at discrete depths indicate that the voids are more extensive at 60 feet bgs. However, the building can be affected by subsurface erosion only when the soils at the near surface begin raveling. The resistivity map at 10 feet bgs shows the weak area that may be related to structural damage within the building. The weak area seems to roughly overlap with the marble band delineated from the boring data.

3.1.5.2 Results from Surface ERI

Surface resistivity imaging was conducted along six transects around the building. The resistivity survey indicates the possible existence of large dissolution features at depth (below 55 feet) such as clay- or water-filled fractures or voids, or severely weathered but saturated bedrock along Lines 3, 4, 5, and 6. On average, the bedrock appears to be buried much deeper northeast of the building. An area free of resistivity anomalies is located northwest of the building. The geophysical interpretations were corroborated with the boring logs of six monitoring wells. Voids up to 2.5 feet were encountered at MW3, MW4 and MW6. The groundwater level data collected in the six monitoring wells after their development indicates that the groundwater flows from west to east underneath the building. In comparison with the hydraulic gradient around WM4, MW5, and MW6, the wider spaced contours underneath the building may indicate a greater hydraulic conductivity of the aquifer. On the other hand, the closely spaced contours around MW4, MW5, and MW6 may indicate the presence of a hydraulic barrier, which prevents the groundwater from flowing further to the east.

3.2 Application of Ground Penetrating Radar to Sinkhole Prediction

3.2.1 Introduction to Ground Penetrating Radar

GPR is often used to obtain a high-resolution cross section of shallow lithologic contacts and subsurface features. Lithologic and hydrogeologic conditions that can be inferred from GPR surveys include sediment thickness, depth to the groundwater table, and breaches in confining beds. Recent technological developments of GPR methods have led to applications in various hydrogeologic settings to predict sinkhole development. The principal factors that affect the performance of the radar system are the electrical properties of the earth material that is being penetrated by the radar energy and the transmitting frequency of the system. Figure 3.9 shows the general GPR response to conductivity and resistivity, fluid properties of subsurface formations, and the transmitting frequency.

The use of high-frequency short-wavelength signal pulses results in greater energy attenuation and, therefore, shallower exploration depths with greater resolution of the signal. The use of low frequency, long-wavelength signal pulses increase exploration depths but reduces the resolution of the signal. Formations with high conductivities, such as clays and ionic contaminants, greatly attenuate radar energy and result in reduced exploration depth. On the other hand, materials with high resistivities, such as dry sand and nonionic contaminants (hydrocarbons), are in favor of GPR signal penetration and result in greater exploration depth.

The GPR equipment produces a graphic display that allows the user to interpret data while in the field. GPR equipment consists of components that are carried either manually or are towed behind vehicles or boats during a GPR traverse. Although dual-antenna arrays are available to minimize unwanted surface noise and better detect small vertical fractures in the subsurface, a single-antenna system is commonly preferred because it is less expensive and easier to maneuver.

GPR data were collected along traverses to predict sinkhole risk in the study area of central Florida. The GPR

Fig. 3.10 Functional operation of the GPR system in the study

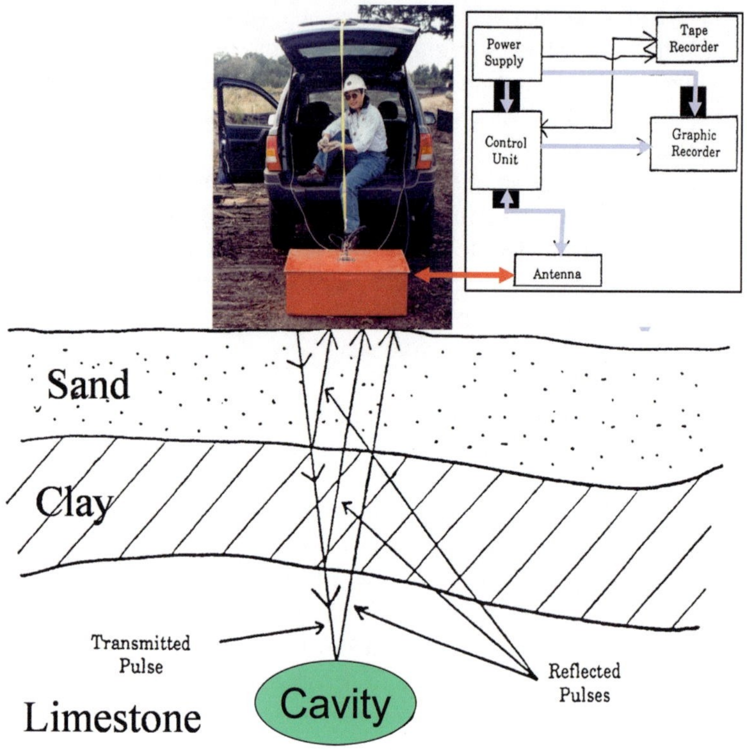

equipment used for this study was a SIR System III with a 100-MHz antenna manufactured by Geophysical Survey Systems, Inc. Radar pulses were generated at a surface antenna and then propagated into the ground. When they encounter a reflection interface, a portion of the radar energy is reflected to the surface and recorded again at the antenna. The wave nature of the reflected energy produces a series of positive and negative peaks (peaks and troughs) which are sensed by the receiver if they exceed a certain threshold. The antenna alternately radiates signal pulses and then receives the reflected signals. A schematic functional operation of the GPR system is shown in Fig. 3.10. The hydrogeological conditions are presented with a three-layer system and a cavity in the Floridan aquifer.

The equipment operated by transmitting short-duration pulses of electromagnetic energy into the shallow subsurface of the terrain being surveyed. Reflected pulses were received by the antenna, processed by the system components, and displayed on a graphic recorder as a continuous time-of-travel and traverse-distance profile that is stored as data on a magnetic-tape recorder. Normally, each reflection produces three detectable surges of alternate polarity which are plotted as three parallel bands on a strip chart, as shown in Fig. 3.11.

Land traverses of GPR equipment required a 2-person field crew, one to drive the vehicle and one to steer the antenna. The manually guided antenna array was towed behind the vehicle by a connecting cable. The optimum traverse speeds used for this study ranged between 1 and 2 mile/h. The truck odometer was used to estimate the traverse distance. Reference ticks were recorded on the recorder chart at 0.1-mile intervals and additional ticks were recorded to mark other surface features.

Knowledge of the local lithologic conditions, equipment settings, transmitting frequency, and antenna array response to the effects of exploration depth and graphic resolution is required to interpret GPR data. Interpretation of GPR data can be simple if stratigraphy is uniform and there are sharp contrasts in electrical properties at reflective surfaces. Interpretation of GPR data is more complicated, however, when lithostratigraphic conditions are complex or when sediments with gradual changes of electrical properties provide weak contrasts in the recorded data. The depth of a subsurface feature or reflector can be estimated if the propagation velocity of a radar pulse is known. The relation between propagation velocity and depth is expressed by the following equation:

$$D = V_m t/2 \tag{3.1}$$

where D is the depth to the reflector in feet; V_m is the propagation velocity in feet per nanosecond; and t is the two-way travel time, in nanoseconds. Table 3.3 presents the propagation velocities for commonly encountered formations. The velocities were theoretically obtained from the dielectric constants of selected materials. These calculated depths of subsurface features are for reference in practice and need to be verified by calibrating the GPR data with other field observations along the GPR traverse. Subsurface properties may be obtained by drilling test borings to obtain lithologic

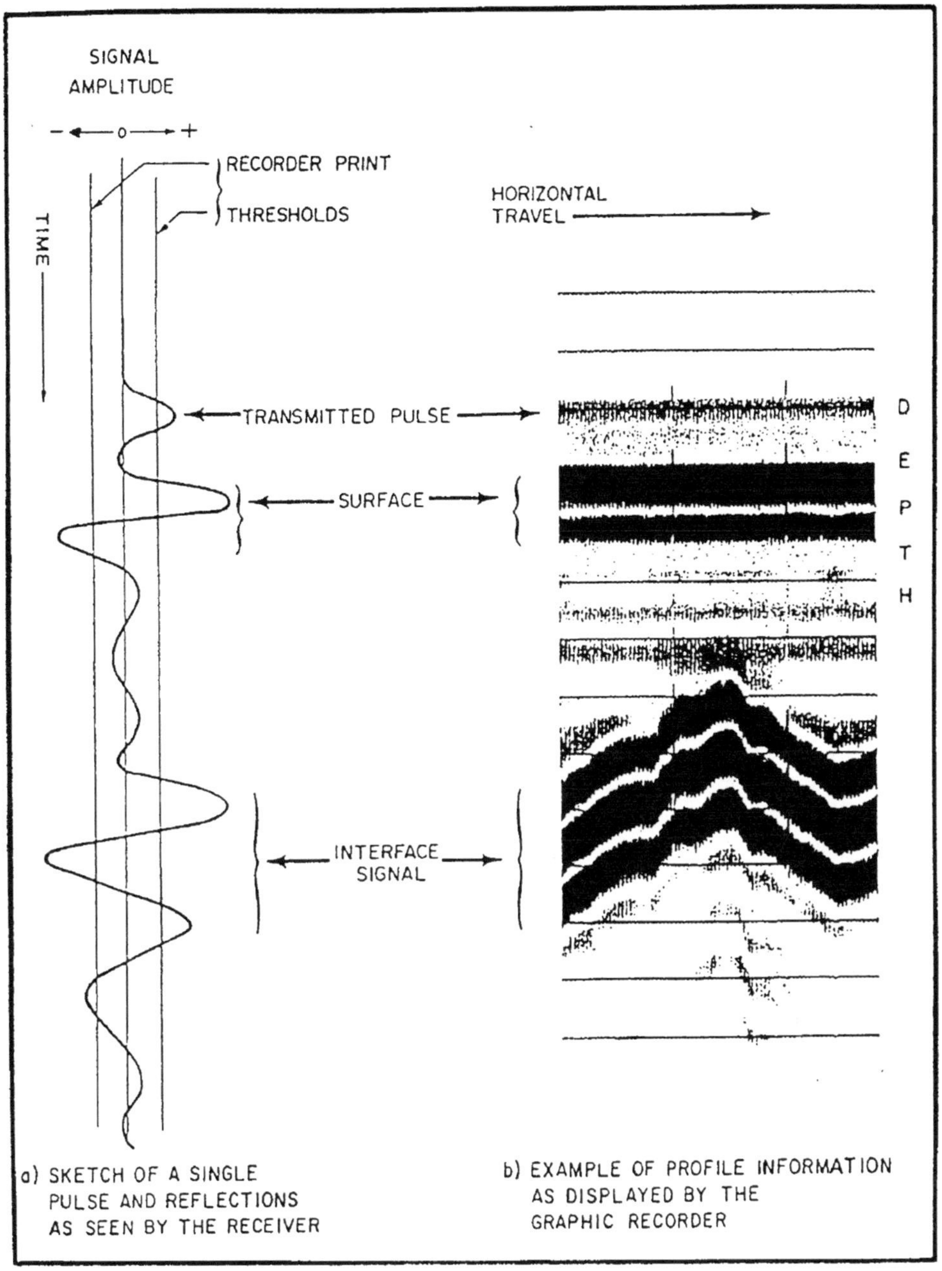

Fig. 3.11 Diagrammatic explanation of the reflected radar wave on the display chart

descriptions, geophysical logging, or measurements of the groundwater table.

3.2.2 Sinkhole Prediction from GPR Reflection Patterns in Central Florida

3.2.2.1 Conceptual Model of Sinkhole Formation in Central Florida

The study site is in central Florida. The major hydrostratigraphic units are the surficial aquifer system, the intermediate confining unit, and the Upper Floridan aquifer. The Floridan aquifer system is the principal aquifer in Florida and is among the most productive in the world. The high productivity of this aquifer is due to the development of secondary porosity caused by dissolution or karst processes. The solution of carbonate rocks by groundwater moving through them is the primary natural factor altering the land surface. Numerous bowl-shaped sinkholes that dot the landscape are a characteristic feature of the study area. An example is shown in Fig. 3.12.

The sinkhole formation processes are schematically shown in Fig. 3.13. The sinkhole collapses occur when there is a breach in the confining/perching strata, and water in the surficial aquifer percolates downward into the

Table 3.3 Radar-pulse velocity in selected materials in the study area (Barr 1993)

Material	Radar-pulse velocity V_m (ft/ns)	Material	Radar-pulse velocity V_m (ft/ns)
Air	0.98	Saturated mixed soil components	0.25–0.44
Freshwater/ Seawater	0.11	Saturated silt	0.31
Dry sand	0.40–0.49	Saturated clay	0.28–0.35
Saturated sand (freshwater)	0.18	Dry sandy on coastal land	0.31
Dry sand and mixed soil components	0.40–0.69	Dry limestone	0.37

Floridan aquifer causing subsurface erosion of the surface sediment and sinkhole development (Beck and Wilson 1988). The groundwater table in the surficial sand aquifer generally lies within a few feet of the land surface throughout the area. Groundwater in the Floridan aquifer is under artesian pressure and will rise in tightly cased wells to a level above the confining clay layer. The potentiometric surface of the Floridan aquifer generally stands 10–20 feet below the water table in the surficial aquifer. Downward leakage of water from the surficial aquifer to the Floridan aquifer, though retarded by the intervening layers of clay and sandy clay, is an important factor in local recharge to the Floridan aquifer. The movement of water from the surficial to the Floridan aquifer is greatly accelerated where the clay is absent or has been breached by subsurface erosion. In addition, the decline in the potentiometric surface of the Floridan aquifer can also trigger collapses.

Sinkhole collapse constitutes a significant geologic hazard in karst because of its inherent suddenness. Calls for an effective predictive method have become increasingly a necessity for sinkhole risk assessment and management. A legitimate karst risk assessment in central Florida includes: (1) detecting areas of down-warped deformation in the mantling sediment and (2) determining whether the groundwater is percolating downward to further undermine the stability of the site. As a relatively inexpensive, nonintrusive geophysical method, the GPR technique has been proven to be useful in detecting subsurface sediment deformation features indicative of sinkhole development, even in areas where surface depressions were not visible.

3.2.2.2 GPR Reflection Patterns of Sinkhole Development

Reflection patterns are useful for the identification of sinkhole development (Barr 1993). The definition of the reflection pattern was based on the interpretation of irregularities in the ubiquitous sand/clay boundary and depressions of the shallow groundwater level in the GPR images. Figure 3.14 shows a GPR profile with interpretation of the hydrogeologically significant features.

The groundwater table and the interface between sand and clay materials shown on the GPR profile have been verified by field-test observations of water levels and by lithologic descriptions of well cuttings or samples from test-hole borings. The GPR profile indicates downwarping or abrupt displacement of shallow strata, overlain by a depressed groundwater table. A breach in the clay layer is obvious and may provide a significant hydraulic connection between surface waters, shallow aquifers, and the underlying Floridan aquifer. Parallel reflections or "transparent" return infers a stratigraphy of sand and clays. The depth of a subsurface reflector was estimated from Eq. (3.1). Although there is no surface depression visible, the GPR

Fig. 3.12 A bowl-shaped sinkhole in central Florida

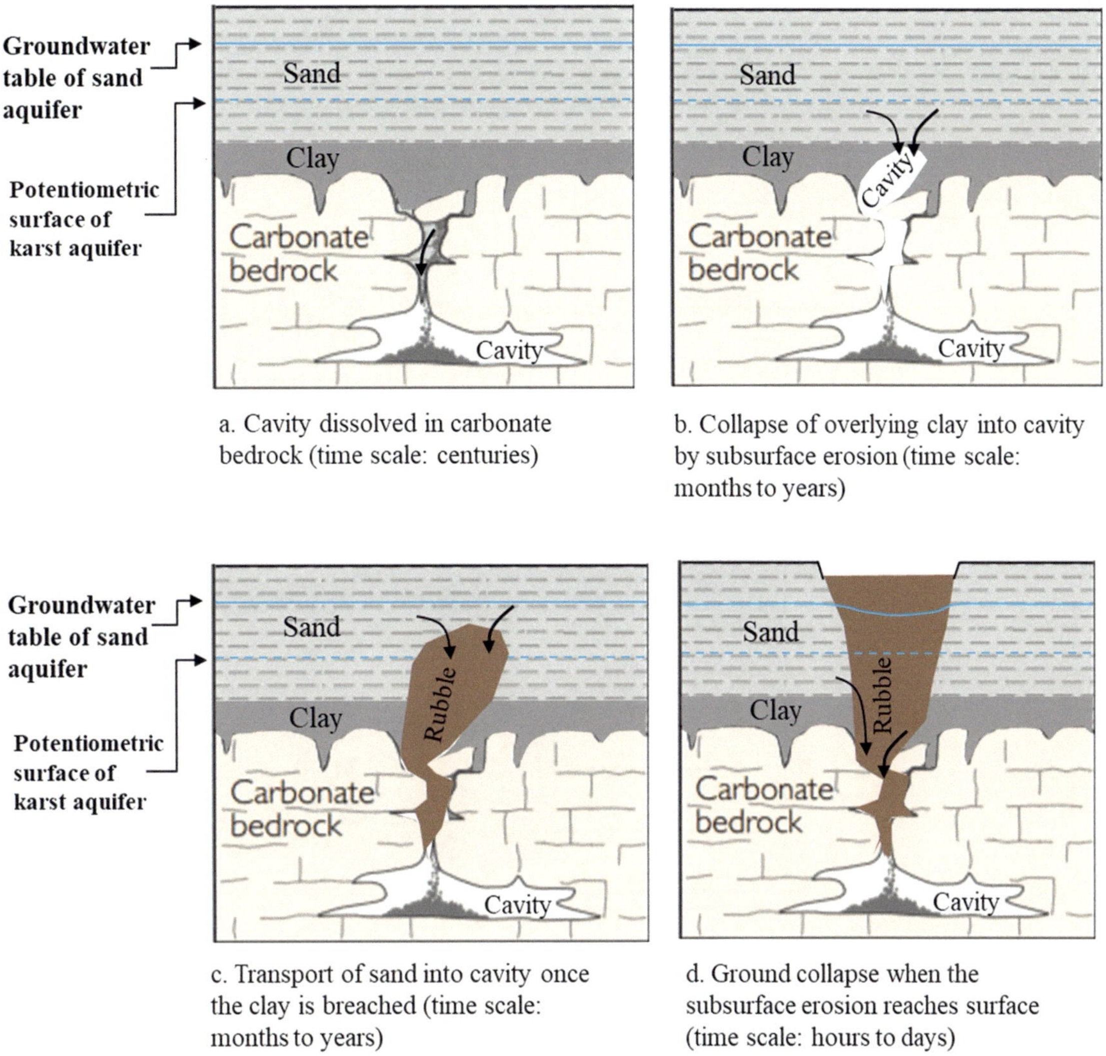

Fig. 3.13 Conceptual site model of sinkhole formation in the study area

profile suggests that sinkhole development is active in such an area. Hydrogeologic conditions and lithology can be inferred from the interpretation of GPR data. Such sinkholes are of most concern because the leaking of shallow groundwater downward into the limestone also facilitates the erosion of the shallow sediment downward into dissolved voids, which is the main mechanism forming cover-collapse or cover-collapse sinkholes, as discussed in the conceptual site model. In addition to periodic GPR surveys to monitor the sinkhole development, engineering measures shall be conducted to ensure ground stability if any infrastructures are constructed.

Figure 3.15 shows another type of GPR profile in the study area. The overlying strata have been disturbed but there is no depression in the groundwater table. This suggests that the clay layer has not been breached. The disturbed reflectors may indicate a zone of collapsed sediments, whereas the collapse is not hydraulically active and is likely in metastable equilibrium. The bottom of this kind of sinkhole may be temporarily blocked by clayey materials. However, this sinkhole may be rejuvenated when the current condition is disturbed. In its present condition, it may nevertheless be prone to slow collapse when loaded.

3.2.3 Sinkhole Prediction from GPR Reflection Patterns in Guilin, China

3.2.3.1 Conceptual Model of Sinkhole Formation in Guilin

The study site is for highway construction in Guilin, China. The area is covered by Quaternary brownish residual red clay layers. The geotechnical properties of the clay are heterogeneous. The soil layer is soft and plastic in bedrock surface lows, often referred to as karst cutters. The

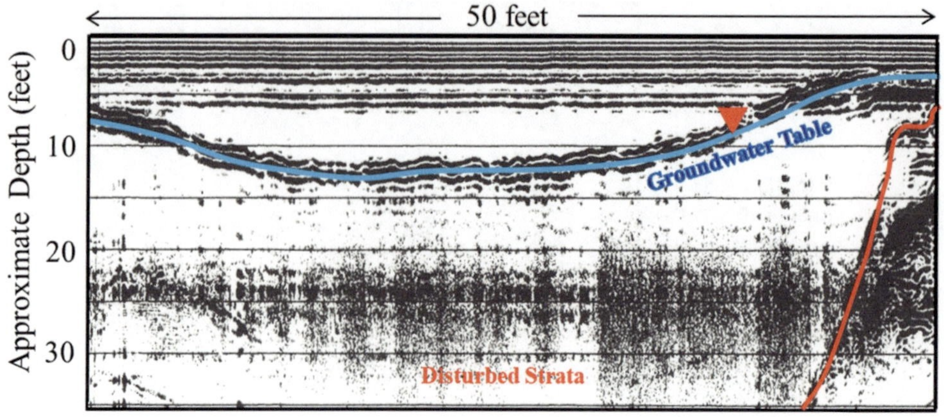

Fig. 3.14 Unmigrated GPR profile with interpretations for a hydraulically active sinkhole

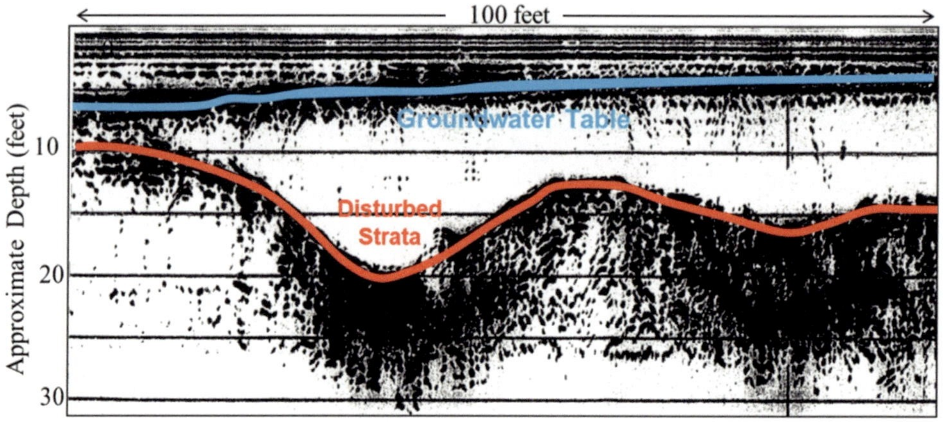

Fig. 3.15 Unmigrated GPR profile with interpretations for a non-hydraulically active sinkhole

thickness of the overburden is less than 5 m in most locations except for the karst cutters where the thickness can be more than 10 m. The clay layer is underlain by Middle-Upper Devonian limestone and dolomite and the Lower Carboniferous limestone. The carbonate bedrocks are outcropped occasionally on both sides of the highway. Because conditions such as carbonate bedrock, geological structures, and climate were in favor of karst development, typical karst features are large karst caves, karst windows, underground rivers, and sinkholes. The size of the collapse ranges from 1 to 2 m, and the depth varies from 0.5 to 4 m. A couple of sinkhole examples are shown in Fig. 3.16.

The study area consists of a hydraulically connected Quaternary porous-medium aquifer and a karst aquifer. The groundwater table is shallow and responsive to precipitation and pumping activities. The average depth of groundwater ranges from 2 to 3 m, while it rises quickly to the surface under intensive rains to form seasonal wetlands. The sinkhole formation processes are schematically shown in Fig. 3.17. Cavities within the bedrock and solution-enlarged fractures or cutters on the bedrock surface are present prior to the highway construction. As shown in Fig. 3.18, soil cavities started in areas where solution-enlarged fractures were present on the bedrock surface. The fact that the groundwater table fluctuated through the bedrock surface also favored cavity growth. The highway construction activities, including surface water drainage modification, water pumping from excavated pits, and heavy equipment uses, destroyed the natural equilibrium and intensified the subsurface erosion processes. The intensified erosion processes accelerated the cavity growth and triggered frequent sinkhole occurrences.

The suddenness of the sinkhole occurrence caused damage to the land, delay in the construction progress, and an increase in construction cost. On several occasions, collapses occurred underneath construction vehicles, causing vehicles and drivers to fall into them. Application of the GPR became necessary to predict sinkhole occurrence so that proactive sinkhole mitigation measures can be implemented. A legitimate karst risk assessment along the highway focused on the detection of areas with sizable cavities in the mantling sediment and monitoring their growth periodically.

3.2.3.2 GPR Reflection Patterns over Soil Cavities

Figure 3.19 presents the field setup of the survey. The GPR equipment used for this study was a SIR 10A with a 100-MHz antenna manufactured by Geophysical Survey

Fig. 3.16 Cover-collapse sinkholes in study area

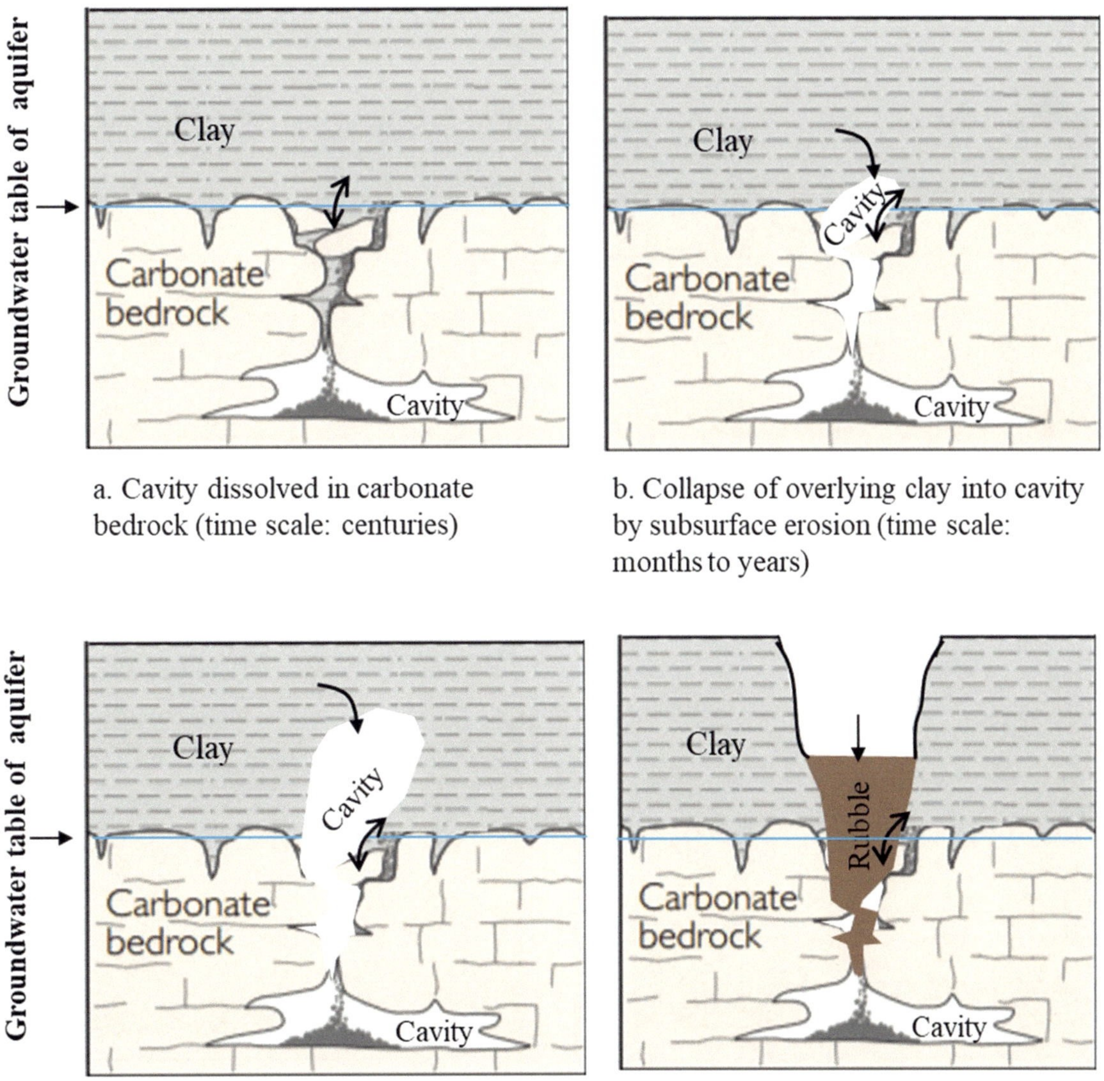

a. Cavity dissolved in carbonate bedrock (time scale: centuries)

b. Collapse of overlying clay into cavity by subsurface erosion (time scale: months to years)

c. Cavity grows as more clay is eroded (time scale: hours to days)

d. Ground collapse when the subsurface erosion reaches surface (time scale: hours to days)

Fig. 3.17 Conceptual site model of sinkhole formation in the study area

Fig. 3.18 Relationship between soil cavity and underlying karst features on the bedrock surface

Fig. 3.19 Functional operation of the GPR system in the study (left: manually pulling antenna; right: data collection system)

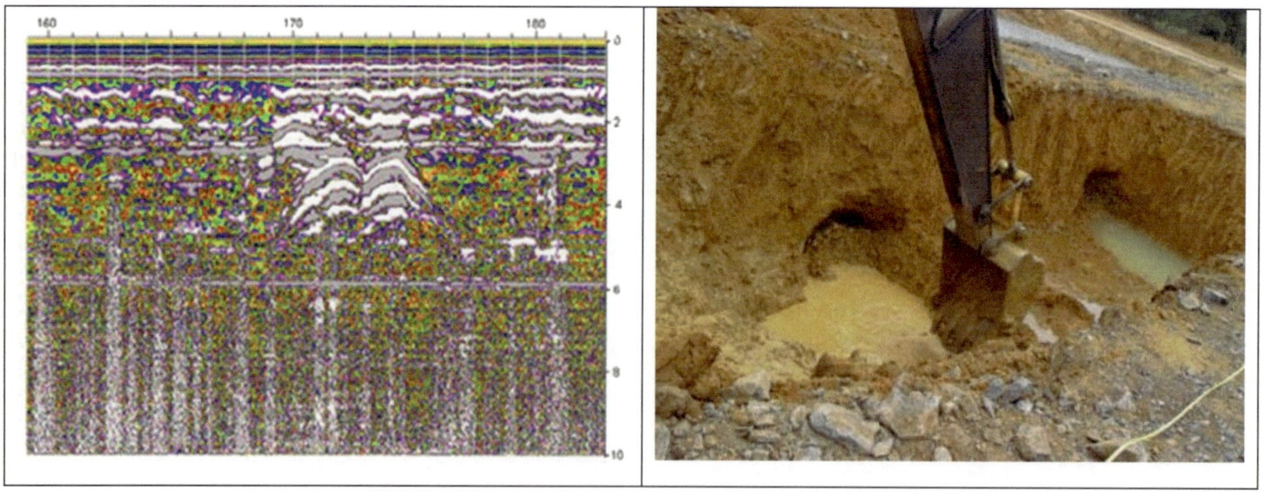

Fig. 3.20 GPR anomalies and soil cavities exposed from Test Pit I

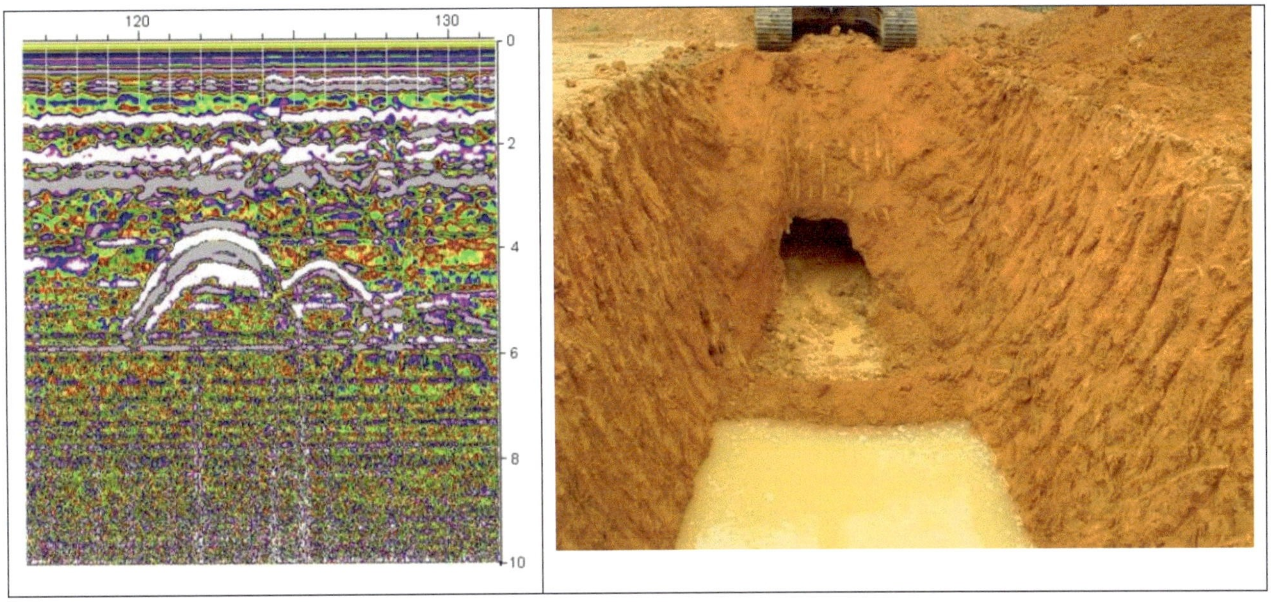

Fig. 3.21 GPR anomaly and soil cavity exposed from Test Pit II

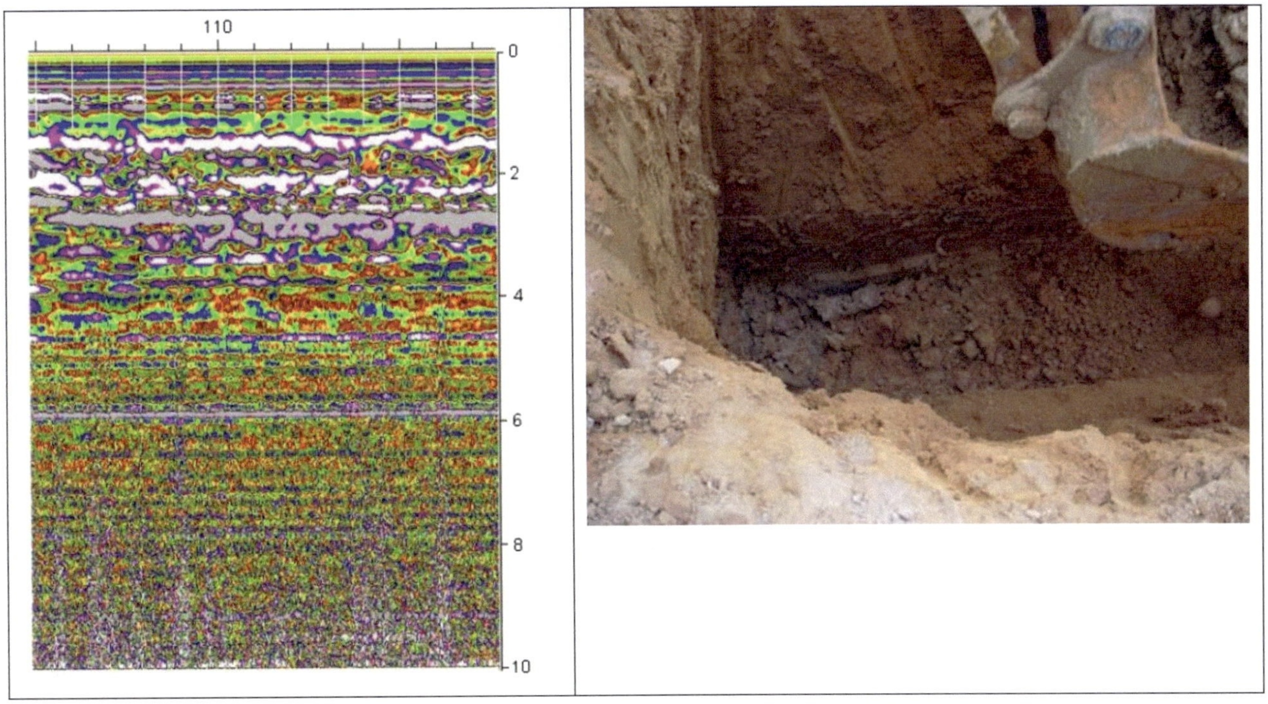

Fig. 3.22 Non-cavity GPR profile and results from Test Pit III

Systems, Inc. The survey was conducted along 12 traverses over a section of highway from K23 + 050 to K23 + 190. Continuous measurements were made with a marker at every 1 m. Figures 3.20 and 3.21 show typical GPR reflection patterns over soil cavities and results from test pits. For comparison purposes, Figure 3.22 shows a GPR reflection pattern where soil cavities are not present.

The GPR surveys identified 33 soil cavities that are distributed in Fig. 3.23. These cavities were either verified by testing pits or treated through grouting or compaction. Three test pits were excavated and are shown in Fig. 3.23. Based on observations from the testing pits, the soil cavities are located where karst features are obvious on the bedrock surface. The largest soil cavity had a diameter of

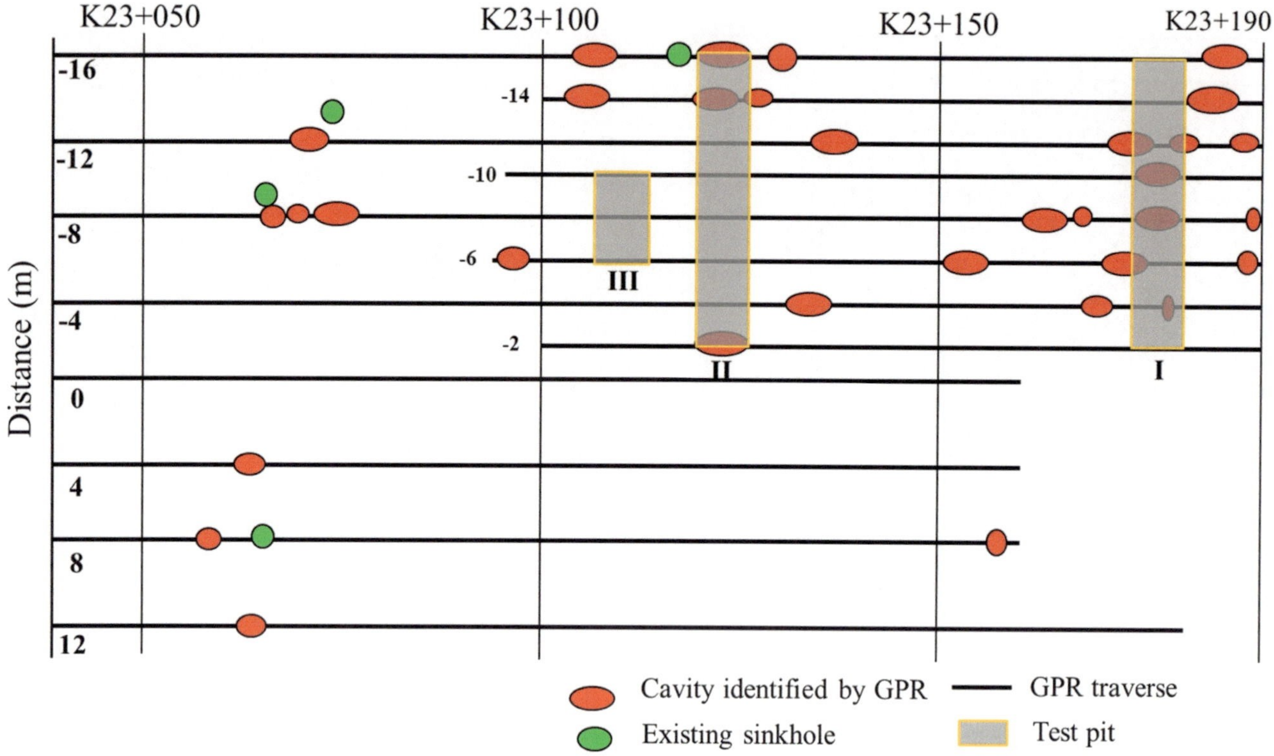

Fig. 3.23 Soil cavities identified from GPR reflection patterns

Fig. 3.24 Comparison of GPR profiles prior to (left) and after soil cavity mitigation (right)

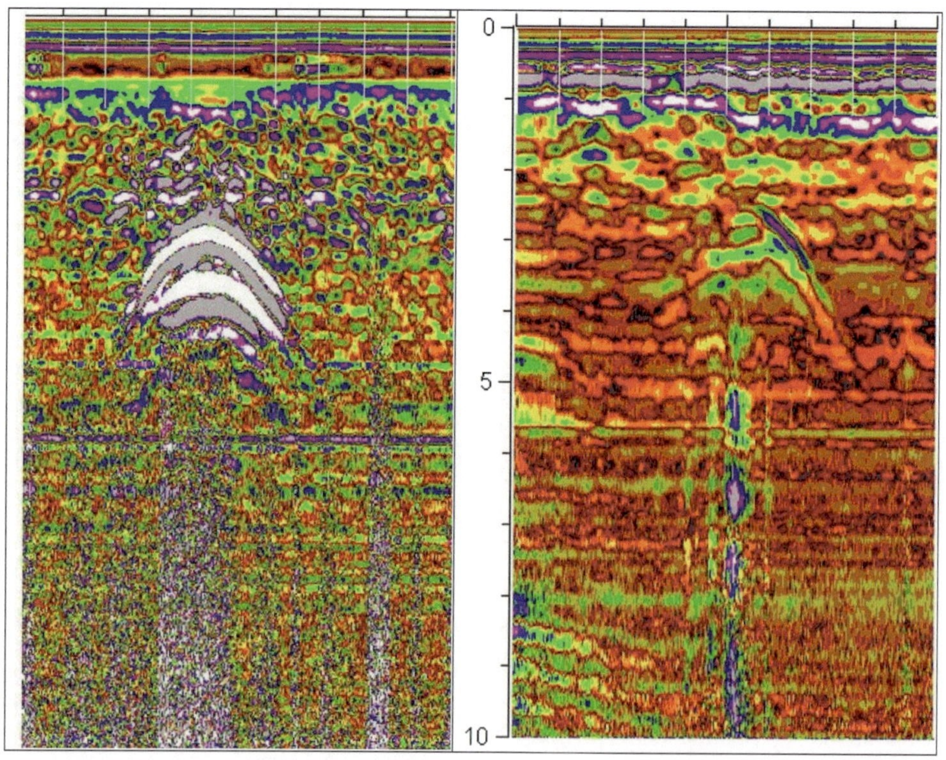

4 m. Loose sediments were accumulated in the lower parts of the soil cavities, whereas the upper parts were open with heights up to 1 m.

Some of the identified soil cavities were mitigated with grouting and compaction. The GPR profiling was used again after the mitigation to verify the effectiveness of soil cavity mitigation methods. Figure 3.24 compares the GPR reflection patterns before and after the soil cavity mitigation. Based on the results from our investigations, GPR has been proven to be an effective technology for sinkhole prediction and monitoring in the study area.

References

Barr GL (1993) Application of ground-penetrating radar methods in determining hydrogeologic conditions in a karst area, west-central Florida, vol 92, no 4141. In: US Geological Survey Water Resources Investigations Report, 26p

Beck BF, Wilson WL (1988) Interpretation of ground-penetrating radar profiles in karst terrane. In: Proceedings of the national water well association second conference on environmental problems in karst terrane and their solutions, Nashville, Tennessee, November 16–18. pp 347–367

4 Monitoring and Early Warning of Karst Collapses Using Brillouin Optical Time Domain Reflectometer

4.1 Basic Principle of BOTDR Technique

Brillouin Optical Time Domain Reflectometer (BOTDR) utilizes Brillouin spectroscopy and optical time-domain reflectometer to measure strain generated in optical fibers as distributed in the longitudinal direction (Guan et al. 2015). Figure 4.1 shows the general working principle of BOTDR monitoring. When a pulse of light travels down an optical fiber, the majority of light travels through but a small fraction is scattered back at every location of the fiber. The frequency of this back-scattered light is shifted from the original input frequency by an amount linearly proportional to the temperature and strain applied at the scattering location. By resolving the back-scattered signal in time and frequency, a complete strain profile along the full length of the fiber can be obtained. There is typically a linear relation between the frequency shift change of Brillouin scattering light in optical fiber and the axial strain received by the optical fiber, and the obtainment of the axial strain of the optical fiber. The theory behind the technology is presented by Horiguchi et al. (1994).

A particular advantage of optical fiber technology comes from the low propagation losses that can be obtained with a single-mode optical fiber. Strains can be measured along the full length (up to 80 km) of a suitably installed optical fiber by attaching a BOTDR analyzer to one end. The fiber can be a simple optical fiber that has an external diameter of 0.9 mm with a single optical fiber placed in the middle of a fiber cable reinforced by steel wires and Kevlar yarns. In either case, the plastic coating and the inner glass core are fixed together so that the strain applied externally is transferred from the coating to the inner core. The technical specifications of the BOTDR analyzer can vary from manufacturer. Table 4.1 lists the major performance indexes of the AQ8603 optical strain analysis meter launched by the Japanese Amtium Company.

Other advantages of the BOTDR technique are as follows.

- Withstands extreme temperatures.
- Impervious to electromagnetic interference.
- No oxidation or problems with water.
- Measures strains in the range of 0.01–1%.
- Optical fiber serves as both a sensing element and an information transmission channel, which will help realize remote monitoring and the creation of alert systems.

4.2 Feasibility Study of BOTDR on Sinkhole Prediction

Figure 4.2a shows the sandbox model of 2 m wide, 2 m long, and 3 m high. Optical fibers were laid in different soil layers at various depths. The strain changes in the fiber were measured as the soil structure was damaged, deformed, and collapsed. Figure 4.2b shows the monitoring results.

The relationship between soil displacement and BOTDR strains can be seen from the model test. The position of peak strain in the optical fibers corresponds to the area of disturbance and cavity growth in the soil. The changes in optical fiber strain in different soil layers indicate the vertical boundaries of the area of disturbed soil. The time series of optical fiber strain shows the ongoing formation of areas of soil disturbance. The model tests show that there is a good relationship between the strain in the optical fiber and the soil deflection when a sinkhole opens (Jiang et al. 2006). The modeling results indicate that BOTDR is a viable method for monitoring and predicting sinkhole collapse or subsidence, especially along linear infrastructure construction projects, such as highways and railways. If used properly, the position and time of sinkhole opening can be forecast by this method.

© The Author(s), under exclusive license to Springer Nature Switzerland AG 2024
X. Jiang et al., *Monitoring and Early Warning Technologies on Karst Lands*, Advances in Karst Science, https://doi.org/10.1007/978-3-031-59045-0_4

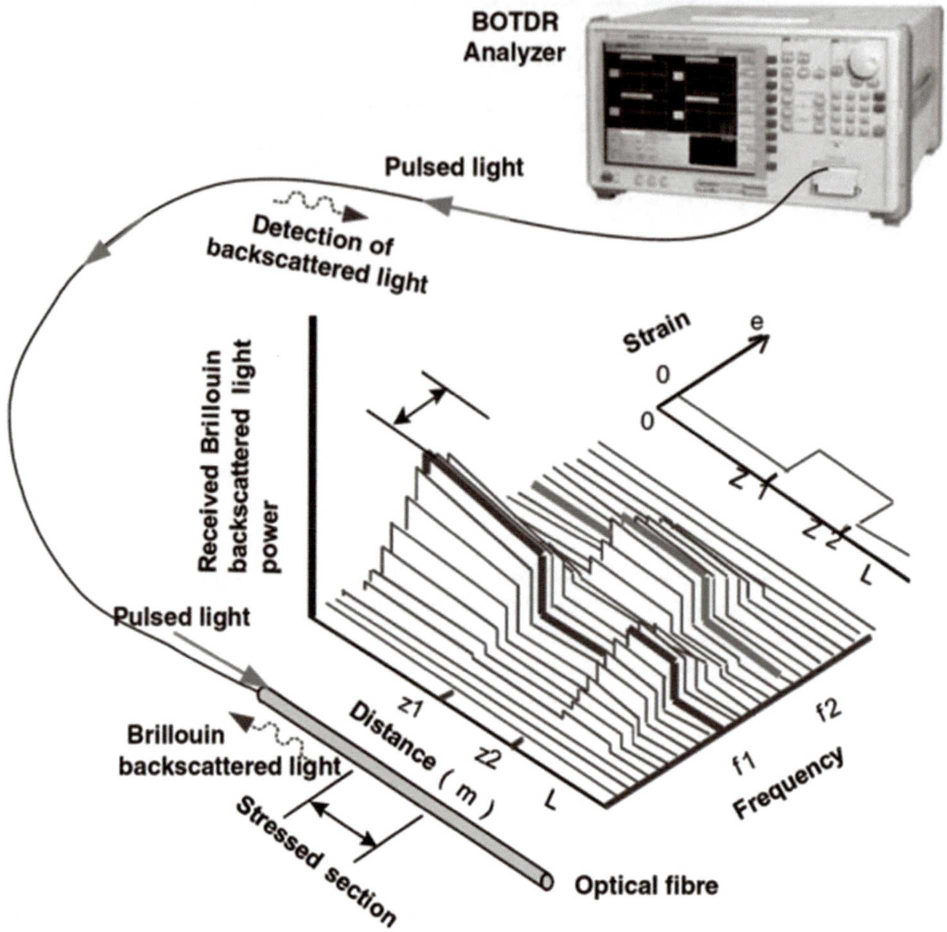

Fig. 4.1 Working principle of BOTDR system

Table 4.1 AQ8603(BOTDR) technical index

Technical index		Optional parameter				
Measured distance (km)		1, 2, 5, 10, 20, 40, 80				
Pulse width (ns)		10	20	50	100	200
Dynamic range (dB)	±0.004%(2 s)	2	6	10	13	15
	±0.003%(2 s)	–	–	8	11	13
Length resolution (m)		1	2	5	11	22
Strain test accuracy		±0.004%(2 s) (± 0.01%)		±0.003%(2 s) (± 0.005%)		
Strain test repeatability (%)		<0.04	<0.02			

4.3 Application of BOTDR Along a Highway in Guilin, China

4.3.1 Site Background

The highway from Guilin to Yangshuo stretches 66 km and encompasses well-developed karst areas. Karst features include karst towers and peaks as well as underground rivers. The groundwater has close hydraulic connections to surface water. The bedrock is mainly Devonian-age dolomitic limestone of the Guilin Group and overlain with the unconsolidated Quaternary soil. Soil composition is heterogeneous with alluvial red clay and usually less than 5 m thick. The depth of groundwater varies from 2 to 3 m bgs, which suggests that the groundwater is usually above the bedrock surface. The soil layer in the saturated zone is soft and plastic. Soil cavities and sinkholes were encountered during the highway construction. Fourteen cavities were found over a stretch of 100 m. Examples of soil cavity and sinkhole are shown in Fig. 4.3.

Although engineering treatment was used in the construction period, new soil cavities develop and grow over time because of the hydrodynamic conditions. Diverting stormwater runoff into existing sinkholes modifies local

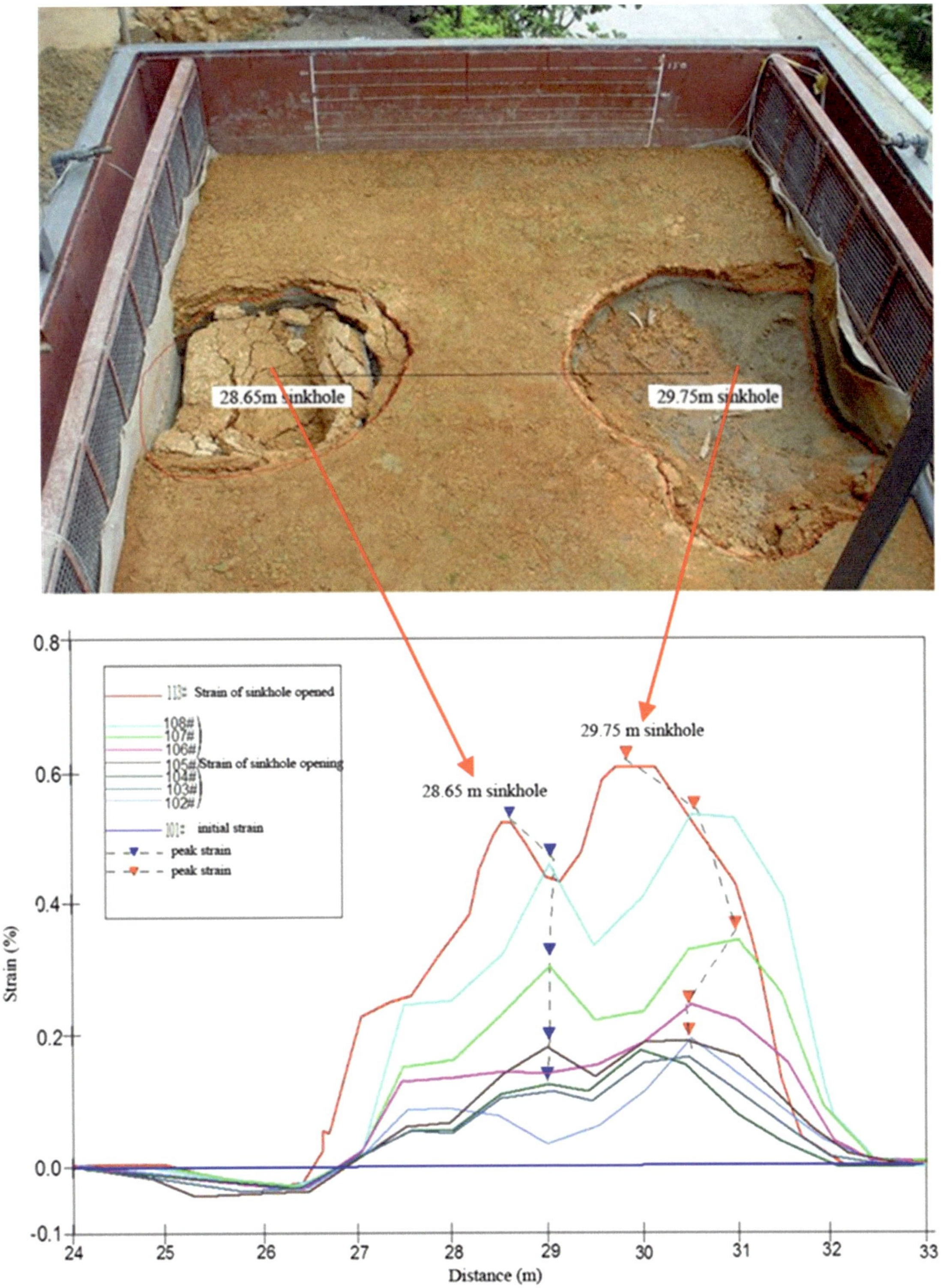

Fig. 4.2 Feasibility study of BOTDR on sinkhole prediction (**a** sandbox model with two sinkholes; **b** measured strain profiles using optical fiber at various depths)

groundwater flow conditions, which may lead to additional sinkhole collapse or roadway subsidence. Sinkhole collapse beneath roadways can be fatal to motorists. Figure 4.4 presents a conceptual site model that demonstrates the interactive relationship between stormwater drainage, existing sinkholes, and the development of new sinkhole/subsidence beneath a highway system. The upward growth of soil cavities and soil displacements in the subsurface typically

Fig. 4.3 Soil cavities (upper photos) and sinkholes (lower photos) encountered during highway construction

Fig. 4.4 Conceptual site model of sinkhole formation under highway

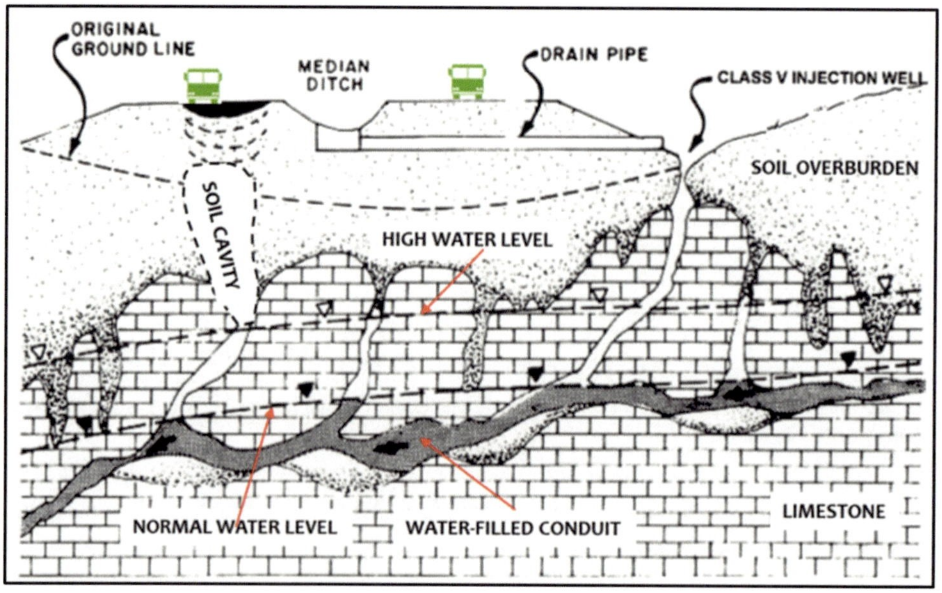

occur prior to the opening of sinkholes on the surface that directly jeopardizes the safe operation of the highway. Monitoring the cavity growing process and soil deformation under the highway roadbed would provide useful data on early warning of collapse risks. BOT DR was used for early warning of collapses over a section of the highway.

4.3.2 Installation of BOTDR for Sinkhole Monitoring

Figure 4.5 shows the area where the BOTDR technique was implemented. The monitoring length is approximately 90 m over a high-risk section of the Guilin-Yangshuo Highway. The excavation and cement mortar bonding optical fiber method were chosen for the fiber installation.

Figure 4.6 shows the installation processes. The fiber cable was placed in excavated channels of approximately 5 m deep and stabilized by pouring in proper cement mortar. Once a cement beam was formed along the optical fiber, the roadbed was covered with geotextile and backfilled. Four parallel optical fibers of 3 m apart (gx1 through gx4 in Fig. 4.5) were installed along the roadbed. The locations and lengths of the optical fibers were recorded and input into the monitoring instrument for data processing.

4.3.3 Monitoring Data Collection and Analysis

Figure 4.7 shows the monitoring data from 2009 to 2011 at optical fiber gx2 for demonstration. The data from the other three fibers are similar. The long-term monitoring data shows that the monitoring data of the optical fiber tends to be stable after the period of construction. The data are relative strain values. The fiber section from 0 to 40 m was reserved. The section from 40 to 60 m was between the monitoring box and the point where the fiber was buried

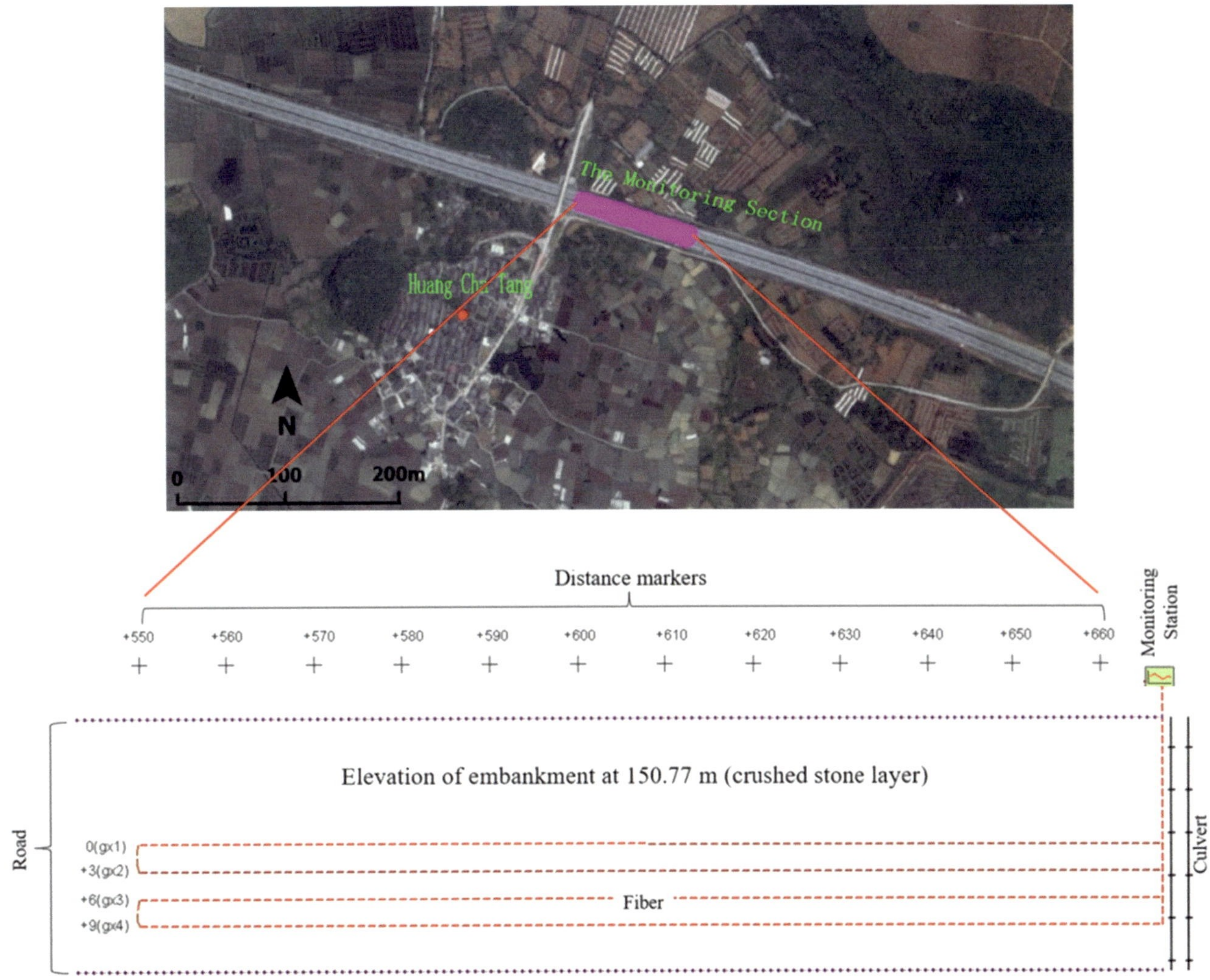

Fig. 4.5 Study site and layout of BOTDR fibers

Fig. 4.6 Installation processes of BOTDR

5 m underground. The section between 60 and 130 m was the key monitoring section. Except for the seasonal change caused by temperature in the reserved section and switch-in section, strain curves are mainly identical in the key section with an overall average strain difference of less than 0.03%, which suggests that there was no karst cave developing under the highway roadbed, and the roadbed was geotechnically safe. Comparison of the data collected in July when

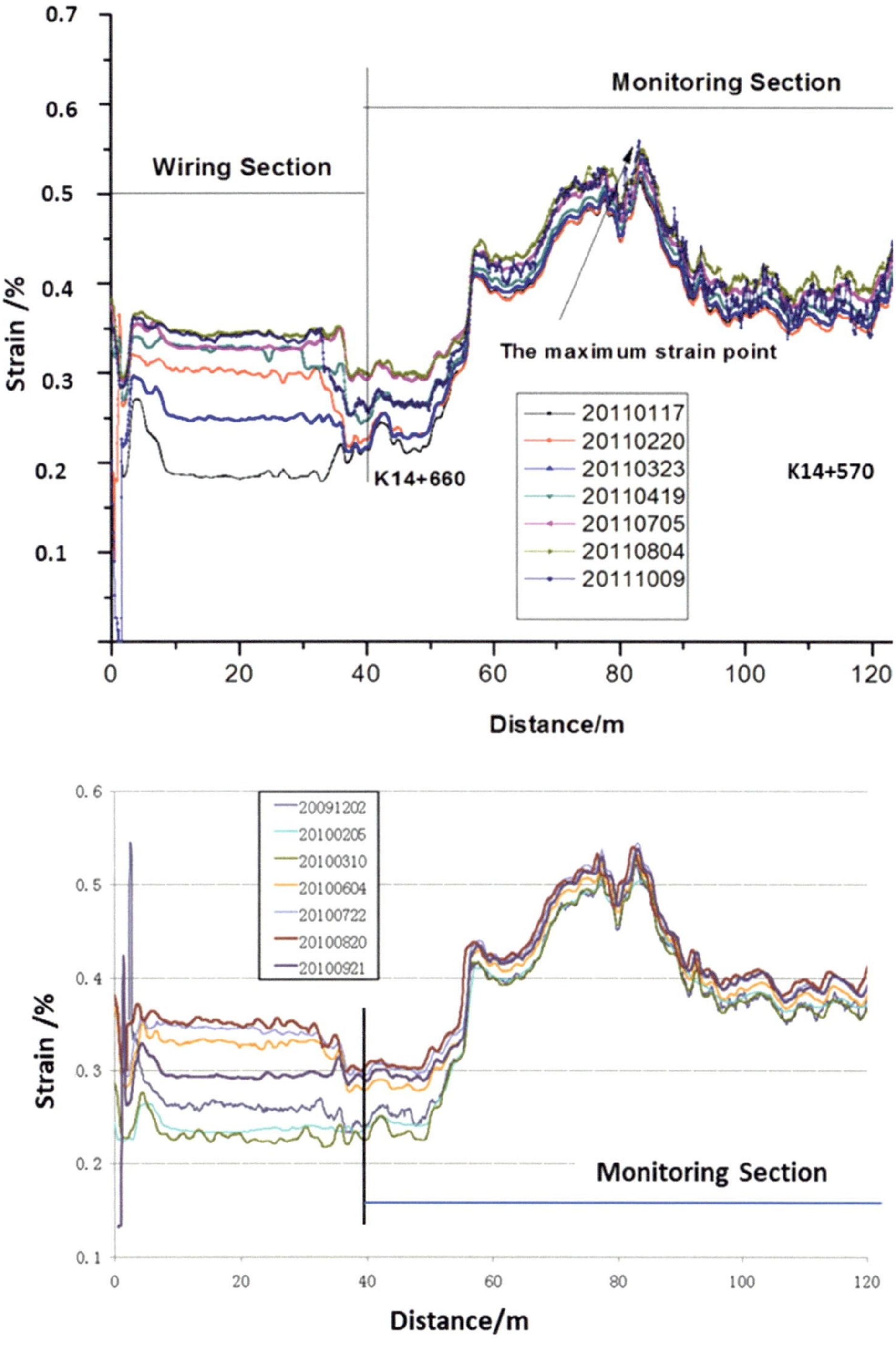

Fig. 4.7 BOTDR data collected at optical fiber gx2

the average temperature was 28 °C with those collected in January when the average temperature was 5.8 °C suggests that the offset of Brillouin scattering was influenced by not only the axial strain of optical fiber but also the temperature in the section on surface. At 5 m below the ground surface, the data shows that the influence of temperature was negligible. A comparison of the data collected during low traffic (12:00–13:00 every day) against the data collected during high traffic (15:00–16:00 every day) suggests that vehicle vibration has little effect on the result of optical fiber strain monitoring.

4.4 Application of Borehole BOTDR in Guangzhou, China

4.4.1 Site Background

Figure 4.8 shows the study site, which is on the west of Guangzhou in south China. The site lithology is composed of unconsolidated overburden with a thickness varying from 18 to 29 m and the Early Cretaceous conglomerates and sandstones of the Dalangshan Formation (K_2dl) and the Carboniferous limestone of the Shidengzi Formation (C_1S). The Dalangshan Formation is characterized with a high content of calcium, and the gravel composition comprises common clastic limestone and calcite. Karst features are well developed in the conglomerates. In addition, the Shidengzi Formation is also a medium-thick-bedded karst layer. The groundwater table elevation varies from 2.5 to 4.5 m above mean sea level, and the range of annual water level fluctuation is between 1.0 and 2.0 m.

Since 2008, more than 30 cover-collapse sinkholes have occurred in the study area. These collapses appear to be related to groundwater level changes, which may be related to pumping conducted as part of the construction of the city subway system. Figure 4.9 shows the groundwater level hydrography in a monitoring well (ZK9 in Fig. 4.8) with the number of sinkholes reported. When the groundwater level at the monitoring well declined to an elevation of 1.88 m, which is about 2.2 m lower than the normal water level, six large sinkhole collapses occurred. When the water level had risen to an elevation of 4.30 m, no further ground collapses occurred. Controlling the groundwater level drawdown would help manage the sinkhole occurrence at the site. However, it would not complete elimination of sinkhole risk. Because all sinkholes occurred in the unconsolidated overburden, monitoring the subsurface soil deformation at critical locations would provide useful information on early warning of sinkhole formation.

4.4.2 Installation of BOTDR in Boreholes

BOTDR systems (YR1, YR2, and YR3) were installed in three boreholes near an active school, as shown in Fig. 4.8. Figure 4.10 shows the installed BOTDR systems. YR1 is located on the north side of the school. The installation depth was 41 m with an optical cable length of 108 m. YR2 is located on the south side of the school with an installation depth of 20 m and a length of optical cable of 64 m. YR3 is located on the south side of the school with an installation depth of 19 m and a length of optical cable of 76 m. The geological formations where the three optical fiber monitoring holes are located are composed of approximately 10 m of fully weathered rock and soil, 14 m of conglomerates, and limestone of the Carboniferous Shidengzi Formation from top to bottom. YR2 and YR3 are arranged on the edge of structural fault F3 at 20 m, close

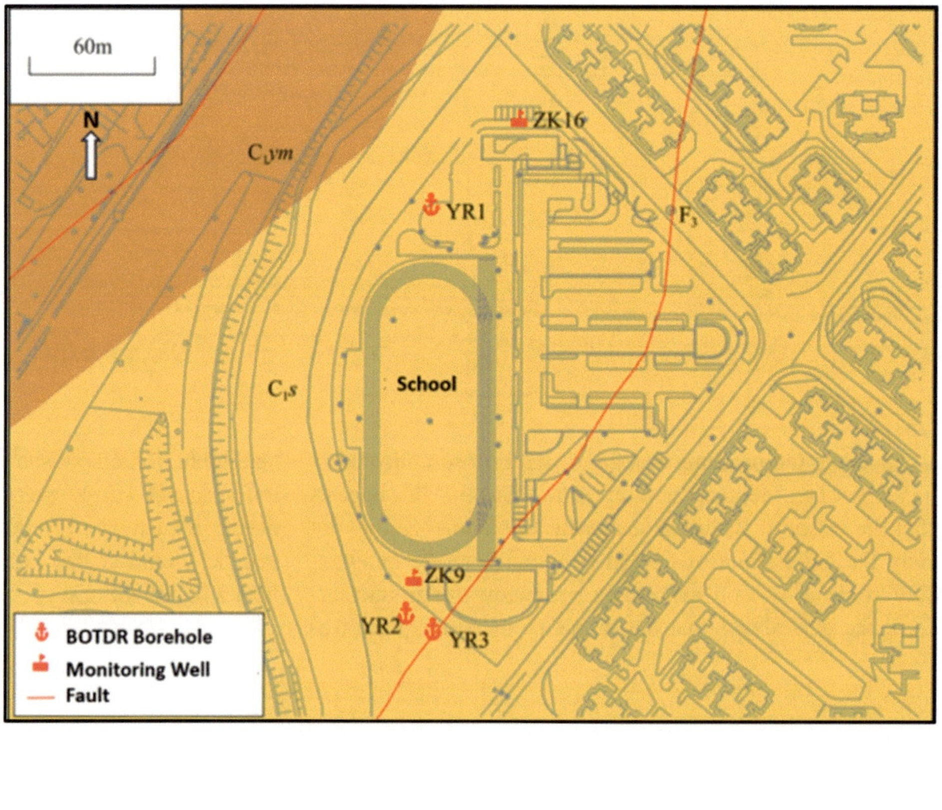

Fig. 4.8 Site location of borehole BOTDR application

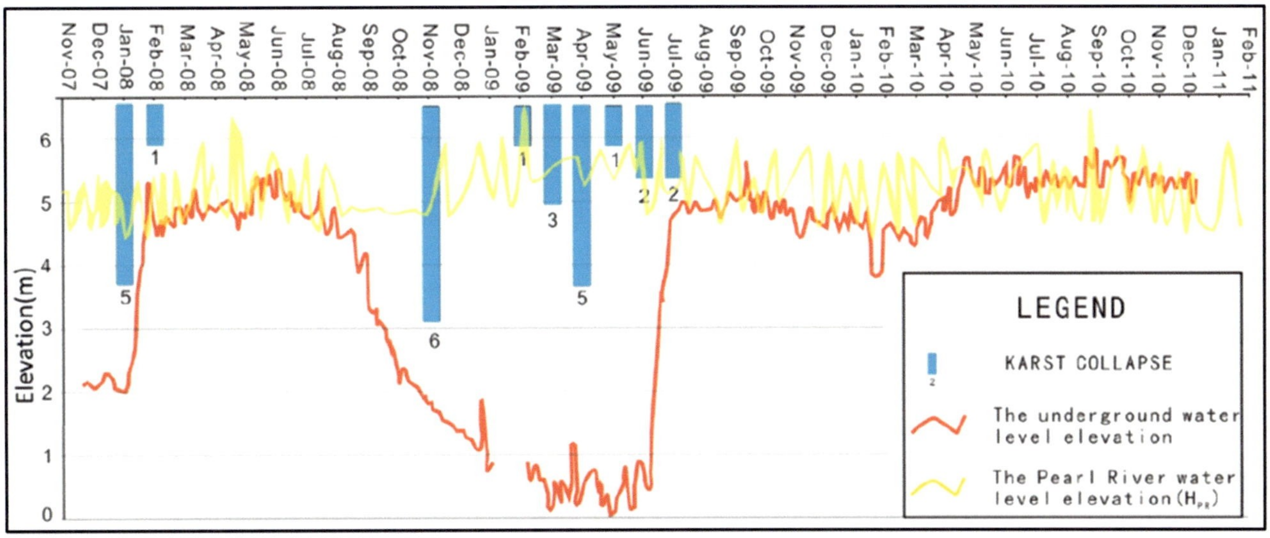

Fig. 4.9 Groundwater level in monitoring well and sinkhole occurrences

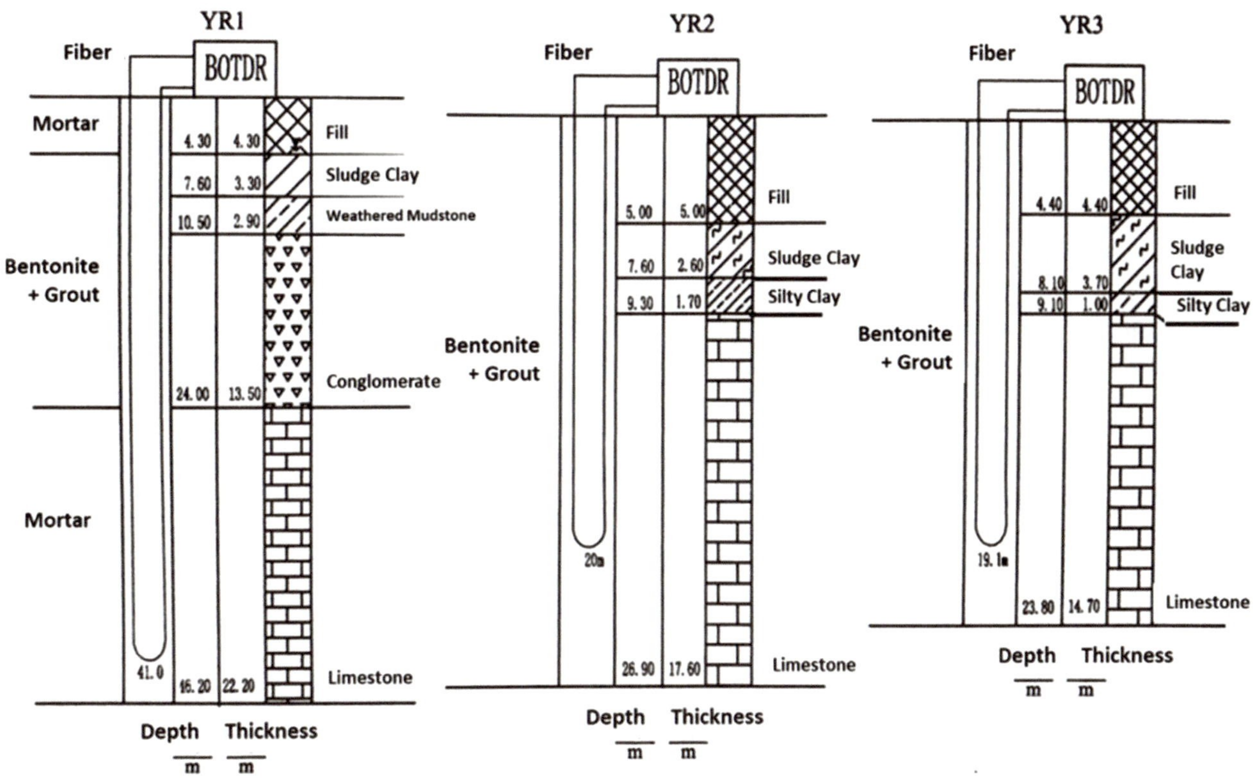

Fig. 4.10 Installation of borehole BOTDR systems

to groundwater/air pressure monitoring well ZK9. YR1 is approximately 100 m away from the fault, 50 m southwest of groundwater/air pressure monitoring well ZK16. Each monitoring well was installed with 4 kinds of optical cables: 5 mm diameter polyurethane optical cable, 2.5 mm communication optical cable, strain relief optical cable, and temperature leather optical cable.

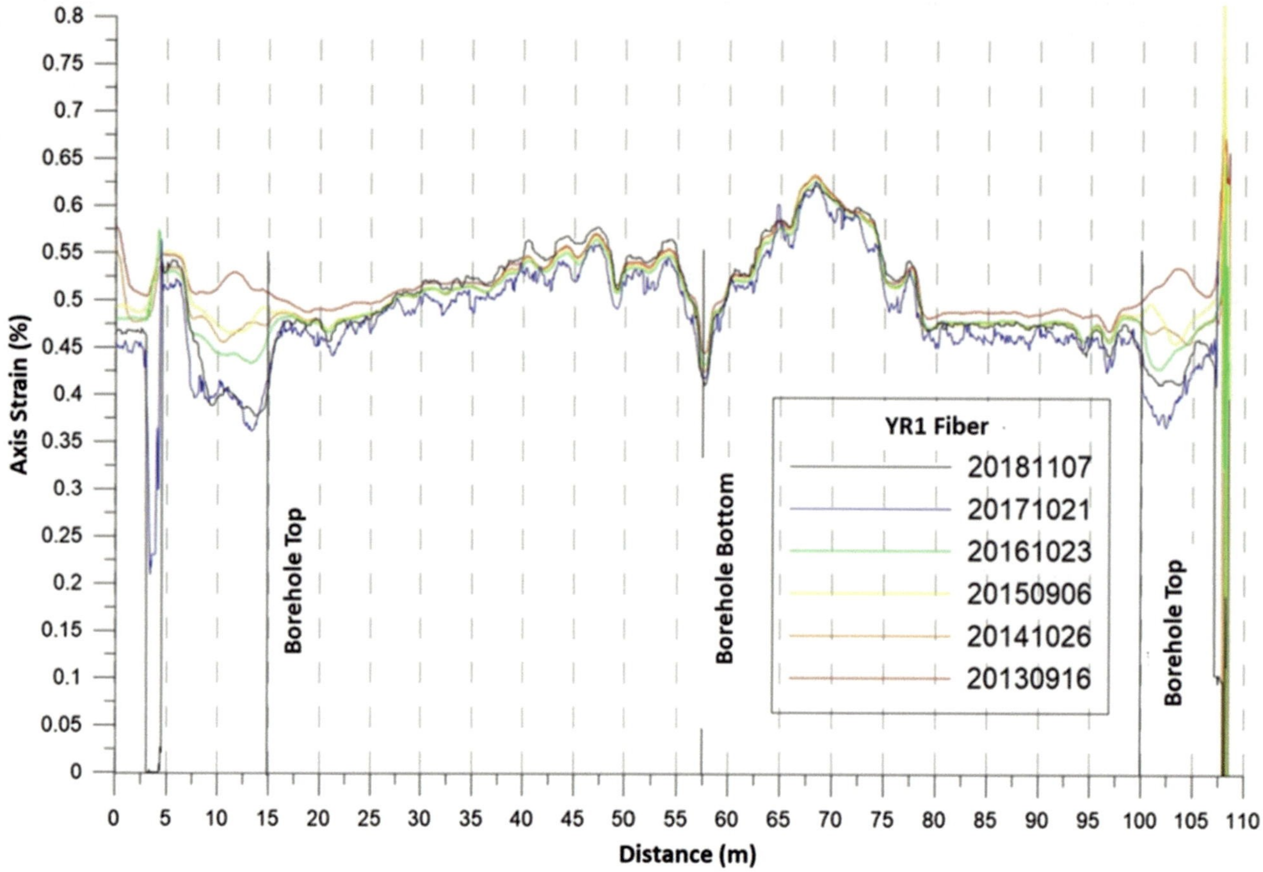

Fig. 4.11 BOTDR data collected in borehole YR1

4.4.3 Monitoring Data Collection and Analysis

Figure 4.11 shows the monitoring data from 2013 to 2018 at borehole YR1 for the communication optical fiber. The data at the other two boreholes are similar for the communication optical fiber. The axial strain at the 40–55 m section of YR1 communication fiber increased significantly on November 7, 2018, which should be caused by the displacement of the optical fiber near 45 m by soil and rock displacements. The strain fluctuations of the fiber near the top of the borehole are relatively large, which could be caused by the temperature difference and the different bending of the fiber coil. The monitoring data from the other three optical fibers did not provide meaningful anomalies.

In summary, the case studies presented in this chapter demonstrate that the BOTDR technique can be a viable technique in sinkhole monitoring and early warning if applied properly. The BOTDR fibers can be installed horizontally in unconsolidated overburden or vertically in a borehole. As a newly developed technique, improvements are expected in the data collection and analysis.

References

Guan ZD, Jiang XZ, Wu YB, Pang ZY (2015) Study on monitoring and early warning of karst collapse based on BOTDR technique. In: 14th sinkhole conference NCKRI symposium, vol 5. pp 407–414

Horiguchi T, Kurashima T, Koyamada Y (1994) 1-m spatial resolution measurement of distributed Brillouin frequency shift in single-mode fibres. In: Proceedings of the NIST/IEEE symposium on optical fibre measurements, Boulder, CO. pp 73–76

Jiang X, Lei M, Chen Y, Ge J (2006) An experiment study of monitoring sinkhole collapse by using BOTDR optical fiber sensing technique. Hydrol Eng Geol 6:75–80

Hydrodynamic Monitoring Technique in Collapse Risk Evaluation at Datengxia Water Conservancy Area, China

5

5.1 Site Background

Datengxia Water Conservancy Project is within the Pearl River Basin of south China. A hydropower reservoir was constructed near Datangxia in the Qianjiang River. The normal storage level of the reservoir is 61.00 m above mean sea level (amsl) with a storage capacity of 2.813 billion m^3, whereas the water level for flood control is 47.60 m amsl. Several karst areas were affected by the project, including the study area discussed in this chapter, Wuxuan Conservation Area (WCA). Figure 5.1 shows the study area and its main characteristics pertinent to karst hydrogeology. The WCA is on the first level terrace on both left and right banks of the Qianjiang River. The banks are steep with the bedrock partially exposed. The terrace elevation is between 65 and 75 m amsl. The highest point of the study area is the peak of Maanshan Mountain at 144.30 m amsl.

Figure 5.2 shows a geological cross section. The overburden consists of the Quaternary residuals (Q_3^{el}) from dissolution of the carbonate rocks and alluvial deposits (Q_4^{pal}). The deposits are unconsolidated and include clay, silt, and clayey sand. The dissolution residuals are across the site, while the alluvial deposits are within 50 m of the river. The thickness of the soil layer is up to 20 m. Aquifer testing suggested that the hydraulic conductivity is between $1.49 \times 10{-}^{6}$ and 1.45×10^{-8} cm/s. Three bedrock formations underlie the area. The Upper Carboniferous System (C_3) is composed of limestone, chert limestone, intercalated chert dolomitic limestone, and dolomite. The Huanglong Formation of the Middle Carboniferous System (C_2h) consists of dark gray medium to thick bedded limestone and dolomitic limestone. The Dapu Formation of the Middle Carboniferous System (C_2d) is composed of thick-bedded dolomite, dolomitic limestone, and medium-thick-bedded limestone. The carbonate rocks of C_2h and C_2d are vulnerable to karst development, and the C_2h formation is more vulnerable to karst development than the C_2d formation because the limestone and dolomitic limestone is purer and thicker. Wells in both C_2h and C_2d formations can produce water at 60–80 m^3/h.

The main karst forms in the WCA include karst collapses, caves, conduits, and springs. There were six sinkholes identified in the study area. Figure 5.3 shows some of the karst features. Karst caves are the general terms used for cavities identified in the study area. There are three karst caves in WCA with entrance heights varying from 0.5 to 4 m. The length of these caves is unknown, but they are accessible up to 5 m by people. The water is visible in the caves and is likely connected to the Qianling River. The Qianjiang River has the lowest discharge datum level of groundwater in the area. Groundwater flows toward the river and discharges through the C_2h dolomitic limestone at a karst spring at the riverside. The discharge point elevation is 35 m, which is above the river water level in dry seasons. The spring becomes submerged when the water level of the Qianjiang River rises. The estimated flow rate varies from 30 to 80 L/s. The average pH is 7.36, and the conductivity is approximately 373 μs/cm.

Forty-nine exploratory boreholes were drilled to understand the karst geology. Approximately 65% of the boreholes encountered cavities with 63 cavities in total. The largest cavity was 14.3 m high, and the smallest was 0.13 m. Approximately 49% of the cavities were less than 1 m in height, as shown in Fig. 5.4a. Sixty percent of the cavities were filled with clay, gravely sand, or breccia. The breccias were composed of a mixture of broken rocks and clay. The borehole data also revealed caves in multiple layers. Figure 5.4b shows the cave distribution with depth based on the 49 boreholes. Fluctuations in the water level of the Qianjiang River controlled cave formation, and a horizontal strong karst development zone has been formed along the Qianjiang River. Because of the limited number of deep boreholes, the lower limit of karst development in the area was not conclusive. However, karst cavities are present at depths of 104 m below the ground surface or at an elevation of – 41 m amsl.

© The Author(s), under exclusive license to Springer Nature Switzerland AG 2024
X. Jiang et al., *Monitoring and Early Warning Technologies on Karst Lands*, Advances in Karst Science, https://doi.org/10.1007/978-3-031-59045-0_5

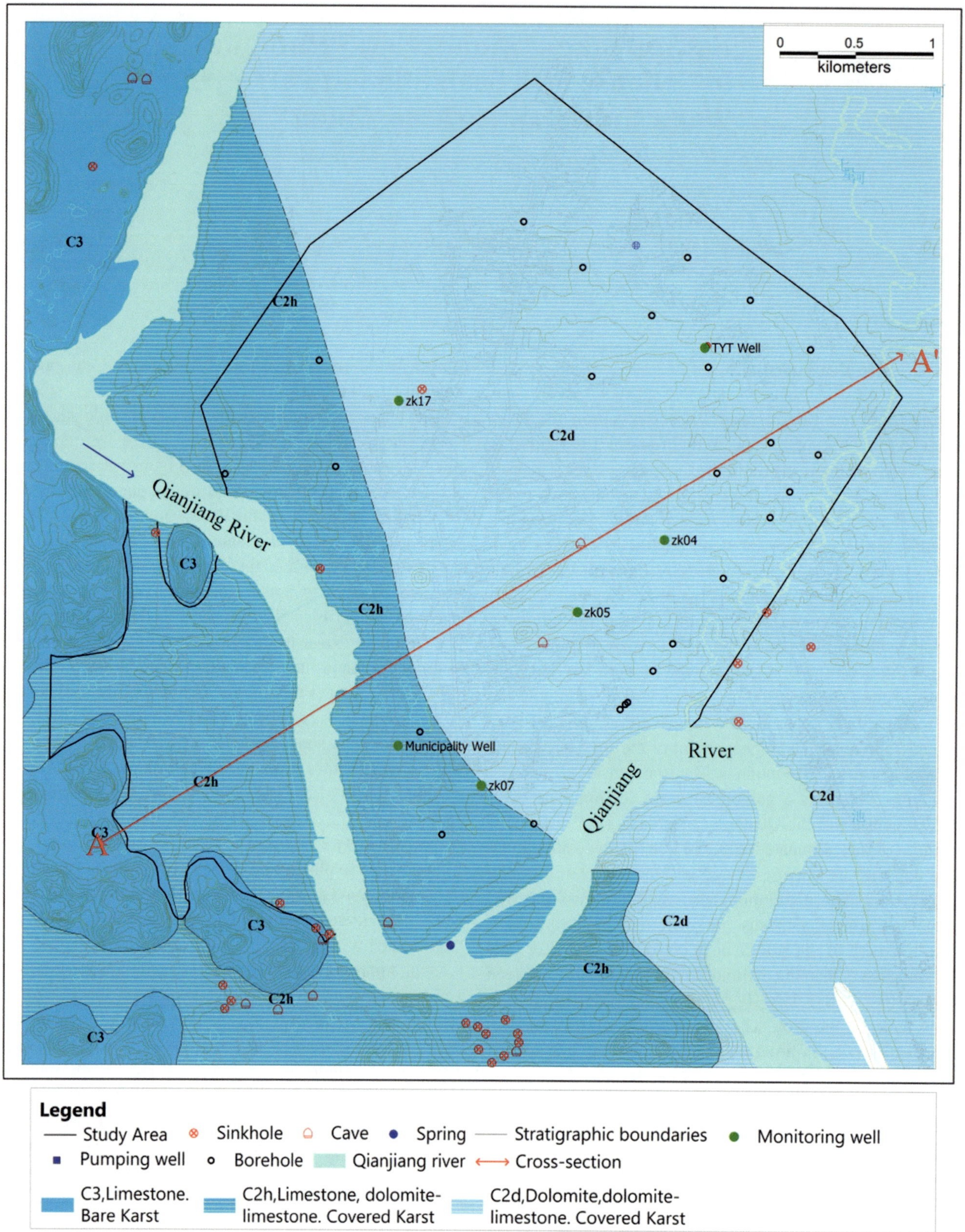

Fig. 5.1 Background setting of study area

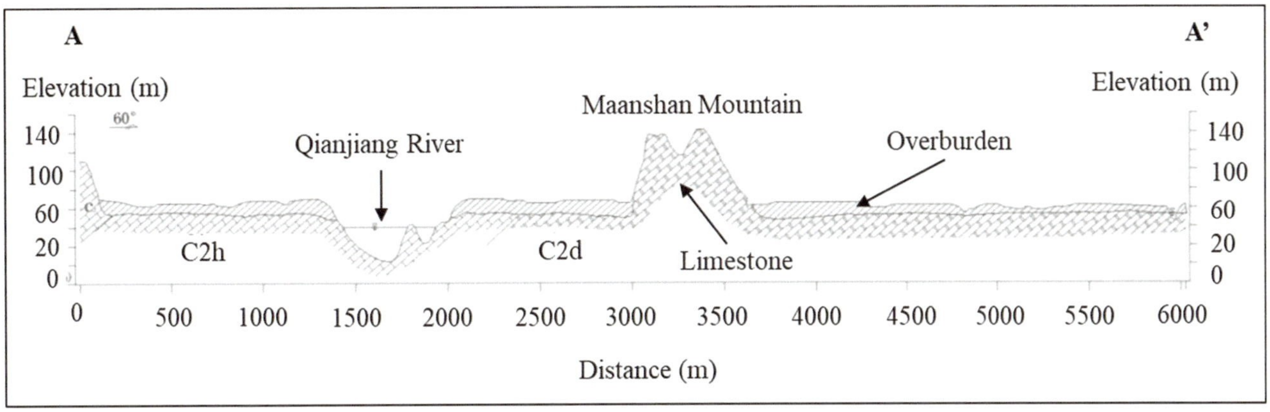

Fig. 5.2 Geological cross section of study site (location of the cross section is shown in Fig. 5.1)

Fig. 5.3 Examples of karst features

Six karst collapses were identified in the study area. Table 5.1 summarizes their main characteristics. The size of these sinkholes varies from less than 1 m to approximately 25 m. The collapse depths were between 5 and 20 m. Their general shapes are circular or elliptic on the plane view. Of the six sinkholes, four occurred in the C_2h formation, and two occurred in the C_2d formation. Five of the six sinkholes are within 200 m of the Qianjiang River. These sinkholes naturally occurred, and the fluctuation of the groundwater level may have been the driving force, whereas dynamics of the karst groundwater are mainly controlled by the water level changes of the Qianjiang River. The frequency and range of water level fluctuation in the river were greater than those of groundwater level. The water level of the

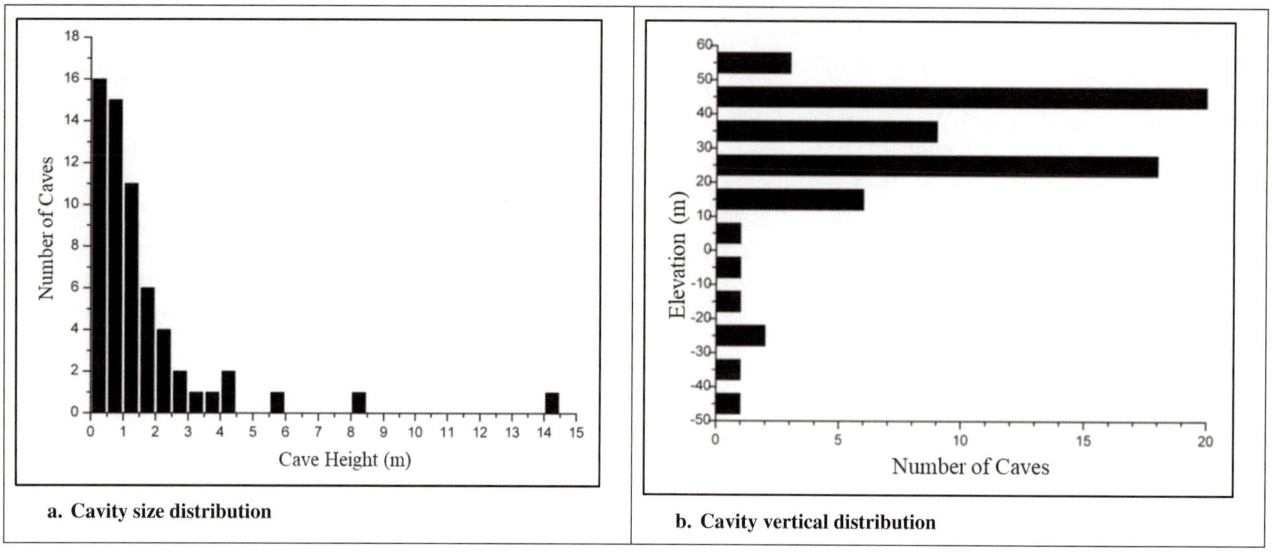

Fig. 5.4 Distribution of karst cavities encountered by boreholes

Table 5.1 Characteristics of sinkholes at WCA

ID	Overburden	Thickness (m)	Bedrock formation	Characteristics		
				Shape	Size (m)	Depth (m)
1	Clay	5–10	C_2d	Circular	3–4 in diameter	0.8
2	Clay	10	C_2d	Circular	0.8	0.3
3	Clay	10	C_2h		2	
4	Clay	10–20	C_2h	Irregular	Long axis = 45; short axis = 20	6 (surface runoff sinking into the swallet)
5	Clay	10–20	C_2h	Circular	25 in diameter	10 (surface runoff sinking into the swallet and discharging to river)
6	Clay	10–20	C_2h	Circular	20	15 (surface runoff sinking into the swallet and discharging to river)

Qianjiang River in the study area had large annual changes of up to 22 m, and the maximum daily increase could reach 7.93 m.

Other factors affecting sinkhole formation in the area were investigated from the sinkhole inventory. In addition to the six sinkholes in the study area, more than 15 sinkholes also occurred immediately adjacent to the study area. Regionally, 102 sinkholes were inventoried, and 73 sinkholes recorded times of collapses. Figure 5.5a shows the temporal distributions of the sinkholes, whereas Fig. 5.5b shows the histograms of sinkholes versus the overburden thickness. Of 102 inventoried sinkholes in the region, more than 65% occurred where the overburden thickness was less than 5 m. Only two sinkholes occurred in overburden soil that was thicker than 20 m. Based on the temporal distribution, approximately 90% of the cover-collapse sinkholes occurred during rainy seasons from March to June.

Laboratory sandbox experiments have demonstrated that erosion and transport of soil particles mostly took place when the water level fluctuated (Zhou 1997). Figure 5.6 shows the effect of water fluctuation on erosion from the experiment using soil samples of 5 cm by 5 cm by 5 cm. The more frequent fluctuation in water level caused a greater amount of soil transport. When the water level kept constant, the soil mass became stable, 4 days after the soil sample was put into the water. Based on the sinkhole characteristics, hydrogeological conditions, and laboratory test results, it was believed that groundwater and its level changes were the main factors governing the sinkhole development in the study area. A groundwater monitoring program was established to evaluate and predict the risk of karst collapses in WCA under routine operating conditions of the reservoir.

5.2 Monitoring of Dynamic Groundwater-Gas Pressures

The dynamic groundwater conditions were studied from five monitoring wells (ZK04, ZK05, ZK07, ZK17, and TYT monitoring well). Unlike groundwater monitoring for

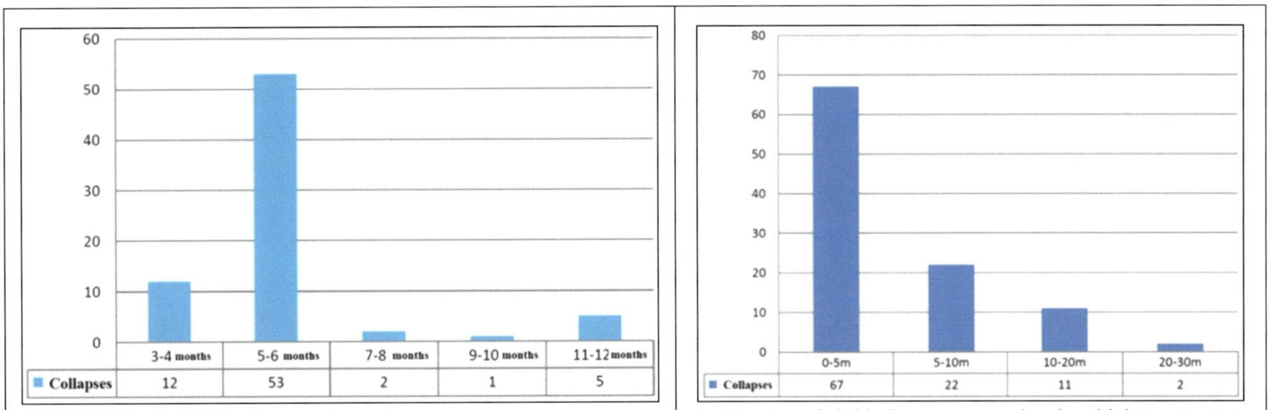

(a) Number of sinkholes by months (b) Number of sinkholes versus overburden thickness

Fig. 5.5 Sinkhole distributions over time and overburden thickness

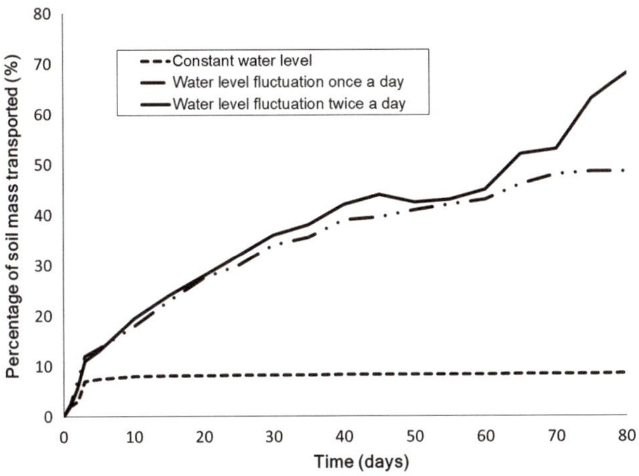

Fig. 5.6 Soil transport intensity in response to water level fluctuations

water resource evaluation, both groundwater and gas pressures should be measured for collapse monitoring. The monitoring wells that housed the pressure transducers were properly sealed, as shown in Fig. 5.7. The monitoring work started in November 2012 and ended in January 2014. A high-frequency interval of 20 min was used for the monitoring. Figure 5.8 presents the monitoring results.

Monitoring well ZK04: During the monitoring period, the water level of monitoring well ZK04 was higher than the water level of the Qianjiang River. The water level of the Qianjiang River was higher than the water level of the ZK04 only once during the flood peak, and the duration was 1 day. The groundwater level generally fluctuated between 45.09 and 52.54 m. There were two anomalies on April 1 and December 28, 2013, when the groundwater level rose by 1.55 m and 5.87 m, respectively, within 20 min. The annual water level varied by 7.45 m. The elevation of the bedrock surface at this monitoring point was 48.99 m. When the rainfall is abundant, the groundwater levels are above the bedrock surface, and when the rainfall is insufficient, the groundwater level drops below the bedrock surface (Fig. 5.9a). The dynamic change in the groundwater level was mainly related to atmospheric rainfall. The groundwater began to respond 24 h after the rainfall, and the groundwater level rose to its peak 5 days later, with a maximum daily variation of 0.93 m. The daily periodic fluctuation of groundwater was not obvious. The change rate of groundwater level was generally between − 0.05 and +0.05 cm/min. The maximum water level decrease rate was 18.73 cm/min. The maximum water level increase rate was 29.34 cm/min.

Monitoring well ZK07: During the monitoring period, the groundwater level of monitoring well ZK07 was higher than the water level of the Qianjiang River most of the time, and the water level of the Qianjiang River was higher than the water level of ZK07 only twice during flood peaks, each time lasting 1–2 days. The groundwater level fluctuated between 41.46 and 45.82 m, with an annual water level variation of 4.36 m. The bedrock surface elevation of the monitoring point was at 47.11 m amsl. The groundwater level had been fluctuating below the bedrock surface (Fig. 5.9b). The dynamic change of the groundwater level was closely related to the atmospheric rainfall. The groundwater began to respond 24 h after the rainfall, and the groundwater level rose to peak 4 days later, with a maximum daily variation of 0.39 m. The daily periodicity of groundwater was noticeable. For example, from March 1 to 9, 2013, the groundwater level rose and fell within a period of 12 h, and the amplitude was approximately 4 cm. The change rate of groundwater level was generally between − 0.1 and +0.1 cm/min. The maximum water level decrease rate was 0.3 cm/min, while the maximum water level increase rate was 0.18 cm/min.

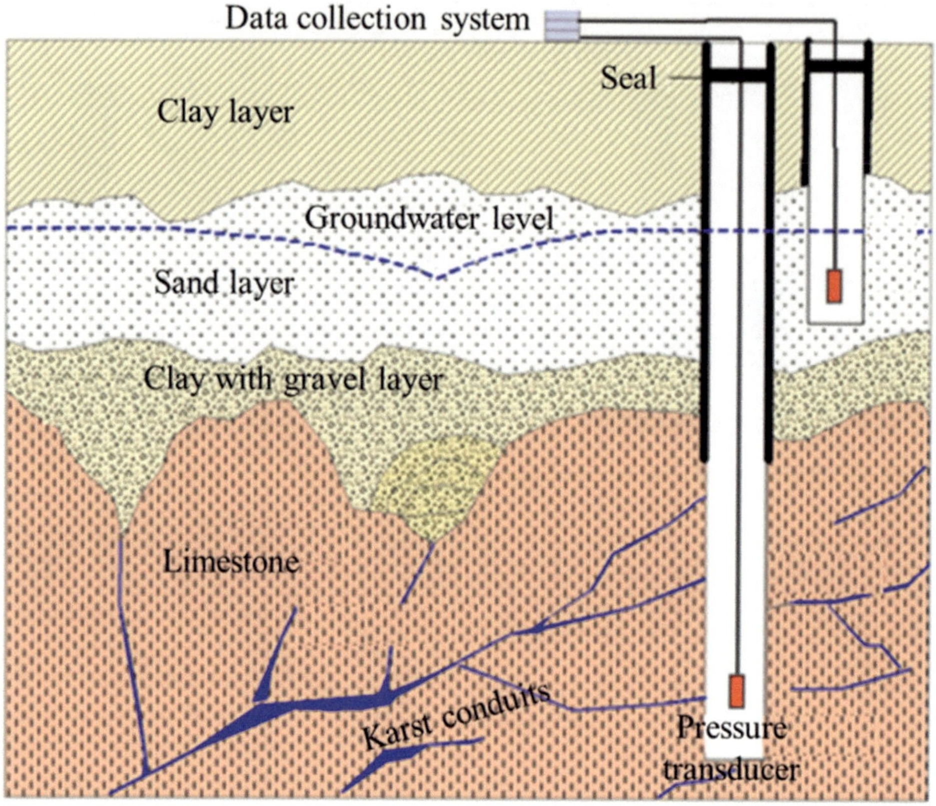

Fig. 5.7 Construction of monitoring wells for groundwater-gas pressures

Monitoring well ZK05: During the monitoring period, the groundwater level of monitoring well ZK05 was higher than that of Qianjiang River, and the water level of Qianjiang River was higher than that of ZK05 only once during the flood peak, and the duration was 1 day. The groundwater level fluctuated between 43.16 and 51.62 m, with an annual water level variation of 8.46 m. The bedrock surface elevation of the monitoring point was at 47.02 m. When the rainfall was abundant, the groundwater level fluctuated above the bedrock surface. When the rainfall was insufficient, the groundwater level dropped below the bedrock surface (Fig. 5.9c). The dynamic change of the groundwater level was closely related to the atmospheric rainfall. The groundwater began to respond 24 h after the rainfall, and the groundwater level rose to the peak 5 days later. The maximum daily variation was 0.8 m. The daily periodicity of groundwater was obvious. For example, from March 1 to 9, 2013, the water level began to drop around 6:00 am and lasted until 7:00 pm, with a drop of about 13 cm. After 00:00, the water level recovered quickly until 6:00 am the next day. The water level recovery time was 12 h. The change rate of groundwater level was generally between − 0.2 and +0.2 cm/min. The maximum water level decrease rate was 0.23 cm/min, and the maximum water level decrease rate was 0.26 cm/min.

Monitoring well ZK17: During the monitoring period, the groundwater level of monitoring well ZK17 was higher than that of the Qianjiang River, and the water level of the Qianjiang River was higher than that of the ZK17 only twice during flood peaks, each time lasting 1–2 days. The groundwater level fluctuated between 43.36 and 50.78 m amsl, with an annual water level variation of 7.42 m. The bedrock surface elevation of the monitoring point was 48.14 m. The groundwater level was below the bedrock surface most of the time (Fig. 5.9d). The groundwater level only rose above the bedrock surface when there was a significant rainfall event. The dynamic change of the groundwater level was closely related to the precipitation. The groundwater began to respond within 24 h after the rainfall, and the groundwater level rose to the peak 5 days later, with a maximum daily variation of 0.32 m. The change rate of groundwater level was generally between − 0.1 and +0.1 cm/min. The maximum water level decrease rate was 0.14 cm/min, and the maximum water level increase rate was 0.12 cm/min.

Monitoring well TYT: During the monitoring period, the groundwater level of monitoring well TYT was higher than that of the Qianjiang River, and the water level of the Qianjiang River was higher than the water level of the monitoring well only once during the flood peak, which

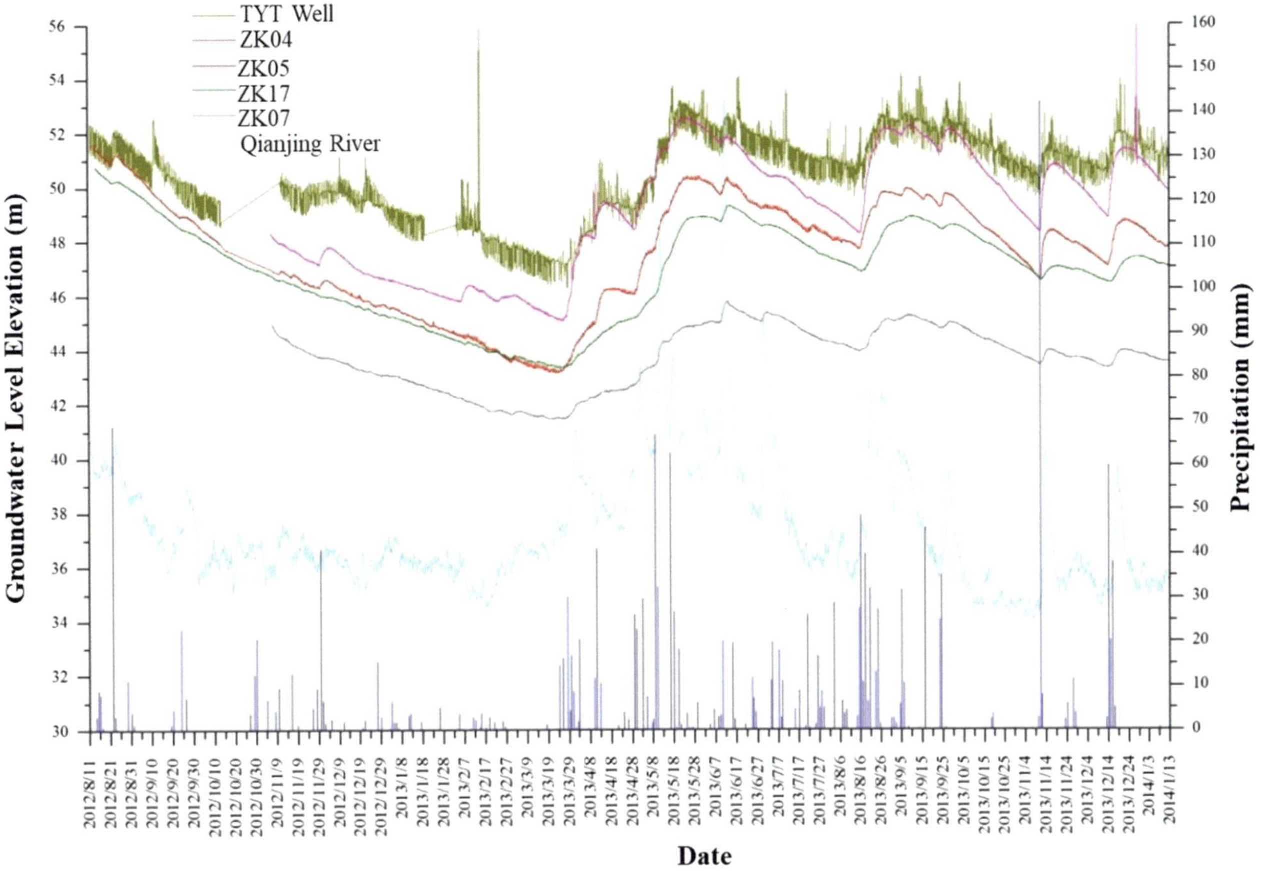

Fig. 5.8 Monitoring results of groundwater-gas pressures

lasted for one day. The groundwater level generally fluctuated between 44.85 and 52.77 m amsl. There was an anomaly on February 14, 2013, and the groundwater level rose by 5.06 m within 20 min. The annual water level varied by 9.57 m. The bedrock surface elevation of the monitoring well was at 50.46 m. When the rainfall was abundant, the groundwater level fluctuated above the bedrock surface. When the rainfall was poor, the groundwater level dropped below the bedrock surface. The groundwater level fluctuated across the bedrock surface many times (Fig. 5.9e). The response of groundwater to rainfall was obvious. The groundwater began to respond 24 h after the rainfall, and the groundwater level rose to its peak 2 days later. Without including the abnormality, the maximum daily variation was 2.48 m. Because of daily pumping of groundwater, the groundwater level fluctuated frequently. The daily cycle was obvious. For example, from March 1 to 9, 2013, water pumping began at around 7:00 am, and the groundwater level dropped until around 12:00 noon. The drop range was 0.8–1 m, and then the water level started recovery. The water level recovery time was 6 h. The pumping activity was either once per day or twice per day, one in the morning and one in the afternoon between 16:00 and 18:00. The change rate of groundwater level was generally between − 5 and +5 cm/min. The maximum water level decrease rate was 19.27 cm/min, and the maximum water level increase rate was 25.32 cm/min.

Table 5.2 summarizes the main characteristics of dynamic monitoring. The range of groundwater level at the monitoring wells varied from 4.36 to 9.57 m. The maximum daily water level range was between 0.39 and 0.93 m. The rate of change in groundwater level was between − 0.3 and 0.3 cm/min. The rate of change could be between − 5 and 5 cm/min under the condition of groundwater mining. The water level of ZK04 and TYZ wells had abnormal readings several times when the groundwater rose several meters in a short time. The maximum daily water level could reach 7.52 m, and the maximum water level change rate could reach 29.34 cm/min. Except that the groundwater level of ZK07 fluctuated below the bedrock surface, the groundwater level of other monitoring points fluctuated up and down the bedrock level. During the drilling, it was

Fig. 5.9 Groundwater levels in reference to bedrock surface

Table 5.2 Summary of groundwater dynamic monitoring

Monitoring wells	Magnitude of annual change (m)	Magnitude of daily change (m)	Maximum decrease rate of groundwater level change (cm/min)	Maximum increase rate of groundwater level change (cm/min)	Relationship between groundwater level and bedrock surface	Relationship between groundwater level and Qianjiang River	Notes
ZK04	7.45	6.11	29.34	18.73	Both above and below bedrock surface	Groundwater level was higher than the Qianjiang River for most of the time	Two anomalous readings with daily change of 0.93 m
ZK05	8.46	0.8	0.23	0.26			
ZK07	4.36	0.39	0.3	0.18	Below bedrock surface		
ZK17	7.42	0.32	0.14	0.12	Both above and below bedrock surface		
TYT	9.57	7.52	19.27	25.32			One anomalous reading with daily change of 2.48 m

found that the soil layer structure at the rock-soil interface was relatively loose and soft, and the epikarst zone was well developed. This soft zone has been interpreted as soil that has sloughed or raveled. These soil particles have the potential to move easily into voids below. Based on this observation, the risk level for soil erosion at ZK07 would be lower than that at the other four monitoring wells. Sinkhole risk still exists in the other four monitoring wells. In areas with a water table higher than the bedrock surface, sinkholes can be induced by lowering the groundwater level, whereas, in areas with water table lower than the soil/bedrock contact, modifications to surface runoff may induce sinkholes. The groundwater level of each monitoring point was mostly higher than that of the Qianjiang River. The water level of the Qianjiang River was higher than the groundwater level of monitoring wells for a short time only during 1–2 flood events.

5.3 Conceptual Site Model of Karst Collapses

Development of the conceptual site model of karst collapses was based on understanding the impacts of groundwater on soil erosion processes in the overburden soil. These impacts include loss of buoyance with water level drop, seepage erosion by water flow, hydraulic fracturing and sloughing by water level fluctuation, and soil strength deterioration by repeatedly wetting and drying. Our studies focused on soil erosion processes under hydrostatic conditions and hydrodynamic conditions.

5.3.1 Study on Soil Erosion Processes Under Hydrostatic Pressures

Fifteen undisturbed soil samples were collected during drilling of the five monitoring wells with three samples per well. These samples were tested for their erodibility using a specially designed device. The sample size was 91 mm in diameter, and the thickness was 30 mm. A 5-mm cut was made at the bottom of each sample to simulate a small cavity. The actual thickness for water seepage was 25 mm.

Figure 5.10a shows a schematic design of the testing device. The water head system consists of a water supply source, a discharge pipe, a discharge valve, an inlet valve, a pulley, and a pulling cord. The water level in the pipe simulates the driving water pressure. The height of the discharge valve controls the height of the water level of the test device and keeps it constant if required. A pulley system was installed on the top of the water supply pipe to adjust the water head. Figure 5.10b is a photo of the constructed testing device.

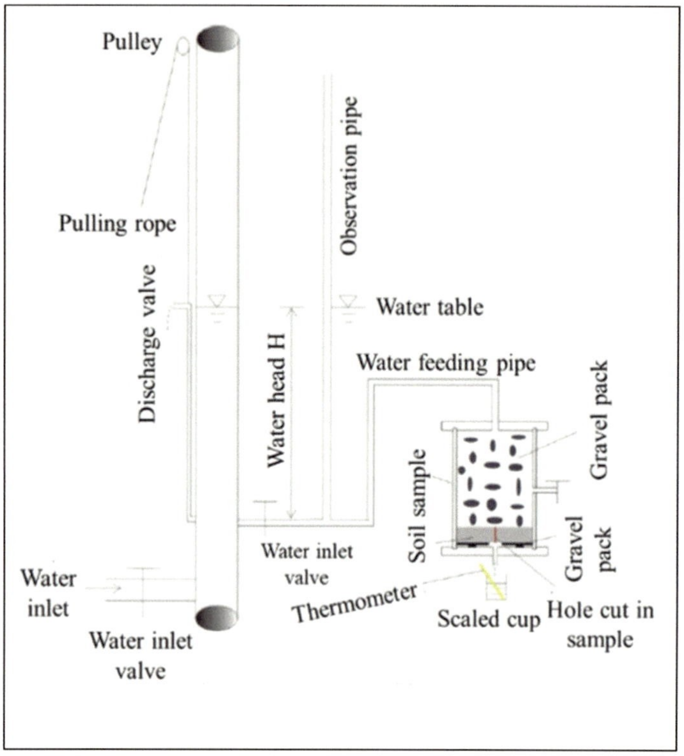

Fig. 5.10 Experimental device of testing soil erodibility. **a** Schematic design; **b** Testing assembly

During the tests, the initial water head was set at 2.0 m. The water head was gradually increased by an increment of 0.25 m. The water pressure at each level was maintained for approximately 8 h. Under each water pressure, the seepage flow was measured three times, and any signs of clay damage were observed. The water heads and measured flows were plotted to show their relationships. One example is shown in Fig. 5.11.

When the water heads varied from 2.0 to 4.0 m, the seepage flow increased gently with the increase of water head, which represented the typical Darcy flow regimes without changes in the porosity. From the water head 4.25 m and higher, the seepage flow showed an upward inflection and increased sharply, suggesting that the porosity structure inside the soil sample changed and the pore size increased significantly under the hydraulic pressures. Soil particles were observed flowing out of the sample assembly, indicative of soil erosion initiation. Based on the testing results, an increase in seepage flow greater than 20% was indicative of porosity changes. Therefore, the water

Fig. 5.11 Relationship between driving water head and seepage flow

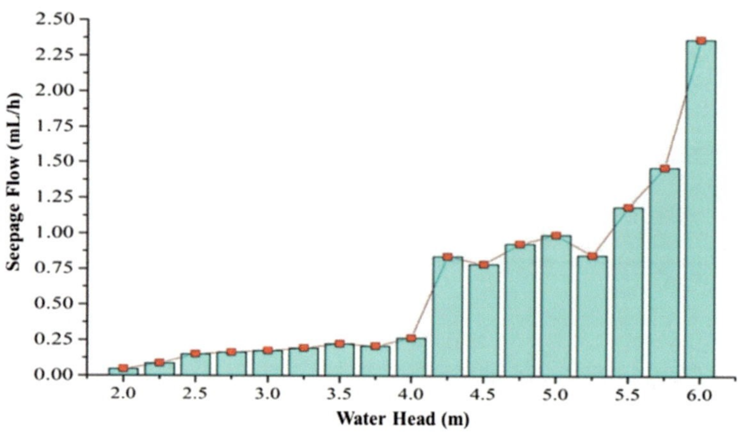

Table 5.3 Testing results of soil sample under hydrostatic pressures

Monitoring well ID	Soil sample ID	Threshold hydraulic head (m)	Average threshold hydraulic head (m)
ZK04	K04-1	4.00	3.75
	K04-2	4.00	
	K04-3	3.25	
ZK05	K05-1	4.50	4.50
	K05-2	4.50	
	K05-3	4.50	
ZK07	K07-1	4.50	4.17
	K07-2	4.00	
	K07-3	4.00	
ZK17	K17-1	4.25	4.17
	K17-2	4.25	
	K17-3	4.00	
TYT	KT-1	2.75	3.25
	KT-2	3.25	
	KT-3	3.75	

head under which the seepage flow increased dramatically was referred to as the threshold water head for the sample test. The testing results are summarized in Table 5.3. On average, the threshold water head ranged from 2.33 to 4.50 m. Figure 5.12 illustrates the water head variations from the bedrock surface at the five monitoring wells. TYT well appeared to be the most vulnerable to soil erosion, whereas ZK04 was probably vulnerable to soil erosion. The other three monitoring wells did not have the hydrostatic conditions that supported the soil erosion.

5.3.2 Study on Soil Erosion Processes Under Hydrodynamic Pressures

Tharp (2003) and Anikeev (1999) investigated the hydraulic fracturing of soil as a mechanism for cover-collapse sinkhole formation under hydrodynamic conditions. Figure 5.13 shows a conceptual model explaining the components that are involved in the erosion processes. A soil void is hypothesized at the soil and bedrock contact. The void is modeled as spherical with spherical symmetry in all boundary conditions to allow a one-dimensional representation of the scenario. When a constant water head drops, hydrofracturing occurs at the radius where pore pressure exceeds the combined radial stress and tensile. Consequently, the annular region inside the radius sloughs off and falls into the void below. The newly exposed face is subject to full head drop. Because the transient pore pressure remains high while the radial stress decreases instantaneously, hydrofracturing and sloughing occur again. Under this erosion mechanism, the spherical shell can peel off gradually until the void reaches the surface and results in a collapse.

In the one-dimensional spherical plastic analysis, stresses are related by the following Mohr–Coulomb failure equation:

$$A\sigma_\theta' - B\sigma_r' = 2c \text{ or } \sigma_\theta' = (2c + A\sigma_r')/B \tag{5.1}$$

where

A $(\mu^2 + 1)^{1/2} - \mu$

B $(\mu^2 + 1)^{1/2} + \mu$

σ_θ' is effective tangential stress, $\sigma_\theta' = \sigma_\theta - P$.

σ_r' is effective radial stress, $\sigma_r' = \sigma_r - P$.

P is potentiometric pressure or pore pressure.

$\mu = \tan\varphi$, where φ is friction angle.

c is cohesion.

Two parameters, soil cohesion c and friction angle φ, function as resistance forces against soil transport. Cohesion is a more important factor and defines the soil tensile or compressive stress. Cohesion is associated with soil composition, cementation, and compaction and is directly related to the vertical consolidation stress, which results from loading of soil column above. Mesri (1975) developed the following empirical equation to calculate the cohesion:

$$c = 0.22\gamma_s d \tag{5.2}$$

where γ_s is the unit weight of soil and d is the initial thickness of soil above void.

The finite-difference equilibrium equation for the void arch is expressed by

$$\Delta\sigma_r'/\Delta r = 2(\sigma_\theta' - \sigma_r')/r \tag{5.3}$$

Substituting Eq. (5.1) into Eq. (5.3) yields

$$\Delta\sigma_r'/\Delta r = [4c/B + 2\sigma_r'(A/B - 1)]/r \tag{5.4}$$

Pore pressure P is required in the calculation of the effective stresses. For the spherically symmetric flow model, the hydrodynamic pressure can be calculated by the following finite-difference equation:

$$\frac{\Delta p}{\Delta r} = K\frac{1}{r^2}\left\{2r\frac{\Delta p}{\Delta r} + r^2\frac{\Delta^2 p}{\Delta r^2}\right\} \tag{5.5}$$

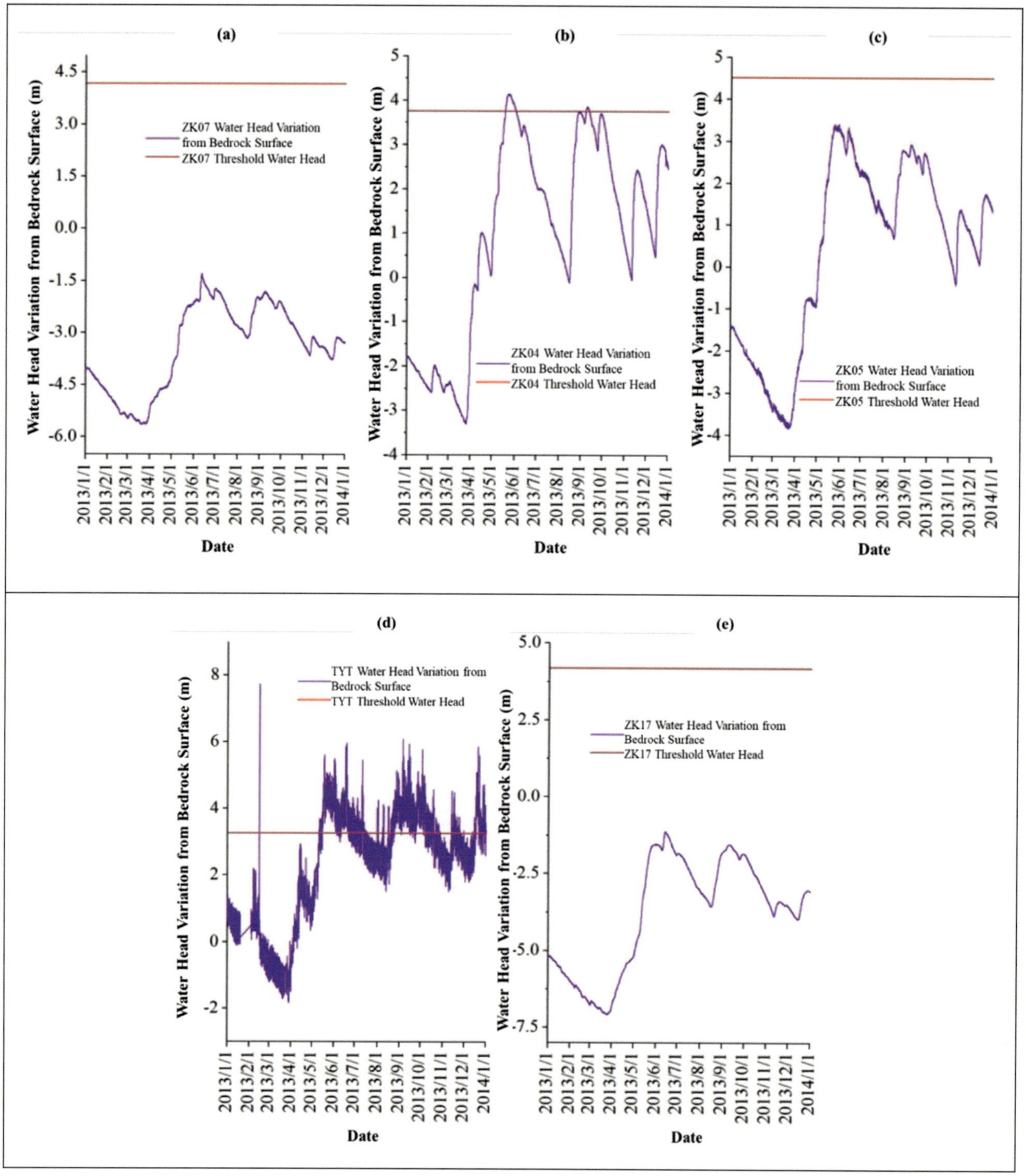

Fig. 5.12 Water head variations from the bedrock surface

Based on Tharp (1997), the non-hydrostatic pressure P can be calculated analytically by the following equation:

$$P = \gamma_w[(h_i - h_w)(1/r_i - 1/r)/(1/r_w - 1/r_i) + h_i] \quad (5.6)$$

where

h_w and h_i are the hydraulic heads at r_w and r_i, respectively.

γ_w is the unit weight of water

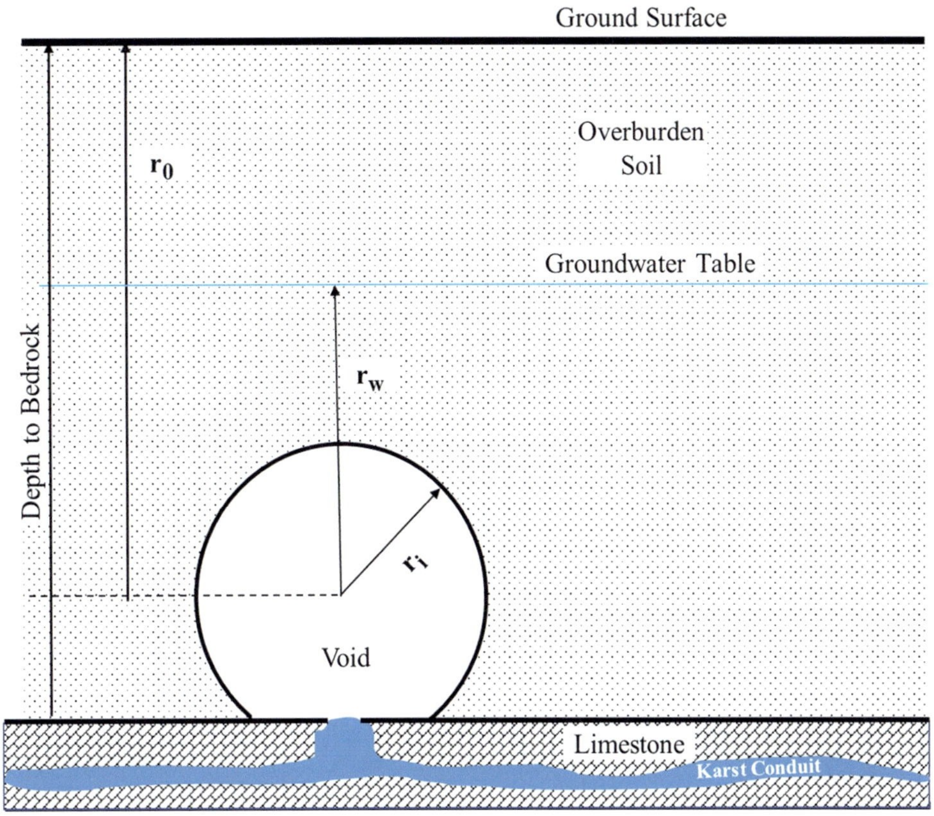

Fig. 5.13 Conceptual model of hydraulic fracturing in overburden soil under hydrodynamic conditions

Assuming a soil depth of 5 m, Tharp (2003) presented the modeling results. Figure 5.14 shows the relationship between groundwater level drop versus rate of level drop under hydrofracturing conditions for various soil depths. The soil depth was used here to reflect the cohesion value as expressed in Eq. (5.2).

Both the groundwater level drop and the rate of the groundwater level drop governed hydrofracturing of soil. This differs from the conventional understanding in which the groundwater level drawdown is the affecting factor, and the rate of groundwater level drop is not accounted for. For soil with smaller cohesion values such as at depths 0.1 or 0.2 m, collapses may occur with a drawdown as small as 0.5 m, while the collapsing process is not sensitive to the rate of groundwater level change. As the soil became more cohesive, the rate of groundwater level change became more important. At a depth of 1 m, a 3-m water level drop is required for hydrofracturing when the rate of water level drop is 0.01 m/h, whereas a 2-m water level drop is required when the rate increases to 1 m/h. At a depth of 2 m, both

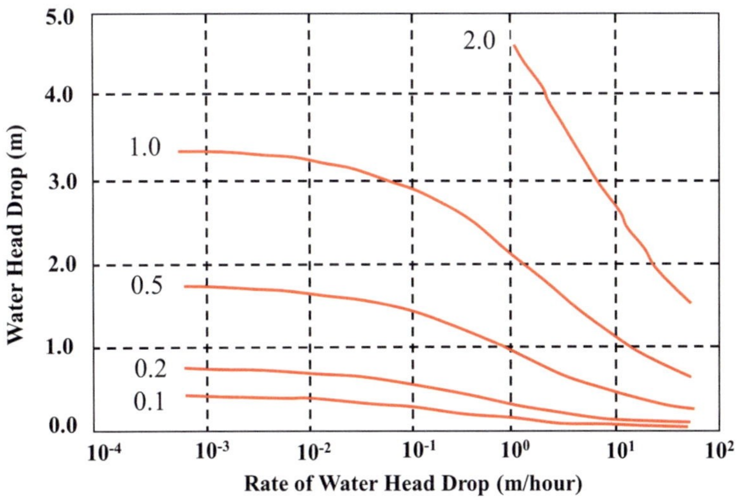

Fig. 5.14 Relationship of water head drop and rate of water head drop when hydraulic fracturing is initiated for variable soil depths (modified from Sharp 2003)

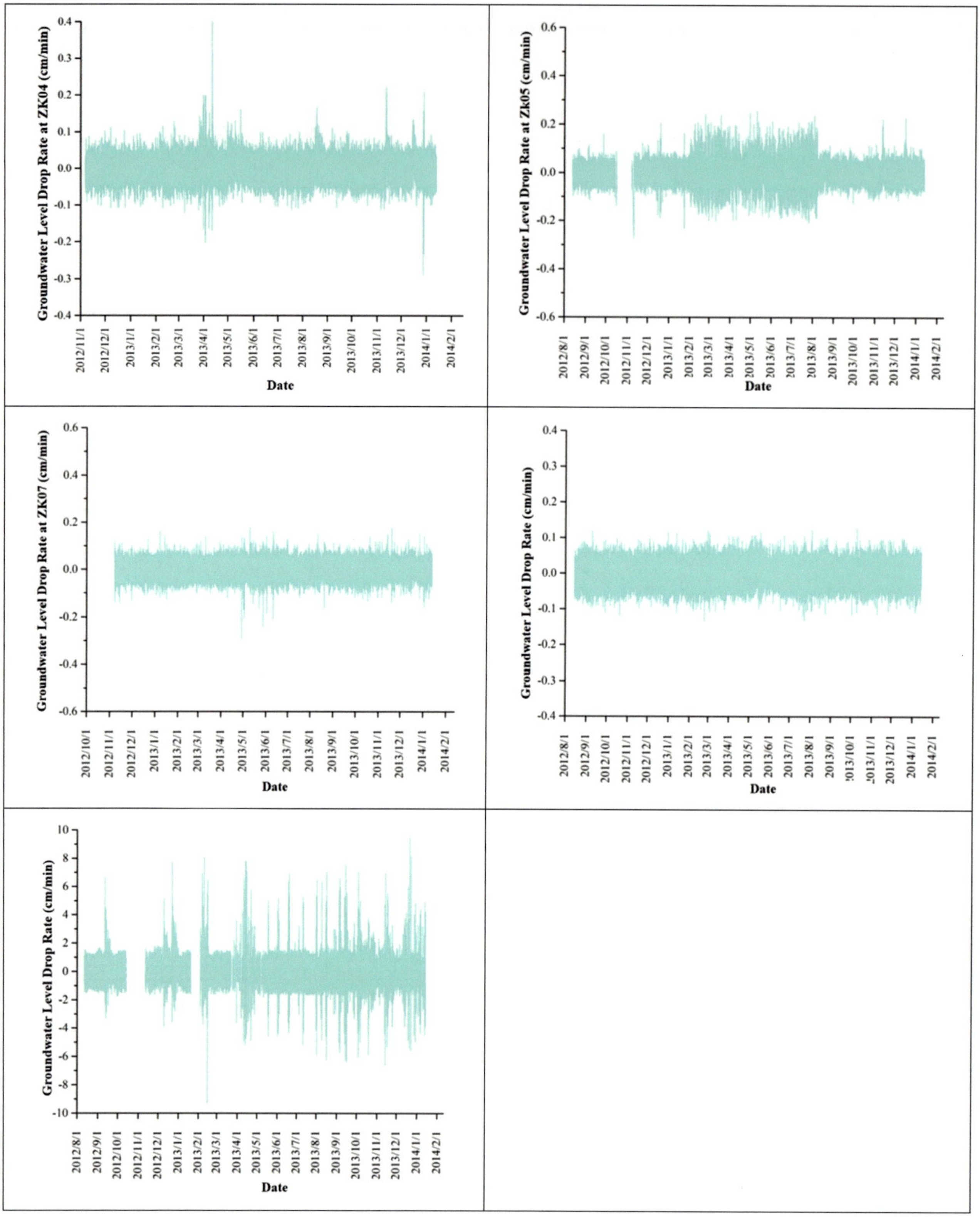

Fig. 5.15 Water level drop rates in monitoring wells

large groundwater level drawdowns and large rates of groundwater level change are required to produce soil hydrofracturing, while the rate of change becomes more important. Although the results shown in Fig. 5.14 were based on a void at 5 m, they were applicable to depths up to 20 m for cohesive soils. For slow drawdowns, the failure conditions are largely independent of the cavity size (Tharp 2003).

Figure 5.15 shows the rates of groundwater level change at the five monitoring wells. The rate varied over two orders of magnitude from 0.03 m/h at ZK04 to 3 m/h at TYT well. The well TYT was the most vulnerable to sinkhole formation under hydrostatic conditions.

5.4 Risk Evaluation of Karst Collapses

Seepage erosion is the primary form of soil loss in the saturated zone under hydrostatic conditions. When the groundwater pressure is greater than the critical water pressure of the seepage erosion, the bottom soil layer is gradually peeled off and lost, potentially forming a soil cavity. The risk level is divided into high, medium, and low. From the perspective of engineering safety, $0.8H$ and $1.2H$ (H is the threshold water head) are used as the boundaries between low and intermediate and between intermediate and high risks. Table 5.4 summarizes the risk classification.

Table 5.5 presents the risk levels of seepage erosion at each monitoring well under hydrostatic conditions based on the groundwater monitoring data. Monitoring wells TYT and ZK04 had an intermediate level risk, while ZK05, ZK07, and ZK17 had low risk.

Hydrofracturing was the primary form of soil loss in the saturated zone under hydrodynamic conditions. The risk level of hydrofracturing was divided into high, intermediate, and low. From the perspective of engineering safety, $0.7R$ was used as the limit of intermediate-high risk, and $0.3R$ was used as the limit between low and intermediate risks. R is the threshold rate in water level drop and can be obtained from Fig. 5.14 or similar modeling results if the soil cohesion is known. Considering the average magnitude of groundwater level fluctuation was similar between these monitoring wells and varied from 3.5 to 4 m, the threshold rate was assumed to be 0.29 cm/min. Based on the principle of small probability events (that is, the frequency of occurrence is very low in a large number of repeated tests), the risk classification is summarized in Table 5.6.

Table 5.7 presents the risk levels of hydrofracturing at each monitoring well under hydrodynamic conditions based on the groundwater monitoring data. Monitoring well TYT had a high risk, while the rest monitoring wells had low risk.

The risk evaluation results provided guidance on water level drop rate in the reservoir during its operations. If the water level dropped too fast, the groundwater level drawdowns might be large enough to cause soil hydrofracturing and thus surface collapses. We recommended that the rate of water level drop of Datengxia Reservoir

Table 5.4 Seepage erosion risk level classification

Risk level	Percentage of data points (%)		
	$<0.8H$	$0.8H$–$1.2H$	$>1.2H$
High			≥ 5
Intermediate		≥ 5	< 5
	< 95	< 5	< 5
Low	≥ 95	< 5	< 5

Table 5.6 Hydrofracturing risk level classification

Risk level	Percentage of data points (%)		
	$<0.3R$	$0.3R$–$0.7R$	$> 0.7R$
High			≥ 5
Intermediate		≥ 5	< 5
	< 95	< 5	< 5
Low	≥ 95	< 5	< 5

Table 5.5 Risk evaluation results on seepage erosion

Monitoring well ID	Threshold water head (m)	Total number of data points	$<0.8H$		$0.8H$–$1.2H$		$> 1.2H$		Risk evaluation
			Number of data points	Percentage (%)	Number of data points	Percentage (%)	Number of data points	Percentage (%)	
ZK04	3.75	26,280	20,446	77.80	5290	20.13	544	2.07	Intermediate
ZK05	4.50	26,278	26,278	100	0	0	0	0	Low
ZK07	4.17	26,290	26,290	100	0	0	0	0	Low
TYT	3.25	25,185	10,477	41.60	9693	38.49	5015	19.91	High
ZK17	4.17	26,282	26,282	100	0	0	0	0	Low

Table 5.7 Risk evaluation results on hydrofracturing

Monitoring well ID	Total number of data points	<0.3*R*		0.3*R*–0.7*R*		> 0.7*R*		Risk evaluation
		Number of data points	Percentage (%)	Number of data points	Percentage (%)	Number of data points	Percentage (%)	
ZK04	31,100	29,779	95.76	1270	4.08	51	0.16	Low
ZK05	35,318	34,577	97.90	732	2.07	9	0.03	Low
ZK07	31,055	30,775	99.10	278	0.895	2	0.006	Low
TYT	34,280	21,059	61.43	5168	15.08	8053	23.49	High
ZK17	37,237	37,156	99.78	81	0.22	0	0	Low

be no greater than 3.0 m/d or 0.08 cm/min to reduce the karst collapse risk in WCA. The hydrodynamic monitoring will continue at the same or high sampling frequencies and be analyzed using the approaches discussed in this chapter.

References

Anikeev AV (1999) Casual hydrofracturing theory and its application for sinkhole development prediction in the area of Novovoronezh Nuclear PowerHouse-2 (NV NPH-2), Russia. In: Proceedings of 7th multidisciplinary conference on sinkholes and the environmental impacts of karst, Balkema, Rotterdam, pp 77–93

Mesri G (1975) Discussion of new design procedures for stability of soft clays. J Geotech Eng Div ASCE 101:409–412

Tharp TM (1997) Mechanics of formation of cover collapse sinkholes. In: Proceedings of 6th multidisciplinary conference on sinkholes and the environmental impacts of karst, Balkema, Rotterdam, pp 29–36

Tharp TM (2003) Cover-collapse sinkhole formation and soil plasticity. In: Proceedings of sinkholes and the engineering and environmental impacts of karst, no 122. Geotechnical Special Publication, pp 110–122

Zhou W (1997) The formation of sinkholes in karst mining areas in China and some methods of prevention. Environ Geol 31(1/2):50–58

6 Detection and Mitigation of Groundwater Contamination from Highway Stormwater Runoff

6.1 Introduction

As a non-point source of water contamination, the impacts of stormwater and highway runoff on surface water have been extensively addressed. However, relatively little attention has been directed toward assessing its impact on groundwater, especially in karst areas. Dissolution of bedrock (usually limestone or dolomite) in karst areas results in a terrane characterized by sinkholes, sinking streams, underground (cave) streams, and springs. The entire landscape formed in soluble rock areas is known as a karst terrane. The term terrane is used rather than terrain to include subsurface features as well as surface features. Groundwater in these settings is more susceptible to contamination because surface water may pass directly into the subsurface with little or no filtration by soil. Because karst groundwater typically flows through relatively large fractures and conduits within the bedrock, it may transport contaminants rapidly from points of recharge (such as sinkholes) to distant cave streams, water wells, springs, and surface streams. This chapter describes a peat filtration system specifically developed for karst areas and its effectiveness in the removal of highway runoff contaminants.

6.1.1 Pollutants from Highway Stormwater Runoff

Highway stormwater runoff is a potential source of pollution. The highway runoff-related pollutants include suspended and dissolved solids, heavy metals, nutrients, bacteria, road salt, herbicides, and hydrocarbons. The predominant contaminants in urban runoff are copper, lead, zinc, chromium, cadmium, nickel, arsenic, cyanide, and asbestos (U.S. Environmental Protection Agency [USEPA] 1983). Stephenson et al. (1999) provided a thorough literature review. Sources of pollutants in road and highway runoff have been investigated by numerous researchers. Major sources of pollutants on highways are vehicles, dust fall, precipitation, and roadway maintenance. Hydrocarbon contamination in highway runoff results from the incomplete combustion of gasoline, oil dripping onto road surfaces and parking areas, and disposal of waste oil at storm drains. Pavement wear and right-of-way erosion contribute to highway runoff pollution, as do highway deicing and right-of-way maintenance activities such as the use of roadside herbicides and road dust suppressants.

Among the factors that affect highway runoff quality are those that affect the sources of pollutants and their transport. In a four-year investigation of factors affecting the quality of highway runoff in the Austin, Texas, area, Irish et al. (1995) identified nine variables—storm duration, storm volume, storm intensity, vehicles during storm, length of antecedent dry period, antecedent traffic count, previous storm duration, and previous storm intensity. Most variations observed in highway stormwater loading in this area may be explained by causal variables measured during the rainstorm event, the antecedent dry period, and the previous rainstorm event.

- The antecedent dry period (ADP) is a commonly cited factor that affects runoff quality. It provides an opportunity for materials to accumulate on the highway surface. The pattern of constituent built-up during that period is an important relationship used in many highway runoff models. Hewitt and Rashed (1992) report a highly significant correlation between the length of the antecedent dry periods and the quantity of dissolved lead and copper removed by runoff events. However, Kerri et al. (1985) reported that there is no statistical significance between the constituent load of a storm and the length of ADP of a storm. Their study established a better relationship with the number of vehicles during the storm. Natural and vehicle-induced winds have been observed to blow materials off the highway during dry weather. Constituents may also be removed during the ADP by volatilization, biodegradation, and chemical decay.

© The Author(s), under exclusive license to Springer Nature Switzerland AG 2024
X. Jiang et al., *Monitoring and Early Warning Technologies on Karst Lands*,
Advances in Karst Science, https://doi.org/10.1007/978-3-031-59045-0_6

- An association of PAHs, total organic carbon (TOC), and extractable organic compounds with suspended solids has been reported. Zawlocki et al. (1980) report that at least 90% of the total aromatic compounds, aliphatic hydrocarbons, and oxygenated hydrocarbons in stormwater runoff are associated with suspended solids. Data presented by Hewitt and Rashed (1992) indicate a correlation between concentrations of PAHs and total suspended solids (TSS) in runoff, suggesting that PAH compounds are adsorbed onto particulate materials.
- Some aromatic hydrocarbons in stormwater runoff appear to be correlated with rainfall intensity rather than TSS concentrations. The concentration of aliphatic hydrocarbons in runoff decreases with each consecutive flush from high-intensity rainfall (Asplund et al. 1980), suggesting that their solubility limits their association with suspended solids. Data presented by Zawlocki et al. (1980) also indicate that most hydrocarbon compounds in runoff remain in solution.
- The effects of precipitation characteristics on the quality of highway runoff have been reported extensively. In a study of highway stormwater runoff in England, Balades et al. (1983) conclude that substantial contaminant loads may result from an individual precipitation event. They state that a few short-duration rainfall events may introduce 30% of the annual highway-related contaminant load.
- Other factors affecting the pollution content of highway runoff include wind and traffic-related air turbulence. The influence of traffic characteristics, such as speed, braking, volume, and density, has been noted by several researchers. In their summary of 15 years of highway runoff research, Smith and Lord (1990) note that impact on water quality is minimal from highways with an average daily traffic of fewer than 30,000 vehicles.

Data from numerous studies (Gupta et al. 1981a, b, c) have suggested that the majority of pollutants are discharged during the beginning of a storm and that the pollutant load decreases with time. This is known as the "first-flush" phenomenon. Accumulated loads of small particle size are flushed rapidly from a highway watershed. However, when rainfall events occur close together with a few dry days for pollutant accumulation, the highway surface may be clean enough to minimize the first-flush effect (Wada and Miura 1984).

6.1.2 Best Management Practices for Highway Stormwater Runoff

Stormwater quality management involves practices designed to reduce the concentration and the mass of stormwater pollutants. The primary objective of highway runoff pollution management is the reduction of the total pollutant loads entering groundwater or surface water. Although all highway runoff contains pollutants, their concentrations do not always constitute a problem for the receiving water. Therefore, runoff management programs must not only identify where the potential highway runoff problems exist; they must also ensure that the treatment is applied to that portion of the runoff with the highest contaminant concentration. In some cases, this may be achieved by treating only the first flush of runoff, rather than the entire volume.

There are many methods for the treatment of highway stormwater runoff. They may be implemented individually or in combination. Some of the most effective stormwater runoff treatment methods are described below:

- Curb Elimination: Curbs can be eliminated from highway designs where room for grassed channels exists. Lack of curbs allows stormwater runoff to flow over the grassed areas, increasing infiltration into the soil.
- French Drains: French drains are pipes that have been placed in a pervious material (usually gravel) for drainage. They are used to drain an area to prevent flooding and route groundwater to a desirable location for treatment or other control measures.
- Temporary Checkdams: Temporary checkdams can be placed on waterways draining land under construction. They can be constructed of sandbags or other impervious materials. Checkdams can be used during construction periods to prevent erosion in unvegetated temporary channels. These channels are temporarily unable to handle design flows and need to be protected until vegetation can be established.
- Dutch Drains: Dutch drains are gravel-filled ditches with an optional drainage pipe in the base. The drains allow infiltration of precipitation "at the source" prior to concentration. This reduces the volume of storm runoff and reduces flood peaks by increasing ground infiltration. Percolation of first flushes can reduce contaminant loads.
- Litter Control: The use of antilitter laws and management programs for solid waste collection reduces solids loading. Reduction of pollutant loads is site-specific. Trash catchers of various screen sizes remove debris such as bottles, cans, leaves, and other refuse. The devices are effective for removing debris, but periodic maintenance is required, especially after storms. This method does not contribute significantly to chemical removal.
- Reduction of Runoff Velocity: Lowering the runoff velocity to a nonerosive level reduces the ability of the flow to transport particulate materials (especially bed load) and encourages sedimentation.
- Grassed Channels: Vegetative stabilization of drainage channels is one of the most effective methods for

controlling sediment contamination. Runoff velocity is reduced by grass channels for velocities of up to about 1.8 m (6 ft) per second (s). Therefore, the site must be correctly graded and stabilized. If used with pervious soils, treatment by infiltration is provided. Removal efficiencies are proportional to the quantity of water infiltrated.

- Overland (Sheet) Flow: Overland flow of runoff through dense grass cover is an effective mechanism for pollutant removal (screening, sedimentation, and infiltration). This technique is usually applicable prior to concentration of runoff in a channel. However, level spreaders can be used to convert channel flow to sheet flow.
- Establishment and Maintenance of Vegetation: On sites that remain bare, erosion can be significantly reduced using temporary mulch and/or seeding. Permanent seeding should be established as soon as possible after final grading.
- Traffic Control on Construction Sites, Placement of Berms, and Crushed Stone at Construction Exits: Construction roads are a particular source of pollution because of their constant use, where feasible, separate routes should be made for construction traffic during dry and wet conditions.
- Swales: Swales are vegetated open channels designed to transport, treat, and store runoff and facilitate infiltration. Cross berms can be installed in swales to enhance infiltration. By constructing a swale with low flow velocity and high infiltration capacity, impurities in runoff waters can be removed by settling, filtration, physical, chemical, and biological processes. It has been shown by Wanielista and Yousef (1993) that shoulder areas of highways are very effective for the removal of hydrocarbons, metals, and solids. By incorporating the hydrologic processes of runoff and infiltration with processes for the removal of impurities, a swale design based on quantity and quality is possible. The water quality improvement can be estimated by the mass of water infiltrated with respect to the mass input to the swale.

Swales also provide erosion control. Strict limitations on water velocities are important for preventing transport and loss of soil. Swale vegetation has been reported to be very effective for the removal of solids and retention of soil (Lord 1987). Contaminants in surface runoff, such as nutrients (nitrogen and phosphorous) and heavy metals (lead, zinc, copper, nickel, iron, cadmium, and chromium), can be reduced and retained at the site. However, regeneration and relocation of loosely bound contaminants may occur at intermediate locations throughout the length of the swale. The grassy cover in swales causes particulate contaminants to be filtered out and settle to the bottom as sediment. Comparison of the effectiveness of swales at two sites has shown better removal efficiencies at the site with a deeper water table and better infiltration. Sedimentation and filtration remove a large fraction of the total heavy metals present. Slow velocities and large contact areas enhance the sedimentation and infiltration processes. The side slopes of swales should be as flat as possible. The placement of berms (swale blocks) across the longitudinal flow path can provide additional infiltration time. Average mass removal rates for both swales are tabulated in Table 6.1 in relation

Table 6.1 Removal of dissolved heavy metals and nutrients in stormwater runoff by swales (Yousef et al. 1985)

Parameter	Average change in total mass (%)		Average change in concentration (%)		Average mass removal rate ($mg/m^2/h$)			
					Range	Average	Range	Average
	Low water table (1)	High water table (2)	Low water table (1)	High water table (2)	Low water table (1)		High water table (2)	
Lead	– 56	– 76	–	– 57	0.2–2.8	1.14	1.3–3.9	2.61
Zinc	– 93	– 77	– 86	– 62	0.7–3.1	1.85	0.8–10.7	5.76
Copper	– 70	– 49	– 17	– 8	0.1–0.8	0.42	0.5–0.6	0.6
Iron	– 89	– 35	– 69	9	4.3–30.0	15.23	1.5–9.2	5.33
Nickel	–	– 71	–	– 51	–	–	0.9–4.4	2.63
Cadmium	–	– 63	–	– 43	–	–	0.2–0.4	0.26
Chromium	– 61	– 50	– 11	0	0.3–3.4	1.85	0.3–1.5	0.9
Orthophosphate	– 61	– 46	– 24	– 9	5.1–56.0	23.8	1.4–25.8	13.6
Total phosphorus	– 63	– 42	– 25	– 3	6.9–61	27.1	1.4–26.4	13.96
Inorganic nitrogen	– 54	– 43	– 28	5	32–104	62.5	4.6–12.9	8.75
Organic nitrogen	– 28	– 41	13	8	9–51	22.8	33.9–38.7	36.3
Total nitrogen	– 51	– 41	– 11	7	41–122	85.3	42.8–46.7	44.8

$mg/m^2/h$ = milligrams per square meter per hour
(1)Average velocity = 0.03 m per second
(2)Average velocity = 0.04 m per second

Table 6.2 Comparison of physical treatment systems

Physical process	Contaminant reduction (%)					
	Suspended solids	BOD	COD	Settleable solids	Total phosphorus	Total Kjeldahl nitrogen
Sedimentation						
Without chemicals	20–60	30	34	30–90	20	38
Chemically assisted	68	68	45	–	–	–
Swirl concentrator/flow regulator	40–60	25–60	–	50–90	–	–
Screening						
Microstrainers	50–95	10–50	35	–	20	30
Drum screens	30–55	10–40	25	60	10	17
Rotary screens	20–35	1–30	15	70–95	12	10
Disc screens	10–45	5–20	15	–	–	–
Static screens	5–25	0–20	13	10–60	10	8
Dissolved air flotation	45–85	30–80	55	–	55	35
High-rate filtration	50–80	20–55	40	55–95	50	21

to contact area and time of flow. Although the results are site-specific, they indicate that longer residence time in a swale and greater contact area (product of wetted perimeter and swale length) result in higher contaminant removal rates. The effectiveness of swales in karst terrane is limited because of the risk of sinkhole collapse associated with the infiltrating runoff.

- Sedimentation ponds: Originally used to control sediment eroded from agriculture lands, sedimentation ponds are now being used in urban environments to remove sediment from runoff to protect wetlands and wet-detention ponds from siltation. The physical configuration of a pond consists of an inlet control, settling volume, debris and sedimentation volume, and an outlet. Dual-purpose sediment ponds are used to manage peak discharge and remove sediment.
- Physical treatment: Physical treatment is similar to point-source treatment. It includes microstrainers, high-rate filtration, and dissolved air flotation. With a loading rate of 815 L per minute per square meter (lpm/m^2), or 20 gallons per minute per square foot (gpm/ft^2), microstrainers can remove 40–50% of the 5-day biochemical oxygen demand (BOD$_5$) and 70% of the TSS. High-rate filtration (980 lpm/m^2, or 24 gpm/ft^2) can remove 40–55% of the BOD$_5$ and 80% of the TSS, while dissolved air flotation can remove 50–60% of the BOD$_5$ and 80% of the TSS at 102 lpm/m^2, or 2.5 gpm/ft^2. A comparison of these physical treatment systems is shown in Table 6.2.
- Terraces and runoff spreaders: Terraces and runoff spreaders increase the time of concentration by increasing the length of the overland flow. The time of concentration is the time required for water to flow to a specified location from the farthest point in a watershed or the time at which the rate of runoff equals the rate of precipitation of a storm. When the time of concentration increases, the hydrograph runoff intensity peak is lowered. Reduction of runoff intensity can be achieved by spreading runoff or by directing it into a system of terraces.
- Peat filters: Peat filters have been used in the wastewater treatment industry for many years. However, their beneficial effects are not entirely understood for stormwater management. This is partially due to the variable nature of precipitation events and the range of temperatures at each site. Peat filtration of highway runoff has been documented. Keith et al. (1995) discussed the installation and contaminant removal efficiency of rock and peat filters in sinkholes receiving highway runoff. Preliminary findings indicate that one peat filter has been removing approximately 80% of suspended particulate materials (measured as TSS and selected total recoverable metals) and about 50% of the dissolved copper and zinc. Rock filters have been removing approximately 33–76% of the TSS and 35–55% of the total recoverable metals. However, rock filters do not appear to remove dissolved metals. While small upgradient detention basins have been installed to capture spills of hazardous materials and to protect the filter systems from sedimentation, the effective life span of the filters remains to be determined.

The performance of a peat and compost filter was demonstrated to be outstanding in removal of total suspended solids, typically providing 90–95% removal (Stewart 1992). The performance for nutrients is less significant. In fact, organic media may be a source of soluble phosphorus and nitrate. Total phosphorous removal ranges up to 49%, while variable removal of metals is typically between 48 and 90%. Removal of oil and gasoline averages about 90% (Claytor and Schueler 1996).

- Catch basins: A catch basin combines a drainage inlet for intercepting stormwater runoff with a storage chamber for particulate material. The finer solids associated with most of the pollutant load are not effectively removed. In addition, most catch basins are not adequately maintained for pollutant removal. The solids accumulated in catch basins can be re-entrained by runoff inflow and can contribute significant pollutant loads to the receiving waters. The original purposes of catch basins were to prevent sewer clogging by trapping materials from gravel roads and to provide a water seal to prevent odor from combined sewers. Catch basins have been advocated as a stormwater management measure for trapping particulate materials, oil, and grease. When properly designed and maintained, catch basins effectively remove coarse solids from stormwater runoff.
- Dry detention basins: Dry detention basins attenuate runoff without a permanent pool of water. Efficiencies are typically half of that obtained using wet ponds. Dry detention basins are used for flood alleviation and drainage structure cost economy. Time of concentration is increased, and peak discharges are decreased. Stormwater runoff peak flow rates are reduced by storing floods and releasing the water from storage at a lesser rate over a longer period. Sedimentation pond sizes can be estimated using settling velocity distributions, rainfall intensity, and watershed characteristics. Many computer models have been developed which can expedite these calculations. This is potentially disastrous to water quality if the designer does not understand the principles (such as detention time) of pollution abatement techniques. Average detention time is typically calculated by dividing the average pond volume by the average effluent flow rate. This generally results in a detention time that is only a few hours, which is inadequate for settling of the smaller fractions of suspended materials associated with pollutants. Materials which are settled out during a runoff event may be washed out by a subsequent event, sometimes resulting in negative pollution abatement performance during the later event.

Based on the results of several field studies (USEPA 1983; Grizzard et al. 1986), it appears that a properly designed extended detention basin can be expected to achieve the following long-term removal rates (expressed as percentages): TSS (50–70), total phosphorus (10–20), nitrogen (10–20), organic matter (20–40), lead (75–90), zinc (30–60), hydrocarbons (50–70), and bacteria (50–90).

USEPA (1986) and DiToro and Small (1979) presented methodologies for planning and design of detention ponds. After the particle size and particle volume distributions associated with the pollutants in the stormwater samples have been determined, the required removal of the various particle sizes needed to achieve the desired water quality is estimated. The hydraulic surface loading rate is determined from the estimated settling velocities for all representative particle sizes. Once an adequate surface loading rate has been determined, the surface area of the pond and its configuration can be designed. The hydraulic surface loading rate can be based on the average flow-through rate in the storage basin (Urbonas and Stahre 1993).

Upon identification of a potential highway stormwater runoff pollution problem, management measures can be implemented to reduce the impact on receiving waters. Effective pollution management measures include vegetative controls, infiltration basins, wet detention basins, wetlands, and other methods. Effectiveness of operation is measured by the reduction of contaminant concentrations and mass. A mass balance is fundamental for analyzing the efficiency of any stormwater treatment system. While the removal efficiencies of stormwater treatment practices vary greatly and the data are site-specific, they are useful as guides. Vegetative controls, wet detention basins, infiltration systems, and wetlands can be used individually or in combination. Significant reductions in pollutant loads from highway runoff can result when the general guidelines are followed (Maestri and Lord 1987).

Treatment systems using swales and soil filtration should be designed to maximize contact with the soil. Because dissolved hydrocarbons appear to be a significant component of highway stormwater runoff (Asplund et al. 1980; Zawlocki et al. 1980), sedimentation and/or filtration may be inadequate for removing all the pollutants from runoff. Adsorption or other removal methods may be required.

The U.S. Geological Survey (USGS) has been collecting data relevant to the design of stormwater runoff treatment systems for several decades, including urban rainfall, runoff, and water quality data. This is compiled into a database, which includes concurrent time-series values of rainfall, discharge, water quality analyses, and watershed characteristics. Its purpose is to present storm loads and characteristic data for most of the stations included in the USGS urban stormwater database. The data can be used to perform statistical analyses of urban stormwater loads based on factors such as region, drainage area size, and land use (Driver and Tasker 1988; Mustard et al. 1987).

- Vegetative controls: Vegetated surfaces can reduce runoff velocity, enhance sedimentation, filter suspended solids, and increase infiltration. Vegetative controls increase pollution removal and decrease soil erosion. They are the primary measures for many highway runoff situations and are very effective when properly installed and maintained. The use of vegetative controls is influenced by the following factors: topography, soils, space, climate, and erosion.

- Wet detention basins: Wet detention basins, or those that maintain a permanent pool of water, have been found to be a highly effective measure for controlling stormwater runoff quality. When space limitations make the use of vegetated ditches impractical, wet detention basins can be utilized. A permanent, underground water storage pool can be used to attenuate runoff if the ground surface is fully utilized. Sedimentation is the principal mechanism for the removal of particulate pollutants. Detention of runoff increases the time of concentration and reduces the maximum runoff discharge rate. Removal efficiency ranges for most well-maintained designs are 25–40% for BOD_5, 30–70% for heavy metals, and about 40–90% for TSS (Wanielista and Yousef 1993). In addition, some basins exhibit substantial reductions in soluble nutrients such as phosphorus, nitrate, and nitrite. This may be attributable to biological activity in the permanent pool.
- Infiltration systems: An infiltration system provides for temporary runoff storage, allowing a large portion of the runoff to recharge groundwater. There are several types of infiltration drainage systems, in which the runoff can be held on or below the ground surface. Runoff is collected in various storm systems prior to being directed into the basin. Infiltration techniques include open basins, trenches, and infiltration wells. Proper design requirements must be met for infiltration systems to provide effective reduction of highway contamination. The requirements include the use of soils and subsoils that are moderately to highly permeable, a water table that is at least 3 m below the bottom of the infiltration point, runoff inflow that is relatively free of suspended solids, and sufficient storage for the runoff event during the infiltration period. Originally designed as management measures to control stormwater runoff or recharge groundwater resources, infiltration systems are being utilized for reducing pollutant loads. Offline retention (exfiltration or other treatment) system efficiencies can be estimated from a frequency distribution for rainfall volumes. A simulation using a mass balance equation for pond volume is needed to design an offline system of adequate size to treat a fraction of the runoff from each storm event. Other equations can be developed to approximate the mathematically complex first flush. Efficiencies vary between 50 and 95% depending on the recharge basin. Periodic inspection of existing sewers must be performed to ensure that storage and treatment are available for stormwater (Wanielista and Yousef 1993).
- Wetlands: Wetlands are areas where the water table is at or near the ground surface, where the land is inundated by relatively shallow water, or where the land supports aquatic vegetation. Wetlands often occur at the interface between terrestrial and aquatic systems. They are generally characterized by high floral productivity and nutrient requirements, high decomposition rates, low oxygen content in sediments and substrates, and large adsorptive surfaces in the substrate. A wetland treatment system is a variation of detention pond methodology because it removes runoff pollutants primarily through sediment retention and vegetative uptake. Wetlands can provide highly effective highway runoff management because they can absorb large quantities of suspended and dissolved materials. Wetland designs differ from conventional detention systems by being shallower, using vegetation as a pollutant removal mechanism, and emphasizing slow-moving sheet flow within the wetland (Wanielista and Yousef 1993). Linker (1989) and Schiffer (1990) also address the use of wetlands for treating urban and highway runoff.

The use of wetlands has been suggested for the removal of nutrients from urban stormwater since the mid-1980s. It has been hypothesized that wetlands can use nutrients in their growth and thereby may reduce concentrations in the water column. Studies (Kedlec and Hammer 1980; Nichols 1983; Watson et al. 1989) have shown that wetlands can be used effectively for wastewater treatment if nutrient loading rates are not excessive.

6.1.3 Impacts of Highway Runoff on Karst Aquifers

Although the impacts of highway stormwater runoff have been investigated by numerous researchers, relatively little attention has been directed toward assessing its impact on groundwater in karst areas. As noted by Barrett et al. (1993), most studies of highway runoff impact on groundwater have been conducted in areas with relatively thick soils, which attenuate many runoff pollutants. However, Betson and Milligan (1985) and Milligan and Betson (1985) note that urban stormwater runoff may have a more significant impact on groundwater quality in karst terranes, where soils may be thin or nonexistent. Moreover, there is little information in the literature to facilitate a direct comparison of highway-impacted and ambient groundwater quality in karst aquifers. The limited information regarding highways in karst areas is reviewed below.

Keith et al. (1995) report the following ranges of concentrations (in µg/l (ppb), except as otherwise stated) for highway runoff contaminants in the karstic Mitchell Plain of Indiana: total chromium (11–94), total copper (13–80), total lead (4.1–89), total nickel (8.9–69), total zinc (100–310), dissolved copper (below detection limit to 9.9), dissolved zinc (4.7–25), and TSS (69–2300 mg/l (ppm)). They also present a preliminary evaluation of the effectiveness of a highway runoff treatment system. In a study of urban stormwater runoff in Tennessee, Betson (1977) reports that water quality was not significantly degraded, compared to that in a rural area. However, he notes that metals associated with eroded soil are important contaminants. His study suggests that many of the detected constituents are derived from the atmosphere and that urbanization does not appear to contribute significantly to water quality degradation, even in areas of karst groundwater recharge. His interpretation is that the karst watersheds in his study were apparently functioning as contaminant filters or sinks. Saleem (1977) reports that chloride concentrations associated with highway deicing vary with soil thickness and season in a karstic dolomite aquifer in Illinois. Bradbury and Muldoon (1992) demonstrate a high degree of correlation between highway locations and zones of elevated chloride levels in the karstic dolomite of Door County, Wisconsin. Werner (1977, 1983) also reports increased chloride concentrations in karst groundwater resulting from highway deicing salts.

Garton (1977) discusses the groundwater impacts associated with construction of an interstate highway through a West Virginia karst watershed. The aquifer supplies springs used by the Bowden National Fish Hatchery. During storm events, clay and silt from the construction site were transported through the cavernous aquifer, greatly increasing turbidity at the springs. More than 150,000 trout were killed during one storm. Additional fish kills resulted from spills of diesel fuel at the construction site, and construction activities have reduced spring discharge by diverting water away from the karst aquifer. Similar effects have been noted at another West Virginia fish hatchery by Werner (1983), who also notes that constructing highways through karst aquifers increases the risk of hazardous materials impacts to groundwater. Hubbard and Balfour (1993) address potential groundwater impacts from the development of a highway through the recharge area of a cave stream in Virginia. Mull et al. (1988a, b) mention the potential for pollution from highway stormwater runoff in a sinkhole plain at Elizabethtown, Kentucky, where Garcia (1992) discusses three karst watersheds associated with municipal supply wells and springs. Based on data from that investigation, highway-associated metals (cadmium, lead, nickel, and zinc) do not appear to be a significant component of stormwater runoff. Barner and Miller (1991) review sinkhole drainage, land use restrictions, sediment control, and the use of detention basins in Springfield, Missouri. Smith (1991) discusses stormwater runoff, sedimentation, and associated pollutants in the karstic Edwards Aquifer of Texas. He notes that road-related runoff has increased levels of lead, total petroleum hydrocarbons (TPH), metals, and oil and grease.

Numerous reports address the environmental and engineering impacts of highways in Tennessee karst areas, including Moore (1984, 1987, 1988) and Moore and Amari (1987). Documented problems include induced collapse and collapse, bridging of caves, and flooding and sedimentation of sinkholes and swallets. Moore (1984) and Moore and Amari (1987) discuss swallet protection in the context of maintaining drainage and flood prevention. Although water quality data are not reported, it is likely that the use of gabions as recommended by the authors has a positive effect on runoff water quality by reducing the amount of sediment and associated contaminants reaching a sinkhole. Kalmes and Mohring (1993, 1995) discuss similar engineering techniques used in Minnesota sinkholes to control erosion and improve groundwater quality in agricultural areas.

Additional publications address engineering considerations associated with highway-induced sinkhole collapse in karst areas of Alabama (Newton 1987) and Pennsylvania (White et al. 1984). Problems of sinkhole collapse and flooding in Tennessee are noted by Kemmerly (1988) and Royster (1984), and karst runoff and flooding issues are reported for Kentucky (Crawford 1988; Currens and Graham 1993; Reeder and Crawford 1988), Yugoslavia (Mijatovic 1987), and Jamaica (Molina and McDonald 1987). Groundwater contamination events are reported from catastrophic sinkhole collapses in Florida retention basins (Jenkins 1988) and Minnesota wastewater treatment facility lagoons (Alexander et al. 1993; Jannik et al. 1991). Retention basin failures are also addressed by Warren (1974). Problems of karst groundwater contamination are compounded by using drainage wells to increase runoff and minimize flooding in areas with inadequate drainage (Hoos 1990). Removal of soil from sinkholes to facilitate drainage increases contaminant levels in groundwater by reducing the natural attenuation of suspended contaminants. Kemmerly et al. (1987) describe a drainage well in a karst area of Tennessee. Crawford and Groves (1984) discuss stormwater drainage wells in karst areas of Tennessee and Kentucky. In addition to a brief treatment of well design, they address problems of groundwater contamination and siltation. Stormwater runoff and drainage wells in the urbanized sinkhole plain of Bowling Green, Kentucky, are also considered in other publications by Crawford (1988).

In a study of urban stormwater runoff in Clarksville, Tennessee, Hoos (1990) notes that concentrations and loads of some contaminants are significantly less in runoff at a drainage well than in the discharge of a downgradient karst spring, suggesting that the drainage well is only one of several pollution sources. However, some roadway-associated contaminants (arsenic, copper, lead, organic carbon, and oil and grease) are contributed to local groundwater primarily by the drainage well. Trace metals are flushed into the aquifer early during each runoff event. Between February and October 1988, calculated metals loads ranged from 13.61 g (0.030 pounds) (cadmium) to 5.4 kg (12 pounds) (strontium), and nitrogen loads ranged from 0.44 kg (0.97 pounds) (nitrate as nitrogen) to 15.4 kg (34 pounds) (organic nitrogen). Beck and Zhou (2002) conducted a more comprehensive study on management of the discharge and quality of highway stormwater runoff in karst areas. This study evaluated the impacts of highway stormwater runoff in karst areas in 15 states, identified sensitive areas that may be affected by highway stormwater runoff, and designed a practical remedial measure applicable to treating highway runoff in karst settings.

6.2 Groundwater Monitoring Under Natural Conditions

6.2.1 Site Description

The study site is at an interchange in Tennessee. The subject interchange was originally constructed in the 1960s and reconstructed in 1981–1982. It has a drainage area of 24 ha (59 acres). Approximately 76,000 vehicles passed through the interchange each day in 1996 and this number increased to 86,000 in 1999. The site is underlain by sinkhole-forming Ordovician limestone of the Holston Formation which has been thrust-faulted over younger sandstones of the Chapman Ridge Formation (Cattermole 1966). Figure 6.1 presents a site map showing the karst features and flow routes of the stormwater runoff.

Stormwater runoff from the interchange drained into a recently collapsed sinkhole (New Sinkhole in Fig. 6.1) by three routes. The most significant is a 107 cm concrete culvert, which received drainage from several inlets in paved and grassy areas throughout the interchange. This pipe is referred to as the "large culvert." In addition, a shallow, 1.8 m wide, concrete-lined ditch (referred to as the "ditch") carried runoff from the highway surface and an adjacent grassy area to the discharge end of the large culvert. From that location, runoff flowed along a 3 m-wide, concrete-lined channel (referred to as the "channel") into the sinkhole, which collapsed beneath the end of the channel. Finally, a 46 cm corrugated-metal culvert transmitted drainage directly from the paved surface in the vicinity of a concrete median barrier. Referred to as the "small culvert", this pipe discharged stormwater from the embankment adjacent to the highway (approximately 4.5 m above the level of the channel). Runoff from the small culvert dropped approximately 1 m to the ground and flowed down a steep, vegetated slope and joined runoff from the large culvert and ditch in the channel before flowing into the sinkhole. This channel ended abruptly in a 4.3 m precipice at the edge of the sinkhole.

Water flowing into the new sinkhole passes through two small karst windows and is subsequently discharged at Holston Spring approximately 128 m from the site. A schematic diagram illustrating the groundwater flow is presented in Fig. 6.2. Dye tracing by Ogden (1994) indicates that this spring discharges groundwater from a long, narrow drainage basin encompassing a variety of urban land uses.

Karst features and their locations

New sinkhole

Fig. 6.1 Testing site on the treatment of highway stormwater runoff

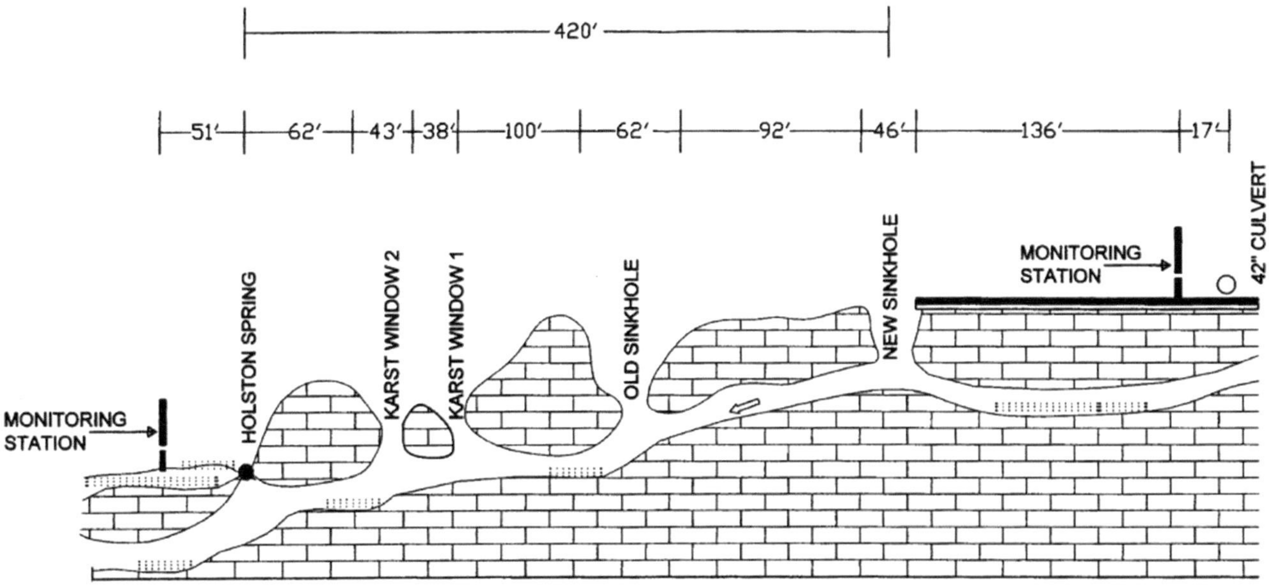

Fig. 6.2 Conceptual groundwater flow model

6.2.2 Runoff Quantity and Spring Discharge Monitoring

The discharge of highway runoff entering the sinkhole and of water resurging from Holston Spring have been monitored continuously since July 1995 to assist with the design of a pilot-scale treatment system and to assess the data from quantitative dye-tracing experiments. Measurement of discharge weighted calculation of pollutant loads before and after treatment. These data are necessary for evaluating the impact of the treatment methodology on the karst aquifer.

Sharp-crested weirs were initially constructed at the spring and sinkhole sites. The weirs were designed and installed to permit cost-effective monitoring of runoff and discharge within a reasonable range of expected hydrologic conditions. The weir installed in the spring was a Cipolletti weir. Compared with a triangular (90° V-notch) weir, a Cipolletti weir can accommodate larger discharge events during heavy storms. The runoff monitoring station at the sinkhole site included a compound weir, combining a Cipolletti with an incised triangular (90° V-notch) weir. This design is very sensitive to small runoff events, but it is also capable of accommodating large volumes of runoff. At each site, a relatively straight section of channel was selected, and the weir was installed perpendicular to the direction of water flow. The constructed weirs have provided essential baseline data for the site. However, they required frequent maintenance, especially in the spring because of chronic leakage. Large tree roots in the stream banks interfered with sealing the edges of the weir and provided preferential pathways for water to bypass it. The pressure associated with the impounded water exacerbated the erosion beneath and beside the weir. Burrowing crawfish may have contributed to the problem as well. Maintenance during the summer consisted of periodically reinforcing the edges of the weir with bentonite and wire mesh. During the winter, sub-zero temperatures and subsequent thawing weakened the bentonite seal, challenging the accurate measurement of the flow.

In 1996, an alternative method for calculating the spring discharge was implemented. Concrete blocks were used to construct a smooth bottom across a section of the stream channel. A current meter was used to measure the stream velocity at regular intervals across the uniform cross section. Discharge for each interval was determined by multiplying the velocity and the cross-sectional area of the interval. Spring discharge was calculated by summing the values obtained for each interval of the stream cross-section. By conducting measurements under a variety of flow regimes and recording water level data, a stage-discharge relationship, or "rating curve", was developed. The rating curve allows the calculation of discharge values from the water level data, which was collected and recorded on a continuous basis.

Maintenance at the sinkhole monitoring station involved removal of silt and debris within the basin detained by the weir. In May 1996, a heavy rainstorm of 0.4 m^3/s or 15 cubic feet per second (cfs) of runoff destroyed the plywood-made weir and washed it and the accessories into the sinkhole. After an extensive search for a suitable monitoring device, a Flo-Tote™ manufactured by Mash-McBirney, Inc., was selected as the alternative. The device was installed into the 107 cm pipe in July 1996 and has been providing reliable data since then.

Preliminary analysis of hydrographs from the sinkhole and spring indicated that a direct connection existed between the sinkhole and Holston Spring. Stormwater runoff draining into the sinkhole from the interchange produced a sharp rise in the water level (and discharge) at the spring, as shown in Fig. 6.3. This increase occurred almost immediately after the drainage began flowing into the sinkhole, which suggests that these features are connected by a phreatic conduit. While the actual stormwater runoff did not arrive at the spring immediately, it displaced water already flowing through the conduit from other parts of the groundwater drainage basin.

The peak discharge values at the spring were sometimes higher than the rate of water entering the sinkhole for the same storm. This confirms that the drainage basin for the spring extends well beyond the site, as indicated by Ogden (1994). However, the hydrographs revealed that storm peaks at the spring were only associated with runoff into the New Sinkhole, suggesting that stormwater entering the aquifer from other sinkholes in the drainage basin was regulated by partial blockage of the conduit and/or its tributaries by urban development. Other evidence supporting this hypothesis included significant flooding problems in the upgradient sinkholes.

6.2.3 Dye Tracing Test

Quantitative dye tracing is useful for investigating aquifer characteristics such as conduit volume and solute-transport characteristics such as velocity, persistence, and dispersion. It involves the introduction of a known quantity of dye and monitoring its concentration through time at a downgradient location. The dye introduced into the New Sinkhole was a 19.8% solution of rhodamine WT (Ci acid red 288) with a density of 1.129 g/cm^3 (or g/mL). Introduction of 100 mL (22.4 g) of dye began at 11:19 h on November 2, 1995. The dye was introduced over a period of approximately 20 min. Dye was introduced with natural stormwater runoff, which was flowing into the sinkhole at a rate of 0.08–0.1 m^3/s (3–4 feet^3/s).

Dye recovery was monitored at Holston Spring, where grab samples were collected and analyzed in the field with a fluorometer. Field analysis was conducted to permit adjustment of the sampling every 30 s (near the peak concentration) to one sample every 15 min (during the recession of dye concentration). Between 11:00 and 17:00, 170 samples were analyzed on-site. However, because field conditions are not ideal for fluorometric analysis, 129 of the samples were also analyzed in the laboratory the following day. This

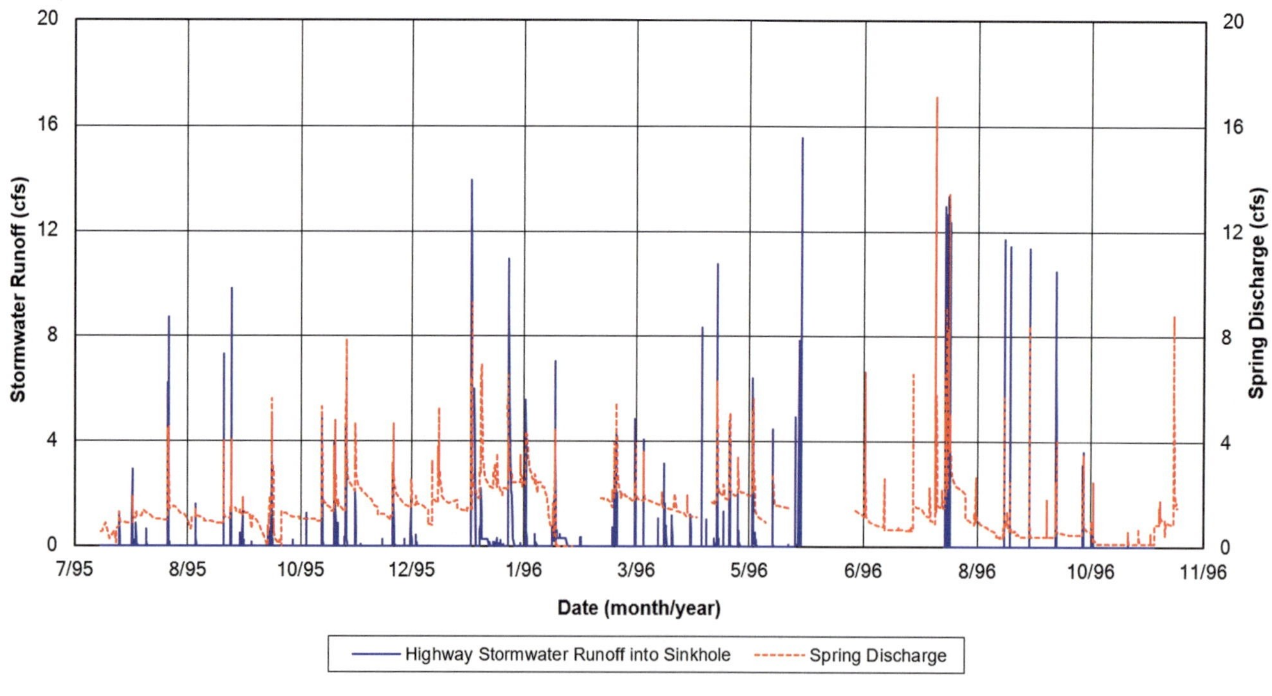

Fig. 6.3 Hydrographs of highway stormwater runoff into sinkhole and discharge from Holston Spring

permitted more precise calibration of the fluorometer and provided time for temperatures to stabilize and turbidity to settle in the samples.

Figure 6.4 shows the dye breakthrough curve, as well as the spring hydrograph for the period of the dye trace. Low levels of dye began to arrive at the spring approximately 20 min after the middle of the dye-introduction interval. The peak concentration (39.9 μg/l or 39.9 ppb) was measured approximately 69 min after dye introduction. The spring water was slightly discolored, but the dull red color was obscured by turbidity resulting from the storm event. Dye concentration had dropped to 0.301 μg/l (0.301 ppb) when the last sample was collected at 17:00—5.5 h after dye introduction. Analysis of a grab sample collected during storm flow almost 5 days later indicated that the rhodamine WT concentration was only 0.034 μg/l (0.034 ppb) above the background level measured at the beginning of the tracer test. Most of the dye had been flushed from the aquifer within 5 days.

Stormwater runoff into the sinkhole and water discharge from the spring were monitored during the dye trace. Staff gauge readings at the spring were recorded periodically to permit verification of the data collected by the water level recorder. Figure 6.5 shows hydrographs for stormwater runoff into the sinkhole and discharge from the spring. The resolution of the spring hydrograph was greater than that for the sinkhole because the data collection frequency of the water level recorder at the spring was increased during the dye trace. Therefore, temporal comparisons between the hydrographs should be made cautiously. The stormwater runoff from the highway interchange reached a peak value of 0.11 m^3/s (3.8 cfs), and spring discharge reached a peak of 0.14 m^3/s (4.8 cfs). Both peaks occurred approximately 40 min after dye introduction.

The increase in discharge at the spring did not necessarily correspond with the arrival of highway runoff from the sinkhole. Rather, increased spring flow might result from the displacement of water already present in the karst conduit by the stormwater runoff draining into the sinkhole. This scenario was supported by two observations: (1) the peak discharge at the spring occurred at approximately the same time as the peak highway stormwater runoff into the sinkhole; and (2) the peak dye concentration did not arrive at the spring for approximately 40 min after the maximum discharge value. This phenomenon was similar to what occurred in a water-filled, U-shaped tube when dyed water

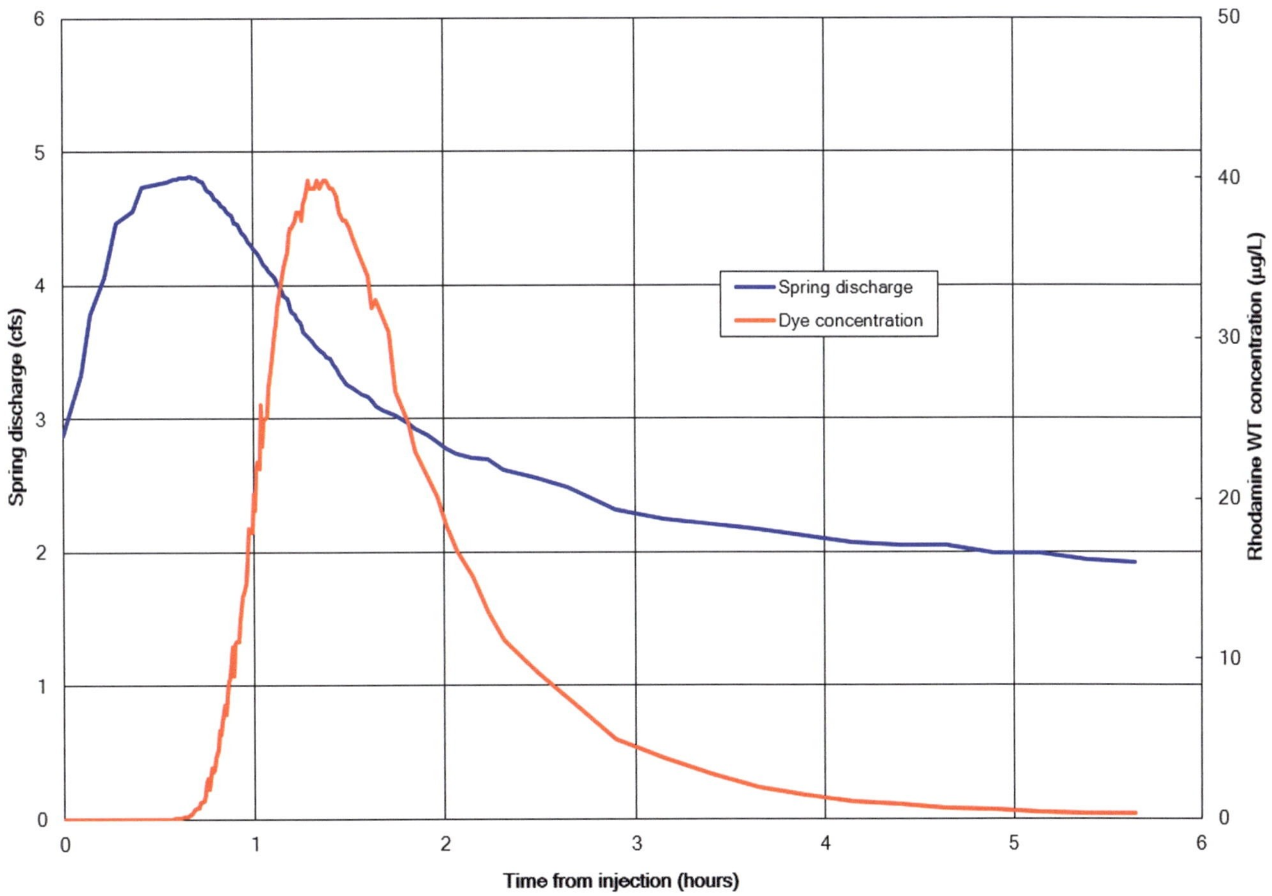

Fig. 6.4 Dye breakthrough curve and hydrograph for Holston Spring during quantitative dye trace

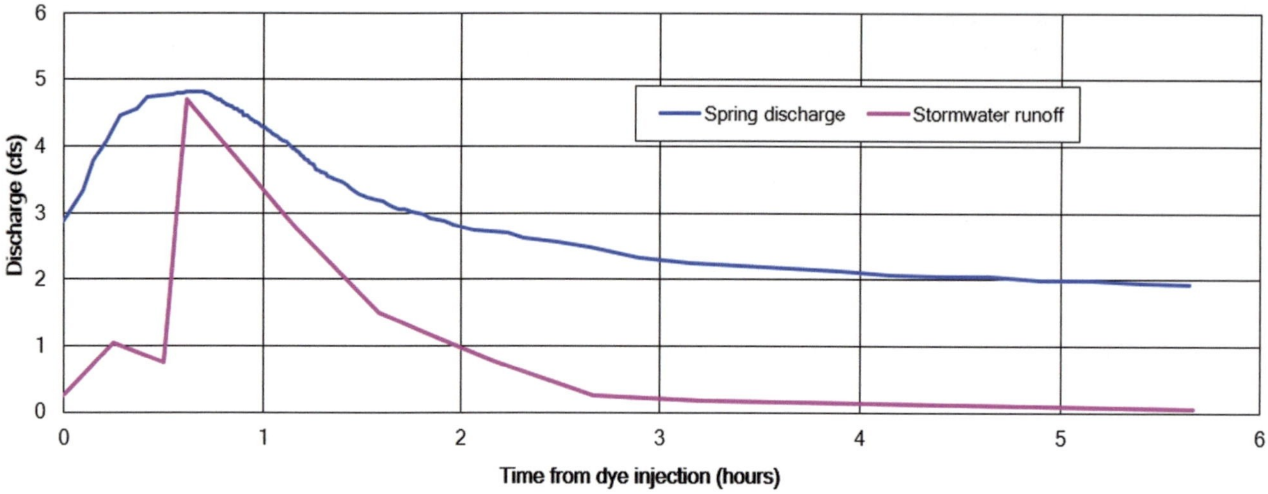

Fig. 6.5 Spring hydrograph and stormwater runoff at the interchange during quantitative dye trace

was poured into one end. The water level rose instantaneously at the other end of the tube, but a longer time was required for the dyed water to flow through the tube.

The dye load (total dye mass recovered) at the Holston Spring was estimated from the dye recovery rate or dye loading curve (Fig. 6.6). Dye loading (recovery rate) was calculated by multiplying the measuring dye concentration and spring discharge for each time interval during the dye recovery. The peak loading for this trace was 4077 μg/s. The calculated dye load was 15.4 g, which was 68.7% of that introduced into the sinkhole. Although recovery of less than 100% of the dye may suggest that some water from the sinkhole did not flow through Holston Spring, it might also be a product of the sampling strategy and non-uniform mixing of the dye across the width of the stream.

The mean travel time for the dye was 1.7 h, with a standard deviation of 0.7 and a skewness coefficient of 0.14. Half the dye (by mass) passed through the spring about 1 h and 40 min after dye introduction. Assuming that the sinkhole and spring are connected by a straight conduit, the mean velocity of dye being transported was 0.024 m/s (0.077 feet/s) or 2.07 km/day (1.26 mile/day). This value should be considered a minimum velocity because it assumed that the conduit was straight. Actual velocity could be as much as 1.5 times that value. Assuming the dye-laden water moved through the conduit as a discrete slug, the conduit volume could be estimated by integrating the spring discharge between the time of dye introduction and its mean travel time, which was best represented by the centroid (mass-weighted mean) of the dye loading curve. Using this method, the volume of the conduit between the sinkhole and spring was estimated to be approximately 592 m^3 (1.1 L, 20,900 $feet^3$).

For each set of flow conditions under which a quantitative dye trace is conducted, interpretation of the results can provide valuable information about the transport of dissolved contaminants. Such information is essential for designing effective sampling strategies for a particular karst aquifer. Sampling a karst spring only during peak discharge would be inadequate for characterizing the range of water quality conditions.

6.2.4 Highway Stormwater Quality

Highway stormwater runoff and spring flow were sampled at the study site during a rainfall event on January 18, 1996. Samplers arrived on site at 13:45 to prepare for the approaching storm. Sample collection occurred between 16:13 and 22:35 during the first pulse of runoff from an event, which continued throughout the night. Precipitation was light and intermittent during several small rain showers between 15:20 and 19:20. A small amount of highway runoff began flowing from the small culvert at 16:10. During this period, four samples (A-1, A-3, A-5, and A-7) were collected from the small culvert. However, there was a minimal amount of runoff, all of which infiltrated the ground before reaching the channel or the sinkhole. Precipitation increased significantly at 19:20, and runoff from the small culvert began flowing into the sinkhole at 19:30. Runoff began flowing from the large culvert at 19:38, but the flow of stormwater was not sufficient for sample collection during the first 3 min. The first sample of that runoff (A-2) was collected at 19:41. After filling the basin at the discharge monitoring station, runoff from the large culvert began flowing through the weir at 19:49 and into the sinkhole soon thereafter. Runoff in the adjacent ditch was not sampled. It contributed a relatively small amount of the total drainage into the sinkhole, and its runoff did not appear to be highly contaminated. Therefore, the contaminant load contributed by the ditch was assumed to be relatively minor.

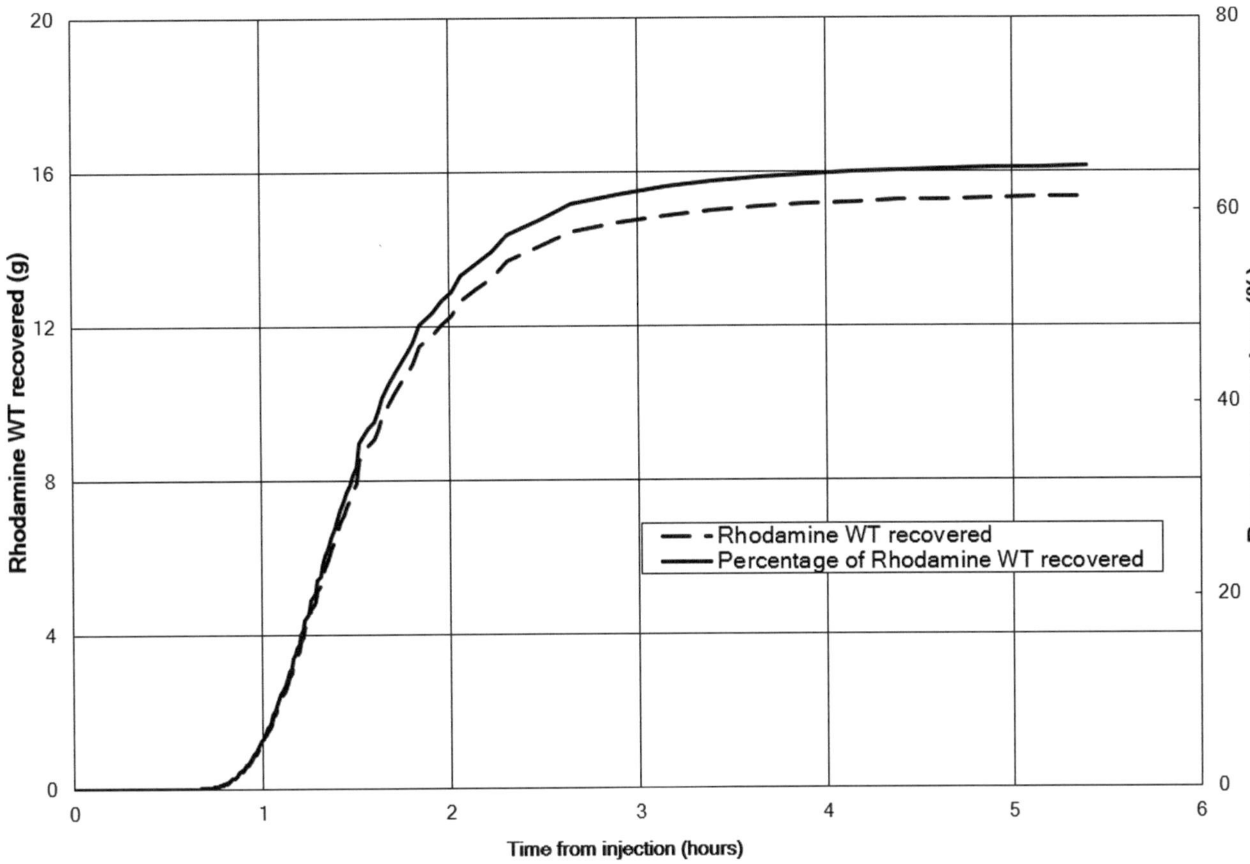

Fig. 6.6 Dye recovery process during quantitative dye trace

Table 6.3 presents the samples collected and analysis results. Beginning with sample A-12 at 20:56; the combined runoff from both culverts was sampled where it flowed into the sinkhole. The final sample was collected at 21:56. Runoff from the small culvert slowed to a trickle soon after precipitation stopped at 21:05, but runoff continued flowing from the large culvert until the subsequent rainfall began at approximately 22:30. A total of 16 samples were collected at the sinkhole site, including six from the small culvert, five from the large culvert, and five from the edge of the sinkhole.

Each sample was collected manually with a dedicated 3.8-L plastic pitcher and distributed into labeled sample containers (with preservatives) as appropriate for each of the following constituents: dissolved Pb and Zn, total Pb and Zn, PAH, TPH, TDS, total suspended solids (TSS), and total volatile solids (TVS). Samples to be analyzed for dissolved Pb and Zn were filtered immediately after collection (and before acidification) using 0.45-μm filters. Because of the high turbidity of the runoff, filters with large surface areas were required to minimize clogging. Following filtration, preservation, and containerization, each sample was recorded on a chain-of-custody form and packaged in a cooler with ice for shipment to the laboratory.

During the runoff event, flow from the large culvert (including the ditch) was monitored by the water level recorder at the weir and confirmed with periodic observation of the staff gauge. The data collection speed of the water level recorder was increased during the sampling event to increase the resolution of the hydrograph. Flow from the small culvert was measured at regular intervals by recording the time required to fill a container of known volume.

Figure 6.7 presents hydrographs for each culvert, as well as for the combined flow from both culverts entering the sinkhole. The maximum runoff was 0.22 m^3/s (8 cfs) from the large culvert including the ditch, which contributed 507 m^3 (17,913 $feet^3$) of runoff to the sinkhole. The peak runoff was 0.01 m^3/s (0.4 cfs) from the small culvert, which contributed 29.6 m^3 (1044 $feet^3$) of runoff to the sinkhole. Runoff from the small culvert accounted for only 5% of the highway drainage flowing into the sinkhole. Even though runoff from the small culvert contained the highest contaminant concentrations, its volume and contaminant load were minor, relative to runoff from the large culvert. The total volume of water flowing into the sinkhole exceeded 536 m^3 (18,957 $feet^3$).

Table 6.3 Highway stormwater runoff and spring water quality

Sample	Time	Concentration (Mg/L)*								
	01/18/96	Dissolved Zn	Total Zn	Dissolved Pb	Total Pb	TPH	PAH	TDS	TSS	TVS
Highway stormwater runoff into sinkhole										
Small culvert										
A-1	16:13	2.40	7.90	BDL	0.374	8.4	0.178	3584	1485	490
A-3	16:29	0.266	1.16	BDL	0.113	4.4	0.155	1772	347	134
A-5	16:40	0.378	0.932	BDL	0.054	3.7	0.140	1772	204	76
A-7	17:00	0.459	1.02	BDL	0.053	3.8	0.136	1820	240	104
A-9	19:56	0.034	1.49	BDL	0.149	126	0.162	236	887	227
A-11	20:26	0.026	0.719	BDL	0.067	8.4	0.131	276	640	173
Large culvert										
A-2	19:41	0.042	0.074	0.005	0.008	0.36	0.012	512	80	80
A-4	19:56	0.028	1.018	BDL	0.172	4.8	0.120	612	653	180
A-6	20:11	0.018	0.791	BDL	0.052	2.7	0.087	308	540	112
A-8	20:26	0.022	0.190	BDL	0.027	1.4	0.041	272	260	107
A-10	20:41	0.009	0.157	BDL	0.024	0.82	0.047	312	84	52
Combined flow into sinkhole from both culverts										
A-12	20:56	0.030	0.097	BDL	0.017	0.90	0.044	328	52	36
A-13	21:11	0.035	0.068	BDL	0.014	0.44	0.034	376	52	52
A-14	21:26	0.034	0.037	0.003	0.010	0.13	0.032	384	105	75
A-15	21:41	0.035	0.037	BDL	0.009	BDL	0.030	384	32	28
A-16	21:56	0.020	0.020	BDL	0.005	0.36	0.028	424	70	60
Holston spring										
B-1	16:17	BDL	BDL	BDL	BDL	BDL	0.001	396	32	32
B-2	19:45	BDL	BDL	BDL	BDL	BDL	0.001	388	24	24
B-3	20:00	BDL	BDL	BDL	BDL	BDL	0.005	372	4	4
B-4	20:15	BDL	BDL	BDL	BDL	BDL	0.002	356	10	10
B-5	20:30	BDL	BDL	BDL	0.003	BDL	0.001	364	8	8
B-6	20:45	0.032	0.032	BDL	0.010	BDL	0.015	296	128	64
B-7	21:00	0.040	0.171	BDL	0.026	1.2	0.038	404	48	36
B-8	21:15	0.037	0.208	BDL	0.031	1.6	0.046	344	336	80
B-9	21:35	0.046	0.161	BDL	0.025	1.2	0.046	364	240	92
B-10	22:05	0.037	0.055	BDL	0.012	0.43	0.029	320	184	96
B-11	2:35	0.031	0.053	BDL	0.007	0.12	0.017	332	<1	<1

BDL—Below detection limit
*mg/l is equivalent to ppm

A background sample (B-1) was collected at Holston Spring at 16:17 before any runoff entered the sinkhole. The second sample (B-2) was collected at 19:45, 15 min after highway runoff began flowing into the sinkhole. Subsequent samples were collected until 22:35, as indicated in Table 6.3. A total of 11 samples were collected at the spring. Compared with the highway runoff entering the sinkhole, water discharged from the spring was relatively clean. Therefore, the plastic pitcher used for sample collection was not dedicated. Rather, the pitcher was rinsed with spring water three times immediately prior to the collection of each sample. All other sample-collection procedures were identical to those used at the sinkhole and described above.

Spring discharge during the sampling event was monitored in the same manner as outlined above for the large culvert. A spring hydrograph for the storm pulse is included in Fig. 6.7. The maximum spring discharge was 0.18 m^3/s (6.5 cfs), and the volume of water discharged was 1257 m^3 (44,373 $feet^3$). Table 6.3 presents analytical results from the first stormwater sampling event. At the sinkhole site, runoff from the small culvert generally contained higher concentrations (but lower loads) of the constituents than runoff from the large culvert. The first sample had the highest concentration of the small-culvert samples. The second sample from the large culvert had the highest concentration from that location.

At the spring, concentrations of dissolved and total Zn, total Pb, TPH, and PAH were very low or non-detectable before 20:30. Dissolved lead was not detected at the spring. Total dissolved solids concentrations remained high during the entire sampling event, with the highest concentration

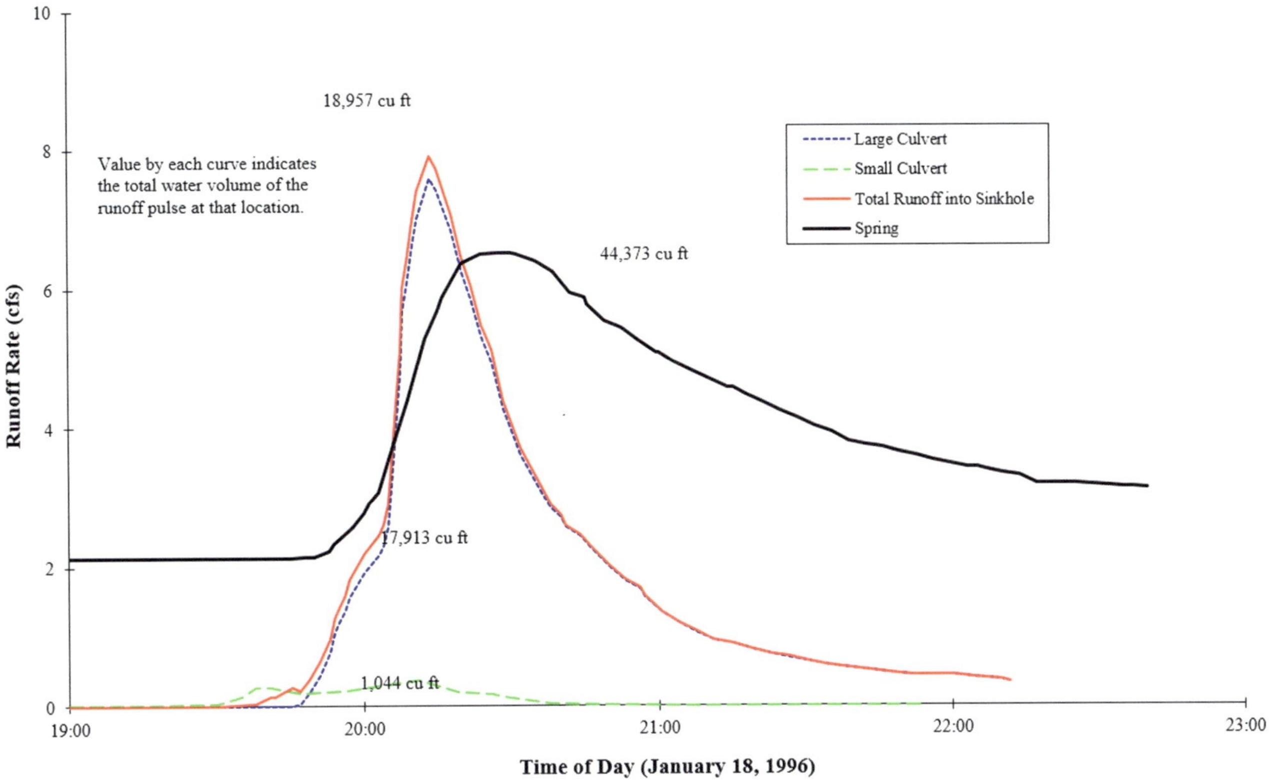

Fig. 6.7 Hydrographs of highway stormwater runoff into sinkhole and discharge from Holston Spring during stormwater runoff sampling

occurring in the first sample. The concentrations of TSS and TVS were identical to each other until 20:30, after which TVS levels were less than TSS levels. In other words, the TSS load was composed entirely of TVS until 20:30, after which TVS constituted only a portion of the TSS load. The peak concentrations for most of the constituents were observed in the sample collected at 21:15.

Contaminant loading rates (mass per unit time) were calculated by multiplying flow rates and concentrations for each sampled interval. The loading curves for each constituent except dissolved lead are shown in Figs. 6.8, 6.9 through 6.10. For most contaminants, peak contaminant loading at the spring lagged the peak at the sinkhole by approximately 1 h. For TDS, however, the peak lagged only 20 min behind. Generally, peak loading rates were lower at the spring than at the sinkhole, except for dissolved Zn and TDS, which were higher at the spring. Loading rates decreased rapidly after peaking at the sinkhole and more slowly after peaking at the spring.

The total contaminant load (masses) flowing into the sinkhole was calculated by integrating loading rates over time for each sampled interval (Table 6.4). The event mean concentration for each analyzed contaminant was also calculated by dividing the total load by the total runoff volume.

Figure 6.11 shows the relative accumulation of contaminant loads at the sinkhole. Analysis of the data from the sinkhole shows that contaminant load is more closely related to runoff volume than contaminant concentration. Although all constituents displayed similar loading characteristics, the cumulative loading rates differed. For example, it took only 30 min for 80% of the TSS, total Zn, total Pb, and TPH loads to flow into the sinkhole; while 40 min was required for PAH and TVS; and 60 min was required for TDS and dissolved Zn. For most contaminants, over 80% of the mass flowed into the sinkhole with only 50% of the runoff flow, which is a characteristic of first-flush phenomenon.

6.3 Performance Monitoring After Installation of a Treatment System

A treatment system should provide a combination of screening, storage, sedimentation, filtration, and adsorption designed to treat the first flush of stormwater runoff. It should be easily installed using readily available materials, resulting in a cost-effective, low-maintenance treatment system. The challenges involved include limited available land area, periodic large runoff events, daily and seasonal temperature extremes, difficult terrain conditions, susceptibility to additional sinkhole collapse, and clogging of the filter media.

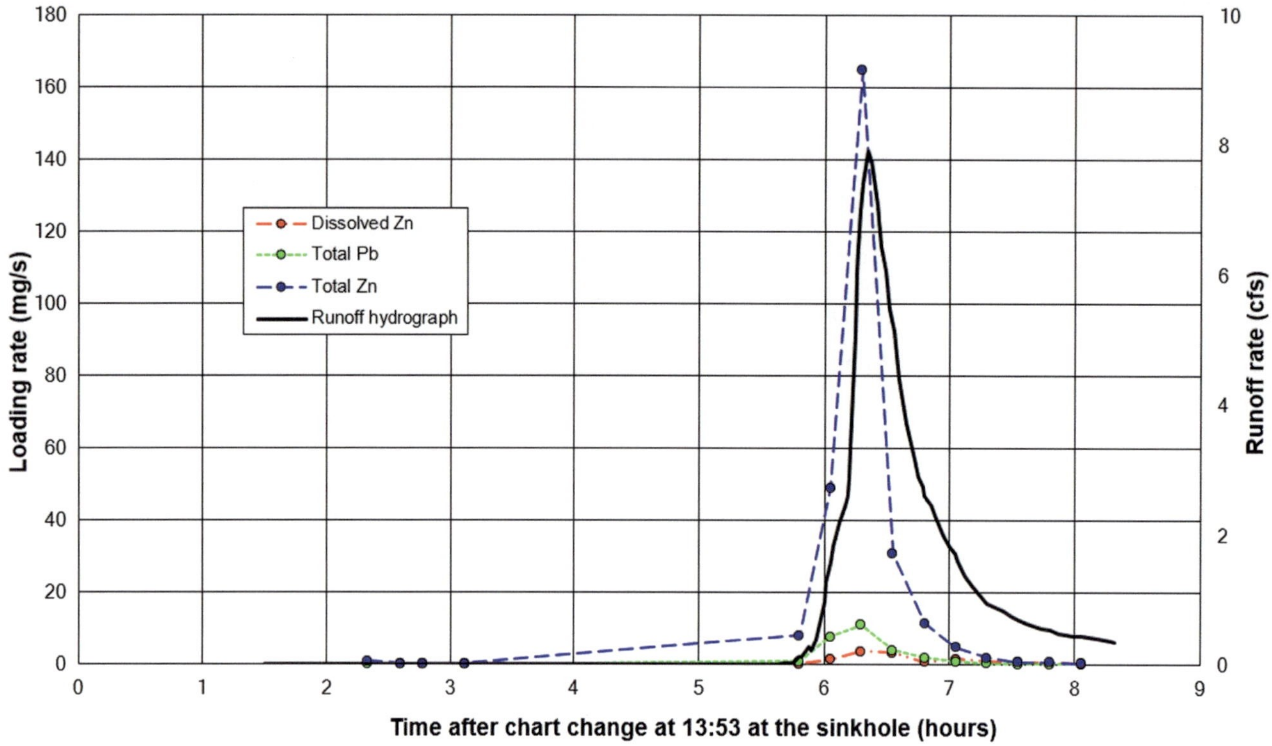

Fig. 6.8 Loading rates of metals for highway stormwater runoff

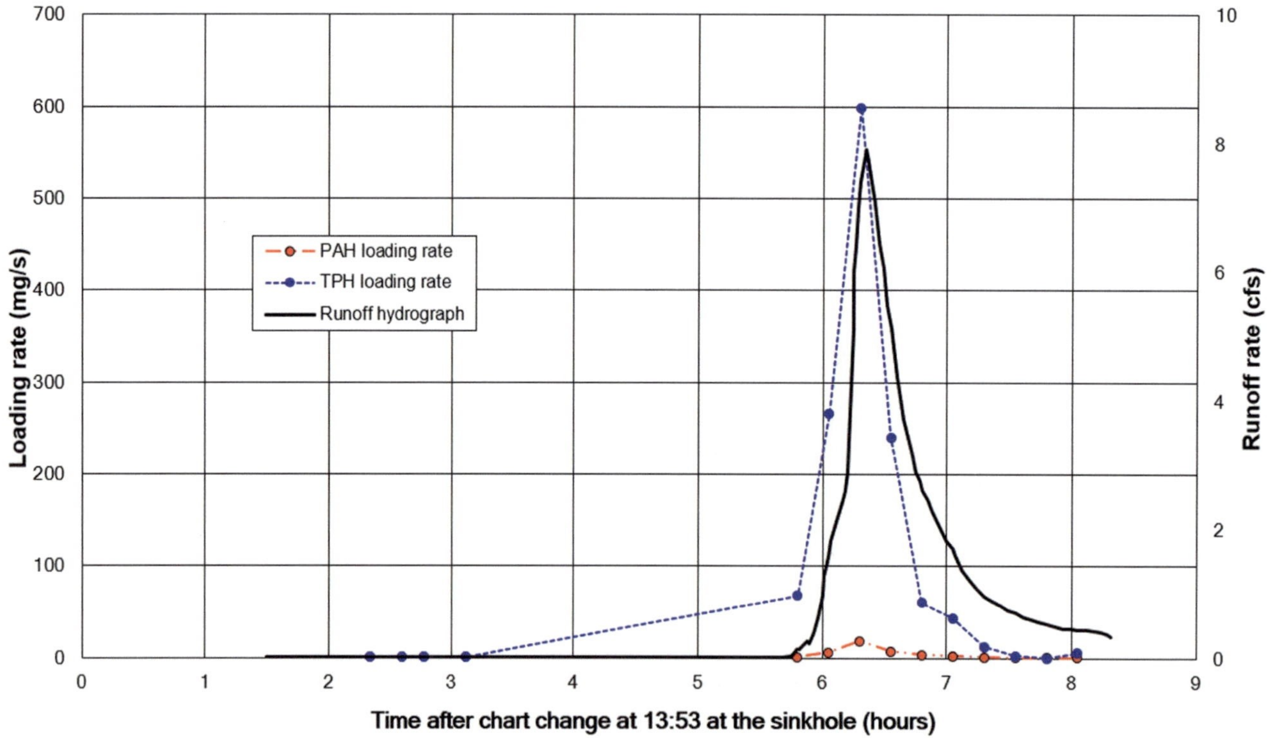

Fig. 6.9 Loading rates of hydrocarbons for highway stormwater runoff

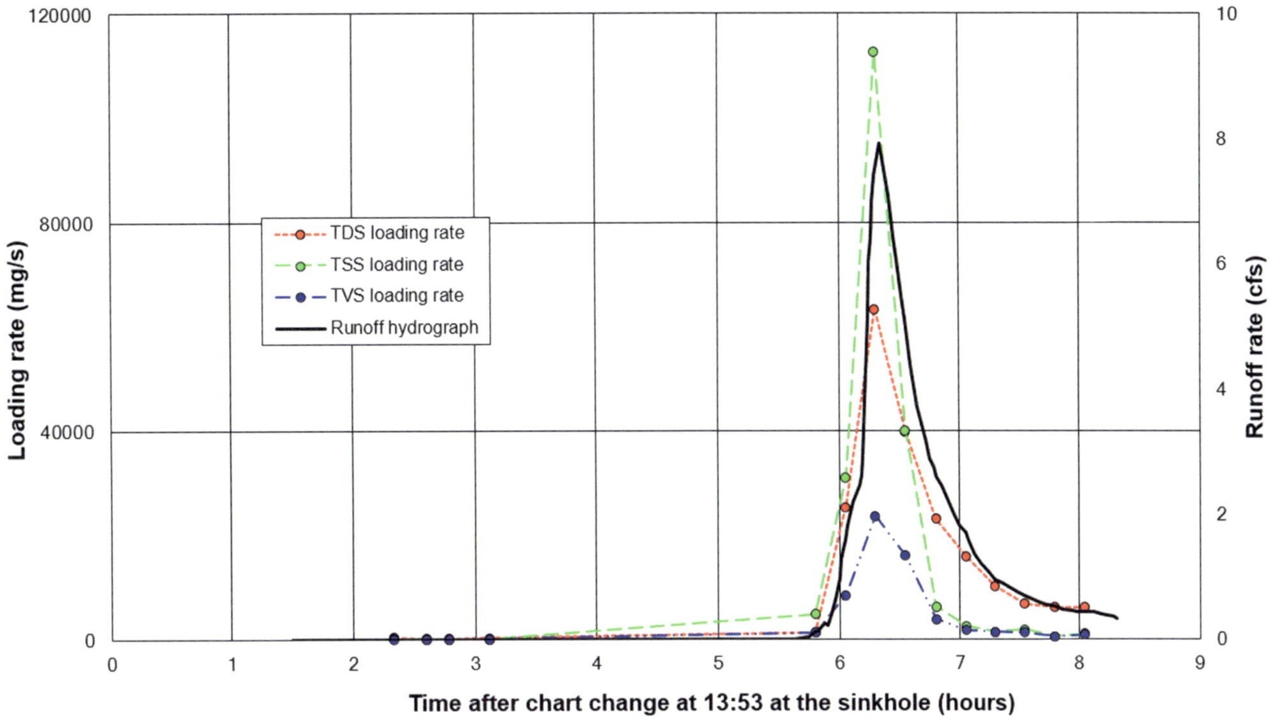

Fig. 6.10 Loading rates of solids for highway stormwater runoff

Table 6.4 Contaminant loads for sampling under natural conditions

	Dissolved Zn	Dissolved Pb	Total Zn	Total Pb	PAH	TPH	TDS	TSS	TVS
Event mean concentration (mg/l)	0.026	0.00	0.554	0.06	0.078	2.89	360	399	114
Load into sinkhole (g)	13	0	281	28	40	1465	182,681	202,228	57,996

6.3.1 Peat Filter

When discussing sand filtration systems, Schueler et al. (1991) listed a peat-sand filter system as a promising technology for the removal of pollutants from stormwater runoff. The use of a peat-sand filtration system to enhance the removal of oil, grease, and other pollutants in urban runoff was proposed in the 1980s (Galli 1990). The filter media are compost made of deciduous leaves. There are three primary pollutant removal mechanisms performed by peat or compost: mechanical filtration to remove sediments, chemical processes to remove soluble metals, and adsorption properties to remove oils and grease. However, filtration systems in karst terranes require not only removal of pollutants but also adequate drainage capacity and structural stability.

The objective of the pilot study was to test the performance of peat moss as a filtration medium in removal of highway runoff contaminants. The batch tests and column tests conducted in the previous phases of the project have demonstrated their ability to remove more than 90% of Cu, Pb, Zn, and oil and grease. Chaney and Hundeman (1979) reported over 95% removal of Cd in peat columns. Peat filters installed in sinkholes in Indiana removed an average of 80% of suspended particulates (measured as TDS and total recoverable metals) and almost 50% of dissolved Zn and Cu (Keith et al. 1995). Due to its strong affinity for oil, peat moss can absorb from five to eight times its weight of oil (McKay 1980). Its average oil removal efficiency could be over 80% (Smith and Marks 1976). The ability of peat as a natural adsorbent for dyes was also demonstrated in a column test in which introduced Rhodamine WT traveled only a few inches before it completely disappeared or was absorbed. Due to both its physical and chemical adsorption/filtration properties, peat has been used in both industrial and domestic wastewater treatment applications (Nichols and Boelter 1982; Lang and Wider 1985).

Peat is a highly organic, and physically and chemically complex material. The type of plants, including mosses, sedges, and other wetland plants, from which the peat originated, and their degree of decomposition heavily influenced both the physical properties and chemical compositions.

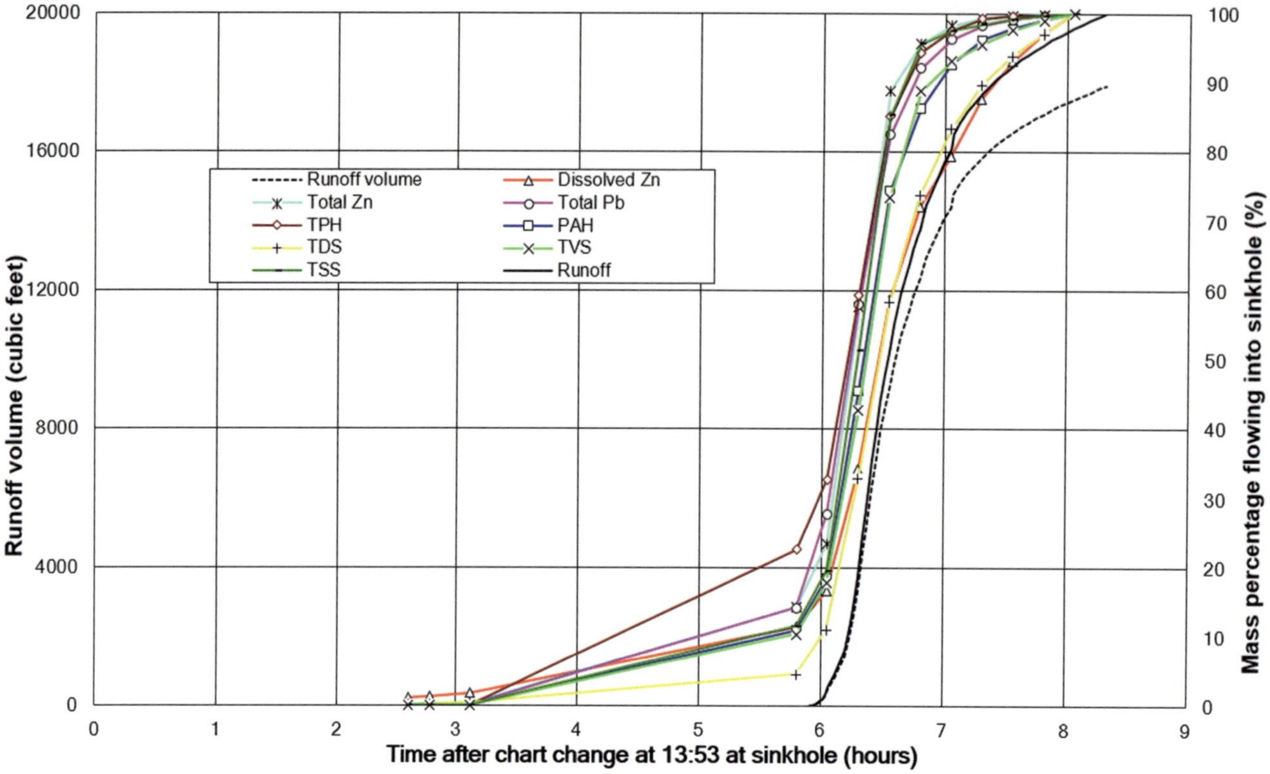

Fig. 6.11 Loading process of highway stormwater runoff into sinkhole

6.3.2 Compression Tests of Peat

The compression characteristics of peat moss were tested to understand its behavior. The test peat was Hyponex Canadian Sphagnum Peat Moss with a packaged density of 320 kg per cubic meter (20 pounds per cubic foot). The density of loose peat ranged from 158.6 to 164.99 kg per cubic meter (9.9–10.3 pounds per cubic foot) (at initial moisture).

The compression testing was performed using a smooth-walled cylinder with a 12.383 cm (4.875 in.) diameter. Peat was loosely placed in the cylinder. A series of increasing loads were applied, and the change in peat thickness was recorded. Three series were conducted, which included (1) loose peat right out of the package with initial moisture content and a starting thickness of 0.91 m (3 feet); (2) loose peat right out of the package with initial moisture content and a starting thickness of 0.305 m (1 foot); and (3) wetted peat with a starting thickness of 0.91 m (3 feet). The results are presented in Table 6.5.

The compressibility of peat is related to the original thickness and the moisture content of the peat. With the same moisture content and the same vertical pressure of 3388.4 kg/m^2 (694 lb/feet^2), the thickness of the peat reduced 0.168 m (0.55 feet) from a starting thickness of 0.91 m (3 feet), while the thickness reduced 0.104 m (0.34) feet from a starting thickness of 0.305 m (1 foot). For the same starting thickness of 0.91 m (3 feet), the peat compressed more with less moisture content when the applied pressure was 2636.5 kg/m^2 (540 lb/feet^2) or less. When the pressure was increased more than 2636.5 kg/m^2 (540 lb/feet^2), the peat with higher moisture content compressed more. At least 1.067 m (3.5 feet) of peat should be used to have the designed 0.91 m (3-foot) peat filter.

Table 6.5 Results of compression tests of Canadian sphagnum peat moss

Vertical pressure (lb/sq-feet)	Peat column height (feet)		
	Loose peat with initial moisture content (10 lb/cu-feet)	Loose peat with initial moisture content (10 lb/cu-feet)	Loose wet peat (23.7 lb/cu-feet)
0	3.00	1.00	3.00
77	2.80	0.85	2.82
154	2.70	0.80	2.75
231	2.65	0.75	2.68
309	2.60	0.73	2.62
386	2.55	0.70	2.58
463	2.50	0.68	2.52
540	2.48	0.67	2.49
617	2.47	0.65	2.45
694	2.45	0.66	2.43

6.3.3 Construction of Treatment System

Figure 6.12 shows the construction processes of the peat filter treatment system. According to the design, the existing "new" sinkhole was filled in with large rubble followed by progressively smaller aggregates. The site was graded to a relatively flat surface, and berms were constructed on the northern side. The treatment system was constructed 3.0–4.6 m (10–15 feet) downstream of the new sinkhole location. Next, the entire water containment area was covered with Bentomat®, a geosynthetic clay liner. The Bentomat® seams were handled by overlapping their adjacent edges. A thin layer of soil was used to cover the liner. The edge of liner was anchored by soils and tied to existing slopes. After the installation of the impermeable liner, the gabion and the baffle wall were constructed. The support wall and the baffle wall were made of Jersey barriers. The compartment between the gabion and the supporting wall contains the peat filter. A slotted pipe was installed at the bottom of peat compartment to collect the treated runoff, which is then discharged to the old sinkhole through a 10.1 cm (4 in.) PVC pipe. Rock fragments (larger than the slot size) were used to cover the pipe to facilitate collection and drainage. Canadian sphagnum peat moss was used to fill in the filter compartment.

Because of its strong hydrophobic property, the peat moss was wetted as it was put in the compartment. More peat was added as the compression process proceeded. Placing the perforated pipe and the rock fragments at the bottom of the filter turned out to be problematic. Because the permeability of the peat was far less than that of the rock fragments, stormwater runoff tended to flow preferentially through the fragments and did not have enough contact time with the peat. This limited the capability of the filter to treat contaminated water, as demonstrated by the results from PAH sampling in the first two sampling events. To overcome this problem, the perforated pipe was later lifted from the bottom and was installed on the top of a 46 cm (18 in.) peat bed. The rock fragments were removed, and the pipe was wrapped in permeable fabric instead. Stormwater runoff was forced to flow through the peat before it discharged into the old sinkhole. This revision ensured the contact of stormwater with the peat. However, there was no water flowing out when the water level in the basin was less than 46 cm (18 in.), which resulted in ponding. The offline detention basin, together with the gaps in the Jersey barriers, allowed the treatment of first flush of highway stormwater runoff. Water flowed over the spillway when the detention basin was full.

6.3.4 Performance Monitoring

Effectiveness of the treatment system was evaluated by collecting stormwater samples at three locations: (A) the large culvert to characterize the original highway runoff; (B) the pre-treatment detention basin to characterize runoff before passing through the peat in the treatment system; and (C) the post-treatment discharge pipe to characterize the treated effluent issuing from the treatment system.

In addition to monitoring the flow rate for runoff from the large culvert and effluent discharging from the treatment system, the following indicator constituents were analyzed to evaluate the effectiveness of the treatment system.

- polyaromatic hydrocarbons (PAHs)
- total (recoverable) petroleum hydrocarbons (TPHs)
- lead (total and dissolved)
- zinc (total and dissolved)
- copper (total and dissolved)
- iron (total and dissolved)
- total dissolved solids (TDS)
- total suspended solids (TSS)
- volatile suspended solids (VSS)
- pH
- specific conductance (SC)
- temperature.

The performance testing was a diagnostic and corrective process. Initially, samples were analyzed only for PAHs, which were selected to provide a simple, cost-effective indication of the system performance. PAH samples were analyzed in the office using immunoassay methodology, resulting in considerable savings over traditional laboratory-based analyses. Only after determining that the treatment system was functioning properly were samples collected and analyzed for the full range of constituents listed above.

6.3.4.1 Sampling and Analysis Plan

Sampling events were selected based on official weather forecasts and availability of personnel and other resources. Online weather-radar summaries were particularly helpful for identifying sampling opportunities. Because of the expense involved in sampling, a decision to mobilize samplers to the site was made only when it appeared that a substantial rain event would arrive at a reasonably predictable time. The use of automatic samplers was an alternative to collect storm-related water samples, but they were precluded by the need to analyze dissolved metals, which requires filtering samples immediately upon collection.

During most sampling events, the field crew arrived on site before the onset of precipitation. During this time, the pH meter was calibrated; data were downloaded from the flow meter and its operation was verified; and devices were set up to filter dissolved-metals samples. Sample collection began at each location as soon as permitted by the flow of runoff. Depending on the intensity of the rainfall, it took approximately 40 min for runoff to flow from the large culvert down the channel and through both rock

Fig. 6.12 Construction processes of peat filter

piles to reach the pre-treatment detention basin. Initially, it took another 10–20 min for stormwater to cover the floor of the basin and begin discharging from the treatment system at the outlet pipe. Raising the drainpipe increased this delay because of the additional water depth required to initiate flow through the peat. The flow rate through the culvert was monitored by the installed Flo-Tote. The flow rate (discharge) of the treatment system was determined by monitoring the time required to fill a shallow, 14.2-L plastic container. Temperature, specific conductance, and pH were measured by placing the meter probe directly into the water being monitored.

Samples were collected in 4-L plastic beverage pitchers. Flat-sided ones were used for collecting shallow runoff directly from the bottom of the large culvert. Rounded pitchers with large handles were used to dip samples from the basin and to collect treated effluent flowing from the discharge outlet. Upon collection, samples were covered with lids and taken to a small moving van for processing. The moving van provided shelter for filtering dissolved-metals samples; transferring samples into labeled containers and preserving them as required for various analyses; and completing field data forms, chains of custody, and other documentation. Immediately after processing, each sample was placed in a cooler with ice.

Although the discharge in the culvert was monitored, accurate estimates of load could not be determined in the tests because of discharge measurement challenges at the outlet pipe. The effectiveness of the filter was evaluated based on the concentration reduction rather than the contaminant load.

6.3.4.2 Sampling Event #1

Thirty water samples were collected in the first sampling event on November 10, 1998, including 10 samples from the culvert, 10 samples from the basin right in front of the gabion, and 10 samples from the outlet pipe discharging the treated water. They were analyzed for PAHs only and the results are shown in Fig. 6.13.

The PAH concentrations of the middle five samples from the culvert were distinctly higher than those in the basin and from the outlet pipe. Comparison of the PAH concentration in the first four or five samples in the basin and from the outlet pipe indicated that the filter seemed to be functioning in terms of contaminant removal. However, starting from the sixth sample, the concentration from the outlet pipe did not differ much from those in the basin. This implied that the treatment system was not functioning properly, and the runoff water was not completely in contact with the peat, or the stormwater was flowing through the system too quickly for contaminants to be removed by the peat. It was hypothesized that this was occurring because the peat was not adequately compacted and saturated. Rudimentary experiments showed that dry peat is hydrophobic, requiring considerable effort to saturate it initially. Based on this analysis, the peat in the filter was dug out and placed back after being wetted and saturated.

6.3.4.3 Sampling Event #2

During the second sampling event on May 5, 1999, 26 water samples were collected and PAHs were analyzed. A valve on the treatment system outlet was also adjusted to slow the stormwater flow through the peat and ensure adequate contact time. Figure 6.14 shows the results of PAH analysis. Once again, the results indicated no improvement in water quality. The stormwater was apparently bypassing the treatment medium. Re-evaluation of the filter design indicated a possibility that water could preferentially flow under the peat along its contact with the underlying material.

Subsequently, all the peat was removed from the filter. They also removed the gravel from around the perforated

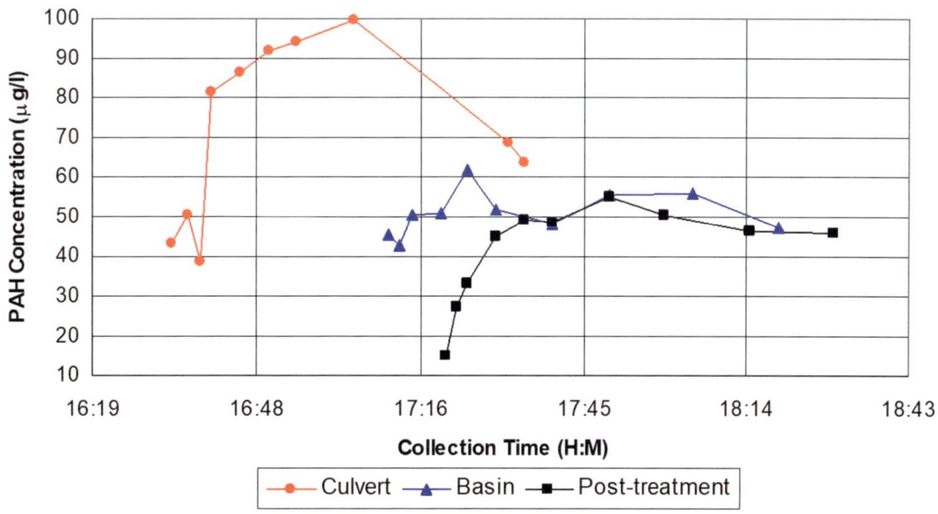

Fig. 6.13 Results of PAH analysis for sampling event #1

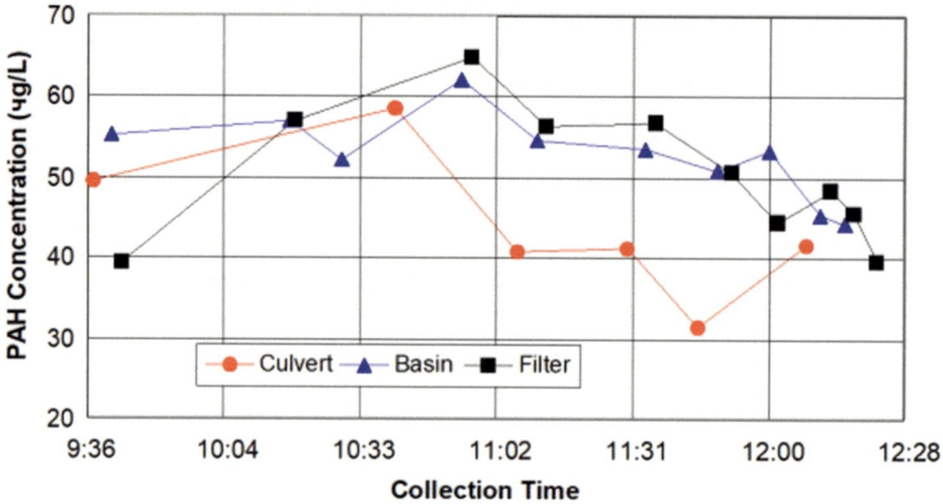

Fig. 6.14 Results of PAH analysis for sampling event #2

pipe and added a 90° elbow and 46 cm (18 in.) of vertical drainpipe. The peat was then saturated and placed back into the filter. As a result of this modification, the perforated collector pipe was no longer on the bottom of the filter but in the middle of the peat moss. The stormwater was then forced to flow through the peat before discharging into the outlet pipe.

6.3.4.4 Sampling Event #3

During the third sampling attempt on September 21, 1999, rainfall was insufficient to produce flow through the treatment system. Because of the higher position of the drainpipe within the peat, it was necessary for 0.61 m (2 feet) of water to build up in the detention basin before effluent is discharged from the system. However, the effect of the detention basin was evaluated by analyzing the samples from the culvert and the basin. There was no substantial change in the PAH concentration for the water from the culvert to the basin, as shown in Fig. 6.15.

The pH, SC, and temperature were monitored in the culvert and basin. The results are listed in Table 6.6. With approximately the same pH for the water in the culvert and basin, the detention basin did not seem to reduce the SC values. Instead, the SC in the basin was greater than that in the culvert, with an average of 243 and 159 μS, respectively.

6.3.4.5 Sampling Event #4

Results of the fourth sampling effort on September 29 and 30, 1999 demonstrated that the treatment system was functioning, as indicated by a reduction in PAH concentration as shown in Fig. 6.16. Except for the #4 sample in the basin that is apparently an outlier, the PAH concentration in the basin ranged from 12 to 17 μg/l (ppb). After the peat filter, the PAH concentrations decreased to 2–3 μg/l (ppb). This is a removal efficiency of well above 80%.

In this sampling event, the pH, SC, and temperature were also monitored at locations where the samples were

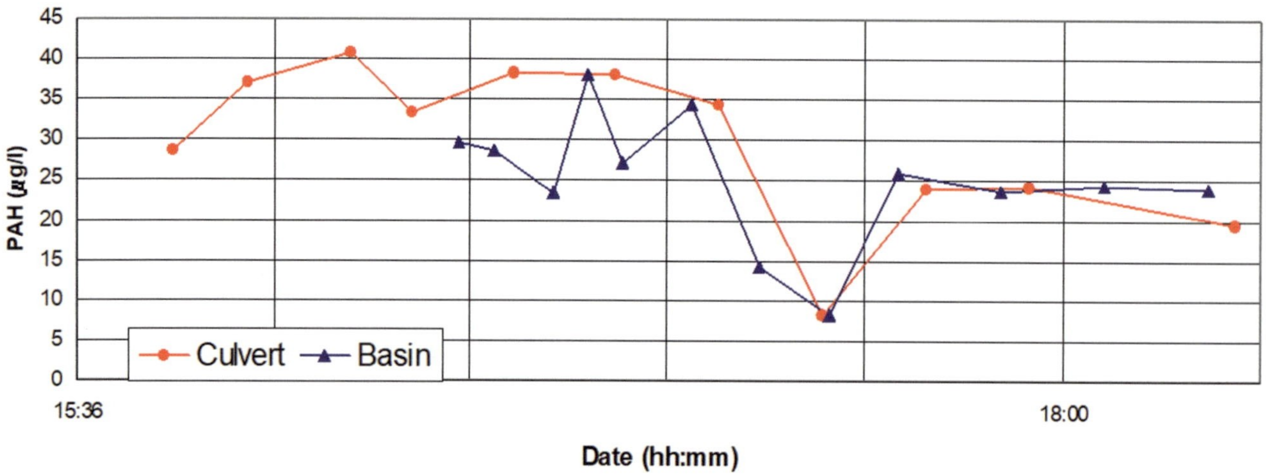

Fig. 6.15 Results of PAH analysis for sampling event #3

Table 6.6 Field parameters during sampling event #3

Culvert				Basin			
Time	pH	SC (μS)	Temp (°C)	Time	pH	SC (μS)	Temp (°C)
15:50	7.65	102.3	17.0	16:32	7.68	399.6	17.2
15:56	7.70	237.2	17.5	16:37	7.78	381.4	17.3
16:01	7.74	199.0	17.5	16:46	7.84	294.2	17.2
16:16	7.80	167.6	17.4	16:51	7.76	299.6	17.2
16:25	7.65	183.3	17.5	16:56	7.92	201.1	17.4
16:40	8.01	152.5	17.3	17:06	7.91	215.3	17.5
16:55	8.03	141.6	17.3	17:16	8.00	181.8	17.4
17:10	8.06	126.3	17.3	17:26	7.99	198.9	17.4
17:25	8.10	126.1	17.3	17:36	7.97	192.9	17.5
17:40	8.09	136.8	17.2	17:51	8.01	183.9	17.5
17:55	8.09	151.8	17.2	18:06	8.01	195.6	17.5
18:25	8.03	183.0	17.3	18.36	8.03	175.9	17.4

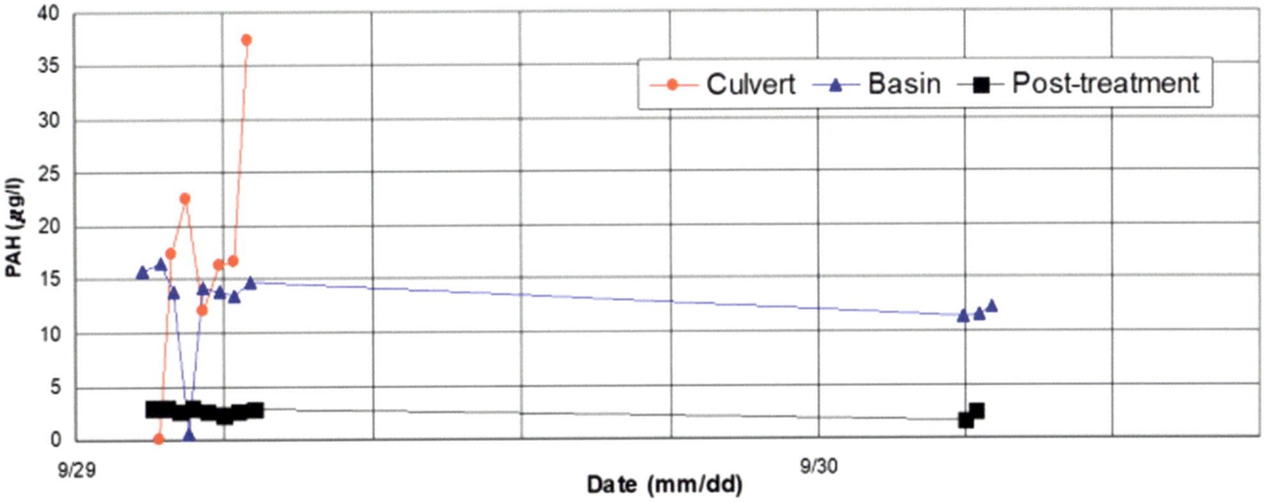

Fig. 6.16 Results of PAH analysis for sampling event #4

collected (Table 6.7). As hypothesized, pH values in the water from the outlet pipe were lower than those from the culvert and basin. The average pH in the treated water was 5.7, while the water prior to the treatment had an average pH value of 7.9. The discharge from the outlet pipe was measured on September 29 using the bucket-and-stopwatch method. The average flow rate was 3.03 L/min (0.8 gpm).

6.3.4.6 Sampling Event #5

Samples collected on April 24, 2000, were analyzed for the entire range of indicator constituents. The results are provided in Table 6.8, and they confirmed that the filter was functioning. All the indicator constituents but TDS showed a decrease in their concentrations after the stormwater passed through the filter. For example, the stormwater had a constant PAH concentration of 0.47 μg/l (0.47 ppb) over the 2.5-h sampling in the outlet pipe, which is much lower than the average PAH concentration of 12 μg/l (12 ppb) in the basin and 21 μg/l (21 ppb) in the culvert. The removal efficiency of the filter for PAHs was more than 95%. The removal efficiency for total lead was promising as well, reducing its concentration from 3.5 μg/l (3.5 ppb) in the basin to less than 1 μg/l (1 ppb) in the outlet pipe. The removal efficiency for TSS and VSS is also promising. The filter had a relatively poor effect on the TPH concentration, reducing the concentration to 0.7 mg/l (0.7 ppm) from 1 mg/l (1 ppm). However, the TDS concentration of the stormwater increased noticeably after flowing through the filter. The TDS concentration in the basin was 122 mg/l (122 ppm), whereas the concentration was 228 mg/l (228 ppm) in the outlet pipe.

Although the concentration of total zinc decreased to 12.3 μg/l (12.3 ppb) from 73 μg/l (73 ppb) in the basin, the analysis results for zinc were confusing. For

Table 6.7 Field parameters during sampling event #4

Date	Culvert				Basin				Post-treatment			
	Time	pH	SC (μS)	Temp (°C)	Time	pH	SC (μS)	Temp (°C)	Time	pH	SC (μS)	Temp (°C)
09/29/1999	11:45	8.34	186.6	20.9	11:47	7.76	160.6	21.9	12:04	5.58	102.1	22.4
	12:19	7.80	195.9	19.6	12:24	7.70	141.1	21.6	12:33	5.71	96.1	22.3
	12:45	7.88	211.8	19.7	12:47	7.59	134.4	22.0	12:59	5.73	94.6	22.3
	13:15	7.89	193.3	19.7	13:17	7.62	129.2	22.1	13:23	5.72	97.0	22.1
	13:45	7.86	220.3	19.6	13:47	7.52	129.1	22.3	13:53	5.73	94.9	22.2
	14:16	7.89	223.0	19.5	14:18	7.50	128.2	22.3	14:24	5.75	97.2	22.2
	14:45	7.84	222.1	19.4	14:47	7.46	127.9	22.0	14:53	5.76	97.3	22.1
	15:15	8.08	112.5	21.6	15:18	7.48	130.8	22.5	15:24	5.75	93.5	22.3
09/30/1999					14:18	7.43	151.7	22.5	14:22	5.77	102.7	21.6
					14:46	7.50	152.3	22.1	14:50	5.73	100.2	20.7
					15:15	7.73	153.0	22.8	15:18	5.69	100.1	21.2

*Degrees C

the treated samples, the total zinc concentrations were *lower* than dissolved zinc concentrations. These values were presented in bold and italic in Table 6.8. A thorough review of field and laboratory procedures did not reveal a cause for this apparent discrepancy. The only difference between sample collection for total and dissolved zinc was that the samples for dissolved zinc were filtered by 0.45 μm filters. Therefore, it was decided that additional samples should be collected and analyzed for both metals to ensure that this was not caused by the filtering procedure.

6.3.4.7 Sampling Event #6

The sixth sampling event occurred on November 8, 2000. Twenty-two samplings were collected in the basin and in the outlet pipe. In addition to zinc and lead, iron was also analyzed. The results are shown in Table 6.9. The majority of the samples in the outlet pipe had lead concentrations under the detection limit. For iron, the filter decreased the total iron concentration from 0.36 mg/l (0.36 ppm) in the basin to 0.14 mg/l (0.14 ppm). However, the filter increased the dissolved iron concentration from 0.05 mg/l (0.05 ppm) in the basin to 0.11 mg/l (0.11 ppm) in the outlet pipe. Again, the total zinc concentrations were *lower* than dissolved zinc concentrations in most of the samples, not just limited to the treated water. As a result of this repeated discrepancy, the sample collection procedures and laboratory analysis procedures were thoroughly reviewed once more. While no satisfactory explanation was found, it was hypothesized that this was caused by laboratory analysis error. The laboratory used the USEPA-approved method for zinc analysis (USEPA 200.7 with detection limit of 0.005 mg/l (ppm)), and the analysis error should be well controlled within a range. This deduction indicated that the discrepancy that was observed in these tests was caused by the low concentrations of zinc and the concentration difference between total zinc and dissolved zinc was less than the laboratory error.

6.3.4.8 Sampling Event #7

To verify the hypothesis that the zinc discrepancy was caused by the low concentration in the runoff, the stormwater runoff was spiked with copper sulfate and zinc chloride and then another round of sampling was conducted on February 20, 2001. These chemicals were added only in the detention basin and the runoff discharged from the outlet pipe would have very low concentrations of copper and zinc. This sampling event was not completely in response to natural precipitation. The runoff generated from the previous day's precipitation was not enough to produce water flow through the filter. To facilitate the flow, several tons of water were artificially dumped into the basin. The laboratory analysis results are listed in Table 6.10. The zinc and copper concentrations were distinctively elevated in the basin. The average total zinc concentration in the basin was 1.21 mg/l (1.21 ppm), and the dissolved zinc concentration was 0.386 mg/l (0.386 ppm). The total and dissolved concentrations for copper in the basin were 2.44 mg/l (2.44 ppm) and 0.102 mg/l (0.102 ppm), respectively.

The discrepancy that was observed in the zinc concentration (the dissolved zinc concentration being higher than the total zinc concentration) occurred again in this round of analysis. However, it occurred only when either the total or dissolved zinc concentration was smaller than 0.12 mg/l (ppm). When the total or dissolved zinc concentration was bigger than 0.12 mg/l (ppm), the zinc analysis results were normal.

This round of sampling and analysis indicated that peat filter could significantly reduce the metal concentrations, and that contaminant removal efficiency is highly dependent on the concentration in the source water. On average, the total zinc concentration was reduced to more than 90%

Table 6.8 Laboratory analysis for samples collected in sampling event #5

Location	Time	PAHs	Lead		Zinc		TPHs	TDS	TSS	VSS
			Dissolved	Total	Dissolved	Total				
	h:m	(µg/l)*	(µg/l)*	(µg/l)*	(µg/l)*	(µg/l)*	(mg/l**)	(mg/l**)	(mg/l**)	(mg/l**)
Culvert	11:35	27.17	BDL	6	69	158	2.57	100	59	23
	11:40	25.90	BDL	6	38	135	2.57	98	40	18
	11:45	24.29	1	5	41	140	1.71	122	58	19
	11:50	18.06	1	4	43	134	1.24	94	24	11
	12:03	24.42	BDL	5	51	140	1.63	109	75	31
	12:12	22.90	BDL	4	42	123	1.79	82	59	23
	12:30	19.38	BDL	3	40	91	1.63	88	31	14
	13:02	17.18	BDL	3	42	86	0.93	146	39	16
	13:33	16.72	BDL	3	34	81	1.16	118	21	13
	14:14	16.63	BDL	3	45	75	0.85	129	38	15
	14:36	27.03	BDL	13	37	161	2.38	66	111	33
	14:59	11.65	BDL	7	36	97	0.7	64	55	21
Basin	11:33	2.75	BDL	2	45	45	0.38	130	10	<1
	11:38	8.05	BDL	3	68	75	0.7	154	23	9
	11:43	5.41	BDL	3	69	69	0.7	128	31	17
	11:48	14.07	BDL	4	43	79	1.32	126	41	21
	12:03	4.96	BDL	2	45	45	0.38	180	29	17
	12:03	12.99	BDL	4	63	88	1.09	94	48	23
	12:18	18.86	BDL	1	46	70	1.24	110	56	23
	12:43	15.31	BDL	3	48	77	0.93	124	24	14
	13:30	14.80	BDL	3	44	87	1.24	108	58	21
	14:00	16.63	BDL	5	65	78	1.24	94	41	18
	14:33	15.73	BDL	6	50	82	1.63	136	33	18
	14:57	16.72	BDL	5	58	73	1.63	118	37	18
	15:26	16.09	BDL	5	45	80	1.09	88	29	15
Post-treatment	14:25	0.47	BDL	1	15	15	0.7	106	8	<1
	14:32	0.47	BDL	2	***36***	***16***	0.49	268	10	<1
	14:41	0.47	BDL	1	***44***	***19***	0.37	238	19	<1
	14:45	0.47	BDL	1	***42***	***14***	0.37	238	18	<1
	14:52	0.47	BDL	1	***37***	***9***	0.29	230	10	<1
	15:00	0.47	BDL	BDL	***44***	***17***	1.09	212	22	<1
	15:08	0.47	BDL	BDL	***32***	***16***	0.93	220	8	<1
	15:15	0.47	BDL	BDL	***31***	***6***	0.49	260	5	<1
	15:18	0.47	BDL	3	***37***	***10***	0.7	254	6	<1
	15:29	0.47	BDL	BDL	***51***	***8***	1.63	228	9	4
	15:40	0.47	BDL	BDL	***39***	***5***	0.93	242	20	<1
	15:51	0.47	BDL	BDL	***33***	***8***	1.32	214	11	<1
	16:02	0.47	BDL	3	***43***	***17***	0.29	252	10	<1

*µg/l is equivalent to ppb
**mg/l is equivalent to ppm
Numbers in bold and italic show discrepancies in zinc analysis

of its previous value, and the dissolved zinc concentration was reduced a little over 70%. The total copper concentration was reduced to 97% with dissolved copper reduction of 69%. It was impossible to quantify the removal efficiency of lead; however, the lead concentration in the outlet pipe was generally lower than that in the basin.

6.4 Summary

The problem of ground and surface water contamination by highway runoff is a common one. However, recharging through sinkholes is more serious as it provides a direct conduit to groundwater and a high likelihood of

Table 6.9 Analysis results for samples collected in sampling event #6

Location	Time	Zinc		Lead		Iron	
		Dissolved	Total	Dissolved	Total	Dissolved	Total
	(hh:mm)	(mg/l)*	(mg/l)*	(mg/l)*	(mg/l)*	(mg/l)*	(mg/l)*
Basin	12:15	***0.073***	***0.041***	ND	ND	0.07	0.43
	12:48	***0.046***	***0.037***	ND	0.001	0.06	0.41
	13:32	***0.052***	***0.035***	ND	0.002	0.06	0.4
	14:01	***0.047***	***0.036***	0.001	0.001	0.05	0.4
	14:30	0.047	0.051	ND	0.001	0.06	0.31
	15:00	***0.054***	***0.034***	ND	0.002	0.05	0.39
	15:00	***0.045***	***0.039***	ND	0.002	0.06	0.29
	15:30	***0.044***	***0.033***	ND	ND	0.06	0.36
	16:00	***0.048***	***0.031***	ND	0.001	0.03	0.32
	16:28	0.047	0.047	ND	0.002	0.05	0.33
	16:58	***0.071***	***0.034***	ND	0.002	0.03	0.34
Post-treatment	12:19	***0.056***	***0.014***	ND	ND	0.08	0.12
	12:52	***0.040***	***0.011***	ND	ND	0.08	0.14
	13:35	***0.057***	***0.017***	ND	ND	0.09	0.12
	14:02	0.022	0.036	0.001	0.001	0.14	0.14
	14:31	***0.048***	***0.021***	ND	0.002	0.1	0.15
	15:00	***0.060***	***0.011***	ND	ND	0.11	0.15
	15:00	***0.055***	***0.014***	ND	ND	0.11	0.14
	15:32	***0.056***	***0.015***	ND	ND	0.12	0.14
	15:59	***0.048***	***0.013***	ND	ND	0.12	0.16
	16:30	***0.034***	***0.021***	ND	0.004	0.14	0.16
	16:58	***0.056***	***0.013***	ND	ND	0.15	0.17

*mg/l is equivalent to ppm
Numbers in bold and italic show discrepancies in zinc analysis

Table 6.10 Analysis results for samples collected in sampling event #7

Location	Time	Zinc		Copper		Lead	
		Dissolved	Total	Dissolved	Total	Dissolved	Total
	(hh:mm)	(mg/l)*	(mg/l)*	(mg/l)*	(mg/l)*	(mg/l)*	(mg/l)*
Basin	15:57	0.437	0.909	0.110	1.18	ND	0.002
	16:31	0.467	1.510	0.120	3.38	ND	0.001
	17:04	0.479	1.060	0.110	2.14	ND	0.002
	17:39	0.279	1.540	0.080	3.54	ND	0.001
	17:39	0.266	1.040	0.090	1.96	ND	0.001
Post-treatment	15:57	0.132	0.161	0.040	0.06	ND	0.002
	16:31	0.122	0.142	0.030	0.11	ND	0.001
	17:04	***0.115***	***0.108***	0.050	0.070	ND	ND
	17:39	***0.102***	***0.079***	ND	0.04	ND	ND
	17:39	***0.091***	***0.075***	ND	0.04	ND	ND

*mg/l is equivalent to ppm
Numbers in bold and italic show discrepancies in zinc analysis

contamination with little chance for natural attenuation in a karst aquifer. The baseline data collection at the study site has demonstrated that karst springs can be seriously affected by highway stormwater runoff. Sequential sampling of the stormwater runoff and spring water indicated similarities in the chemograph of the spring and the stormwater runoff. The peak concentration of the indicator constituents at the spring tended to be attenuated because of dilution, and it lagged approximately 45 min behind the peak concentration of the stormwater runoff. However, brief high amplitude increases in certain pollutants in the receiving water may have lasting detrimental impacts on cave biota and water quality. Extensive literature review indicated that construction of a filtration system is probably one of the best management practices in karst terranes to remove most contaminants before they are transported into the aquifer. An environmental impact assessment should be conducted from karst perspectives when planning land use in these sensitive areas.

Another specific problem addressed by this study was the evaluation of the effectiveness of a peat-filtration system for the treatment of highway stormwater runoff. The laboratory testing and other studies have shown that peat moss is an effective adsorption filter for the removal of dissolved metals and organic materials. The removal efficiency of peat could be as high as 95% for PAHs and oils and greases and 90% for heavy metals. However, performance testing revealed several problems, which would probably not occur in well-controlled laboratory column tests. For one, the construction cost and potential maintenance effort precluded the installation of the filtration system into the sinkhole, although the system had been well-designed. Other problems were associated with the characteristics of peat.

- The hydrophobic property of peat made it difficult to wet and thoroughly saturate.
- The thickness of peat was greatly reduced when compressed and wetted.
- The low hydraulic conductivity of peat can cause preferential flow of stormwater through other media.

The price paid for learning peat behavior in the field was several revisions of the filter design and re-construction of the filter. During the first four rounds of sampling, only PAH samples were collected and analyzed to check the filter function. The fifth round of sampling collected samples for most of the indicator constituents, and the results from it formed the basis for evaluating the effectiveness of the filter. The last two rounds of sampling were conducted to determine the reasons for discrepancies in zinc analysis revealed in the fifth sampling event.

The performance monitoring of this study indicated that the filter had an excellent capability to remove PAHs and reduce PAH concentration as much as 95%. However, it did not exhibit a strong capability to remove TPHs. The TPH concentration was reduced to 0.7 mg/l (ppm) from 1 mg/l (ppm) in the basin. While the removal efficiency for dissolved lead could not be quantified in the test, the removal efficiency for total lead was approximately 70%. The removal efficiency of the filter for TSS and VSS was also promising. However, the TDS concentration in the stormwater increased noticeably from 122 mg/l (ppm) to 228 mg/l (ppm) after flowing through the filter.

Although the concentration of total zinc decreased to 12.3 μg/l (ppb) in the outlet pipe from 73 μg/l (ppb) in the basin, the analysis for zinc was confusing. In some of the samples, the total zinc concentrations were *lower* than dissolved zinc concentrations. Further sampling was conducted to determine the causes of this discrepancy; however, no definitive answer was forthcoming. The most likely reason was that the laboratory analysis error was greater than the difference between the dissolved and total zinc concentrations. Laboratory analysis of the last round of samples indicated that the removal efficiency of heavy metals was dependent on the concentration in the source water. If the source water has a higher metal concentration, the filter would have higher removal efficiency. Otherwise, the removal efficiency might be limited. At a concentration level of less than 2 mg/l (ppm) for total zinc, the test indicated a removal efficiency of 90% for total zinc and 70% for dissolved zinc. The filter had similar removal efficiency to copper.

References

Alexander EC Jr, Broberg JS, Kehren AR, Graziani MM, Turri WL (1993) Bellchester Minnesota lagoon collapse. In: Beck BF (ed) Applied Karst geology, proceedings of the fourth multidisciplinary conference on sinkholes and the engineering and environmental aspects of karst, Panama City, Florida, pp 63–72

Asplund RL, Ferguson JF, Mar BW (1980) Characterization of highway runoff in Washington State. Washington State Department of Transportation Report WA-RD-39.6

Balades JD, Cathelain M, Marchandise P, Peybernard J, Pilloy JC (1983) Chronic pollution of intercity motorway runoff waters. Water Sci Technol 17(6–7):1165–1174

Barner WL, Miller P (1991) Case histories of several approaches to stormwater management in urbanized karst terrain, Southwest Missouri. In: National Ground Water Association, Proceedings of the third conference on hydrogeology, ecology, monitoring, and management of ground water in karst terranes, Nashville, Tennessee, December 4–6, p 741–759

Barrett ME, Zuber RD, Collins ER, Malina JF Jr, Charbeneau RJ (1993) A review and evaluation of literature pertaining to the quantity and control of highway runoff and construction. Technical Report CRWR 239 (First Edition), Center for Research in Water Resources, University of Texas at Austin

Beck BF, Zhou W (2002) Management of the discharge and quality of highway stormwater runoff in karst areas. U.S. Department of Transportation Federal Highway Administration, February, 230 p

Betson RP (1977) The hydrology of karst urban areas. In: Dilamarter RR, Csallany SC (eds) Hydrologic problems in karst regions. Western Kentucky University, Bowling Green, pp 162–175

Betson RP, Milligan JD (1985) The effects of carbonate geology on urban runoff. In: Perspectives on nonpoint source pollution, proceedings of a national conference, Kansas city, Missouri, May 19–22

Bradbury KR, Muldoon MA (1992) Hydrogeology and groundwater monitoring of fractured dolomite in the Upper Door Priority Watershed, Door County, Wisconsin. Final report submitted to the Wisconsin Department of Natural Resources by the Wisconsin Geological and Natural History Survey, 74p

Cattermole JM (1966) Geologic map of the John Sevier quadrangle, Knox County, Tennessee. U.S. Geological Survey, Map GQ-514, scale 1:24,000

Chaney RL, Hundeman PT (1979) Use of peat moss columns to remove cadmium from wastewaters. J WPCF 51(1):17–21

Claytor RA, Schueler TR (1996) Design of stormwater filtration systems. The Center for Watershed Protection, Silver Spring, MD

Crawford NC (1988) Karst ground-water contamination from leaking underground storage tanks_Prevention, monitoring techniques, emergency response procedures, and aquifer restoration. In:

National Water Well Association, Proceedings of the second conference on environmental problems in karst terranes and their solutions, Nashville, Tennessee, November 16–18, pp 213–226

Crawford NC, Groves CG (1984) Storm water drainage wells in the karst areas of Kentucky and Tennessee. U.S. Environmental Protection Agency and Center for Cave and Karst Studies, Bowling Green, Kentucky, 52p

Currens JC, Graham CDR (1993) Flooding of Sinking Creek, Garretts Spring karst drainage basin, Jessamine and Woodford Counties, Kentucky. In: Beck BF (ed) Applied karst geology, proceedings of the fourth multidisciplinary conference on sinkholes and the engineering and environmental aspects of karst, Panama City, Florida, pp 145–56

DiToro DM, Small MJ (1979) Stormwater interception and storage. J Environ Eng Div ASCE 105(EE1)

Driver NE, Tasker GD (1988) Techniques for estimation of storm-runoff loads, volumes, and selected constituent concentrations in urban watersheds in the United States. U.S. Geological Survey Open-File Report 88-191, 80p

Galli J (1990) Peat-sand filters, a proposed stormwater management practice for urbanized areas. Department of Environmental Programs, Metropolitan Washington Council of Governments, Washington, DC

Garcia R (1992) Quality of storm-water runoff in three watersheds in Elizabethtown, Kentucky. U.S. Geological Survey, Water-Resources Investigations Report 92-4078, 27p

Garton ER Jr (1977) The effects of highway construction on the hydrogeologic environment at Bowden, West Virginia. In: Dilamarter RR, Csallany SC (eds) Hydrologic problems in karst regions. Western Kentucky University, Bowling Green, pp 439–449

Grizzard TJ, Randall CW, Weand BL, Ellis KL (1986) Effectiveness of extended retention ponds. In: ASCE, proceedings of an engineering foundation conference on urban runoff quality_impact and quality enhancement technology. New England College, Henniker, New Hampshire, June 23–27

Gupta MK, Agnew RW, Kobriger NP (1981a) Constituents of highway runoff, Volume I, state-of-the-art report: FHWA/RD-81-042. Federal Highway Administration, Washington, D.C., 111p

Gupta MK, Agnew RW, Meinholz TL (1981b) Constituents of highway runoff, Volume II, procedural manual for monitoring of highway runoff: FHWA/RD-81-043. Federal Highway Administration, Washington, D.C

Gupta MK, Agnew RW, Gruber D, Kreutzberger W (1981c) Constituents of highway runoff, Volume IV, characteristics of runoff from operating highways: FHWA/RD-81-045. Federal Highway Administration, Washington, D.C

Hewitt CN, Rashed MB (1992) Removal rates of selected pollutants in the runoff waters from a major rural highway. Water Res 26(3):311–319

Hoos AB (1990) Effects of storm-water runoff on local ground-water quality, Clarksville, Tennessee. U.S. Geological Survey, Water-Resources Investigations Report WRI-90-4044, 57p

Hubbard DA Jr, Balfour WM (1993) An investigation of engineering and environmental concerns relating to proposed highway construction in a karst terrane. Environ Geol 22:326–329

Irish LB Jr, Lesso WG, Barrett ME, Malina JF Jr, Charbeneau RJ, Ward GH (1995) An evaluation of the factors affecting the quality of highway runoff in the Austin, Texas Area. Technical report CRWR 264, Center for Research in Water Resources, Bureau of Engineering Research, College of Engineering, The University of Texas at Austin, Austin, Texas

Jannik NO, Alexander EC Jr, Landherr LJ (1991) The sinkhole collapse of the Lewiston, Minnesota wastewater treatment facility lagoon. National Ground Water Association, Proceedings of the third conference on hydrogeology, ecology, monitoring, and management of ground water in karst terranes, Nashville, Tennessee, December 4–6, pp 715–724

Jenkins DT (1988) Development of storm water management criteria for sensitive karst areas in North-central Florida, U.S.A. In: National Water Well Association, Proceedings of the second conference on environmental problems in karst terranes and their solutions, Nashville, Tennessee, November 16–18, pp 333–343

Kalmes A, Mohring E (1993) Development and construction of sinkhole treatments to improve water quality and control erosion. In: 26th annual [Minnesota] water resources conference

Kalmes A, Mohring E (1995) Sinkhole treatment to improve water quality and control erosion in Southeastern Minnesota. In: Beck BF (ed) Karst GeoHazards: engineering and environmental problems in karst terrane, proceedings of the fifth multidisciplinary conference on sinkholes and the engineering and environmental aspects of karst, Gatlinburg, Tennessee, pp 265–272

Kedlec RH, Hammer DE (1980) Wetland utilization for management of community wastewater: 1979 operations summary. Houghton Lake Wetland Treatment Project, NTIS PB80-170061

Keith JH, Bassett JL, Duwelius JA (1995) Modification of highway runoff quality by sinkhole drainage structures, Highway 37 improvement project, Lawrence County, Indiana. In: Beck BF (ed) Karst GeoHazards: engineering and environmental problems in karst terrane, proceedings of the fifth multidisciplinary conference on sinkholes and the engineering and environmental aspects of karst, Gatlinburg, Tennessee, pp 273–284

Kemmerly PR (1988) Influence of urbanization on sinkhole development and sinkhole flooding in karst terrain, Clarksville, Tennessee. In: National Water Well Association, field guides for field trips #2 and #3, second conference on environmental problems in karst terranes and their solutions, Nashville, Tennessee, pp 109–157

Kemmerly PR, Weber LC, Higgins CS Jr (1987) Multi-disciplinary analysis and design of class V drainage wells for a large commercial development. In: Beck BF (ed) Karst hydrogeology: engineering and environmental applications, proceedings of the second multidisciplinary conference on sinkholes and the environmental impacts of karst, Orlando, Florida, pp 311–320

Kerri KD, Racin JA, Howell RB (1985) Forecasting pollutant loads from highway runoff. Transp Res Rec 1017:39–46

Lang GE, Wider RK (1985) Modification of acid mine drainage by sphagnum-dominated wetlands and the water quality. In: Proceedings of the conference on wetlands of the Chesapeake: protecting the future of the bay, Easton, MD, pp 176–184

Linker LC (1989) Creation of wetlands for the improvement of water quality: a proposal for the joint use of highway right-of-way. In: Hammer DA (ed) Constructed wetlands for wastewater treatment, pp 695–701

Lord BN (1987) Nonpoint source pollution from highway stormwater runoff: science of the total environment, vol 59. In: Highway pollution, proceedings of the second international symposium, London, England, pp 437–446

Maestri B, Lord BN (1987) Guide for mitigation of highway stormwater runoff pollution: science of the total environment, vol 59. In: Highway pollution, proceedings of the second international symposium, London, England, pp 467–476

McKay G (1980) Filtration and separation: peat: an adsorbent/filtration medium for wastewater treatment: Water Serv 357–359

Mijatovic BF (1987) Catastrophic flood in the polje of Cetinje in February 1986, a typical example of the environmental impact of karst. In: Beck BF (ed) Karst hydrogeology: engineering and environmental applications, proceedings of the second multidisciplinary conference on sinkholes and the environmental impacts of karst, Orlando, Florida, pp 299–303

Milligan JD, Betson RP (1985) The effects of carbonate geology on urban runoff, part 2, water quality aspects. In: Perspectives on nonpoint source pollution, proceedings of a national conference, Kansas City, Missouri, May 19–22

Molina M, McDonald F (1987) Sinkhole management and flooding in Jamaica. In: Beck BF (ed) Karst hydrogeology: engineering and environmental applications, proceedings of the second multidisciplinary conference on sinkholes and the environmental impacts of karst, Orlando, Florida, pp 293–298

Moore HL (1984) Geotechnical considerations in in the location, design, and construction of highways in karst terrain: the Pellissippi Parkway Extension, Knox-Blount Counties, Tennessee. Sinkholes: their geology, engineering and environmental impact. In: Proceedings of the first multidisciplinary conference on sinkholes, Orlando, Florida, October 15–17, pp 385–389

Moore HL (1987) Sinkhole development along 'untreated' highway ditchlines in East Tennessee. In: Beck BF (ed) Karst hydrogeology: engineering and environmental applications, proceedings of the second multidisciplinary conference on sinkholes and the environmental impacts of karst, Orlando, Florida, pp 115–118

Moore HL (1988) Treatment of karst along Tennessee highways. In: Sitar N (ed) Geotechnical aspects of karst terrains: exploration, foundation design and performance, and remedial measures, proceedings of ASCE national convention, Nashville, Tennessee, May 10–11, vol 14, pp 133–148

Moore HL, Amari D (1987) Sinkholes and gabions: a solution to the solution problem. In: Beck BF (ed) Karst hydrogeology: engineering and environmental applications, proceedings of the second multidisciplinary conference on sinkholes and the environmental impacts of karst, Orlando, Florida, pp 305–310

Mull DS, Liebermann TD, Smoot JL, Woosley LH (1988a) Application of dye-tracing techniques for determining solute-transport characteristics of ground water in karst terranes. U.S. Environmental Protection Agency, 103p

Mull DS, Smoot JL, Liebermann TD (1988b) Dye-tracing techniques used to determine ground-water flow in a carbonate aquifer system near Elizabethtown, Kentucky. U.S. Geological Survey, Water-Resources Investigations Report 87-4174, 95p

Mustard MH, Driver NE, Chyr J, Hansen BG (1987) Urban-stormwater data base of constituent storm loads; characteristics of rainfall, runoff, and antecedent conditions; and basin characteristics. U.S. geological survey water-resources investigations report 87-4036, 328p

Newton JG (1987) Development of sinkholes resulting from man's activities in the Eastern United States. U.S. Geological Survey Circular 968, pp 7 and 15

Nichols DS (1983) Capacity of natural wetlands to remove nutrients from wastewater. J Water Pollut Control Fed

Nichols DS, Boelter DH (1982) Treatment of secondary sewage effluent with a peat-sand filter bed. J Environ Qual II(1):86–92

Ogden AE (1994) Investigation of sinkhole flooding problems in the Chilhowee Park area, Knoxville, Tennessee. In: Prepared for City of Knoxville Department of Engineering by Ground Water Consulting Services, Cookeville, Tennessee, 18p

Reeder PP, Crawford NC (1988) Urban storm water runoff removal, Bowling Green, Kentucky. In: Proceedings of the second conference on environmental problems in karst terranes and their solutions, Nashville, Tennessee, November 16–18, vol 2, pp 69–95

Royster DL (1984) Use of sinkholes for drainage: construction and difficult geology: karstic limestone, permafrost, wetlands, and peat deposits. Transp Res Rec 978:18–25

Saleem ZA (1977) Road salts and quality of groundwater from a dolomite aquifer in the Chicago area. In: Dilamarter RR, Csallany SC (eds) Hydrologic problems in karst regions. Western Kentucky University, Bowling Green, pp 364–368

Schiffer DM (1990) Wetlands for stormwater treatment: FL/DOT/SMQ/90-377. Florida Department of Transportation, Office of Materials and Research, Gainesville

Schueler TR, Kumble PA, Heraty MA (1991) A current assessment of urban best management practices, techniques for reducing non-point source pollution in the coastal zone, review draft. Anacostia Restoration Team, Department of Environmental Programs, Metropolitan Washington Council of Governments, Washington, DC

Smith HB (1991) Erosion and sedimentation control methodologies for construction activities over the Edwards Aquifer in Texas. In: National Ground Water Association, Proceedings of the third conference on hydrogeology, ecology, monitoring, and management of ground water in karst terranes, Nashville, Tennessee, pp 725–740

Smith DL, Lord BN (1990) Highway water quality control: summary of 15 years of research. Transp Res Rec 1279:69–74

Smith EF, Marks HB Jr (1976) The use of modified forms of peat as an oil coalescer. Journal of Environ Sci Health 12:727–734

Stephenson JB, Zhou WF, Beck BF, Green TS (1999) Highway stormwater runoff in karst areas—preliminary results of baseline monitoring and design of a treatment system for a sinkhole in Knoxville, Tennessee. Eng Geol 52(1–2):51–59

Stewart WS (1992) Compost storm water treatment system. W & H Pacific Consultants, Portland, OR, Final Report

U.S. Environmental Protection Agency (USEPA) (1986) Methodology for analysis of detention basins for control of urban runoff quality: EPA 440/5-87-001

Urbonas B, Stahre P (1993) Stormwater: best management practices and detention for water quality, drainage, and CSO management. Prentice Hall, Englewood Cliffs, New Jersey, p 449

USEPA (1983) Results of the nationwide urban runoff program, volume I: final report. Water Planning Division WH-554, NTIS PB84-18552, Washington, DC

Wada Y, Miura H (1984) Quantification of accumulated loads on road surfaces and their runoff characteristics. Water Sci Technol. River Basin management, proceedings of a conference, York, England, July 4–8, 1983, vol 16, no 5–7, pp 463–472

Wanielista MP, Yousef YA (1993) Stormwater management. Wiley, New York, 579p

Warren WM (1974) Retention basin failures in carbonate terranes. Water Resour Bull 10(1):22–31

Watson TJ, Reed SC, Kedlec RC, Knight RL, Whitehouse AE (1989) Performance expectation and loading rates for constructed wetlands. In: Constructed wetlands for wastewater treatment—municipal, industrial and agricultural. Lewis Publishers, Chelsea, Michigan

Werner E (1977) Chloride ion variations in some springs of the Greenbrier limestone karst of West Virginia. In: Dilamarter RR, Csallany SC (eds) Hydrologic problems in karst regions. Western Kentucky University, Bowling Green, pp 357–363

Werner E (1983) Effects of highways on karst springs: an example from Pocahontas County, West Virginia. In: Dougherty PH (ed) Environmental karst. GeoSpeleo Publications, Cincinnati, Ohio, pp 3–13

White EL, Aron G, White WB (1984) The influence of urbanization on sinkhole development in central Pennsylvania. In: Beck BF (ed) Sinkholes: their geology, engineering, and environmental impact, proceedings of the first multidisciplinary conference on sinkholes, Orlando, Florida, pp 275–281

Yousef YA, Wanielista MP, Hvitved JT, Harper HH (1985) Fate of heavy metals in stormwater runoff from highway bridges: science of the Total Environment, 33:233-244

Zawlocki KR, Ferguson JF, Mar BW (1980) A survey of trace organics in highway runoff in Seattle, Washington: Report WA-RD-39.9, prepared for Federal Highway Administration for University of Washington, Department of Civil Engineering, p 44

Early Detection of Contaminant Release from Waste Disposal Facilities

7

Groundwater monitoring is to detect the nature and magnitude of changes in groundwater quantity and quality because of natural processes or human activities. Achieving these goals is often challenged in karst terranes with peculiar characteristics including multiple porosity elements ranging from small pore spaces, solution-enlarged fractures/conduits to caves, and a spectrum of flow regime from laminar flow to turbulent flow. Figure 7.1 shows a schematic conceptual site model of a relatively isolated karst system, which includes a hypothetical waste disposal facility labeled as SITE. The groundwater flow and associated contaminant transport are controlled by discrete conduits and dissolution-enlarged fractures. While the small pores of the rock matrix function as storage, most karst groundwater and contaminants move through discrete conduits, underground rivers, and discharge at springs. The dynamics of groundwater flow make karst terranes susceptible to sinkholes, vulnerable to contamination, and unamenable to groundwater remediation. Consequently, construction of disposal units is typically not recommended in highly developed karst terrains.

7.1 Overview of Detection Monitoring Methods

7.1.1 Monitoring Strategy

Karst land covers approximately 13% of the world's ice-free land area and provides water to as much as 25% of the world's population (Ford and Williams 2007; Hollingsworth 2009). Construction of disposal facilities in such terranes is not totally avoidable. If a new solid waste disposal unit is proposed, the proponent must demonstrate that the proposed location in the karst terrain will not cause any degradation to the local groundwater resources and the environment. Meeting this requirement requires groundwater monitoring and early detection of any release. Figure 7.2 presents general procedures for establishing a groundwater monitoring program. An effective monitoring program in a well-developed karst aquifer shall include not only monitoring wells but also springs or underground rivers that are hydraulically connected to the investigation site (Quinlan 1990; Quinlan et al. 1991; Schindel et al. 1996; Ewers 2016). For a particular site, the monitoring program shall be tailored to the established conceptual site model and the project objective.

Construction of a representative conceptual site model is the basis for early detection of groundwater contamination from waste disposal facilities in karst terranes. Reliable monitoring of ground water in karst terranes can be done, but it is not cheap or easy (Quinlan et al. 1991). Design of a detection monitoring program in karst terranes addresses the following elements that are not typically as significant in other terranes:

- First, one needs to demonstrate that the groundwater system at the study site is monitorable. Both springs and wells are potential monitoring locations if they are connected to the groundwater system, and they are not impacted by any other disposal facilities.
- Second, due to dynamic responses to recharge events, particularly discharge and chemical constituents at karst springs, multiple-parameter, long-term, and high-frequency monitoring may be required to collect background data. Sampling and analysis plans should be designed to reflect the unique characteristics of the monitoring locations. Characterization of the natural variations in water quality may require sampling efforts under different flow conditions.
- Third, evaluation of the potential impact of waste disposal units on the groundwater system requires an effective statistical evaluation program. Due to heterogeneity of karst aquifers, intra-locational comparison is more practical, although inter-locational comparison is the preferred and more accepted strategy in porous-medium

© The Author(s), under exclusive license to Springer Nature Switzerland AG 2024
X. Jiang et al., *Monitoring and Early Warning Technologies on Karst Lands*, Advances in Karst Science, https://doi.org/10.1007/978-3-031-59045-0_7

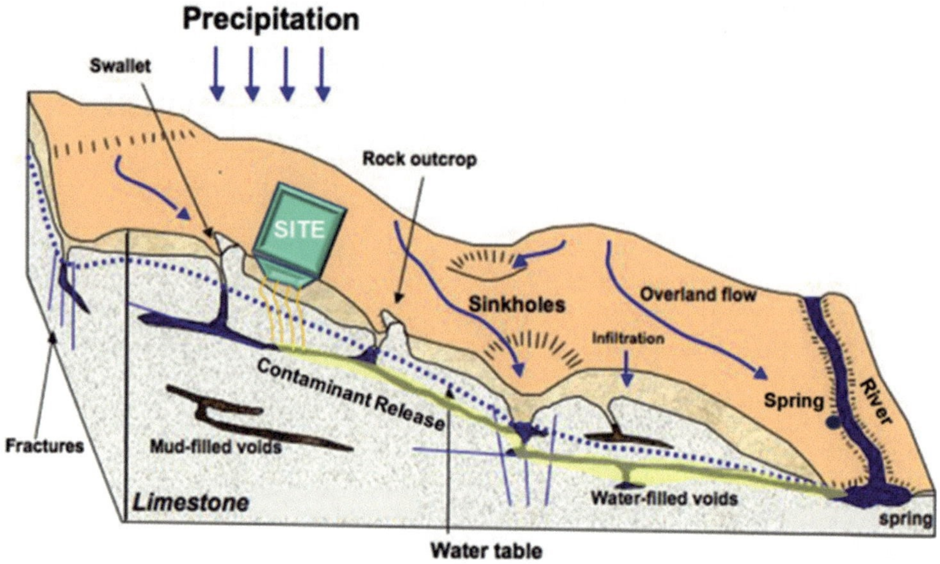

Fig. 7.1 Schematic conceptual site model of a karst aquifer

types of aquifers. Sufficient groundwater monitoring data prior to construction of waste disposal units is required to develop the intra-locational statistical evaluation.

In the case study presented in this chapter, procedures to address these above-mentioned challenges were presented for two springs. Seven dye tracing tests, two spring instrumentations, nine background sampling, flow-weighted concentrations, and an innovative statistical evaluation method were presented to illustrate the detection monitoring program. These procedures were developed to evaluate potential contaminant release from a solid waste disposal facility constructed in a relatively isolated karst terrane. Although the specific procedures may not be readily duplicated, the overall technical approaches discussed herein will shed light on groundwater monitoring programs in other karst areas.

7.1.2 Monitoring Locations

Under USEPA's Resource Conservation and Recovery Act, for example, a minimum of four wells, one upgradient from the unit, and three downgradient from the source are required to monitor the release of contaminants (USEPA 2004). The upgradient well is intended to sample background concentrations of the relevant constituents presumably unaffected by the contaminant source, while the downgradient wells represent points of compliance, where concentrations are established to determine whether contaminants are migrating from the regulated unit. While this design would typically be effective in porous media where contaminants tend to spread out in blob-like plumes, contaminants in well-developed karst aquifers can pass undetected by even closely spaced wells.

Figure 7.3 shows schematically groundwater flow and associated contaminant transport pathways in a karst system. The path highlighted in yellow represents the likely migration route of contaminants if a release or leak occurs at SITE. When the probability of monitoring wells encountering the conduits is small the monitoring wells, designated by red circles, are not capable of detecting release of contaminants from the waste disposal facility or landfill. On the other hand, springs are preferred monitoring locations, and it is perceivable that any release of contaminants would move to and discharge at the springs. From a remediation perspective, springs are exposure points of contaminated groundwater to receptors such as human beings or plants. Therefore, the springs can be remedial action targets to eliminate the exposure pathway.

However, use of springs as either monitoring locations or remedial targets shall be based on a thorough understanding of the springs including their characteristics and relationship with the entire drainage basin. This chapter presents a case study to demonstrate how springs helped accomplish project goals if used properly and the challenges of using springs as compliance monitoring points.

In many karst aquifers, springs are an important component of conceptual site models, and identification of springs has been an essential part of site characterization. Because springs can provide data of aquifers in scales typically larger than monitoring wells, they have been considered as more effective monitoring points for detection of contaminant release (Zhou et al. 2002; Quinlan et al. 1991).

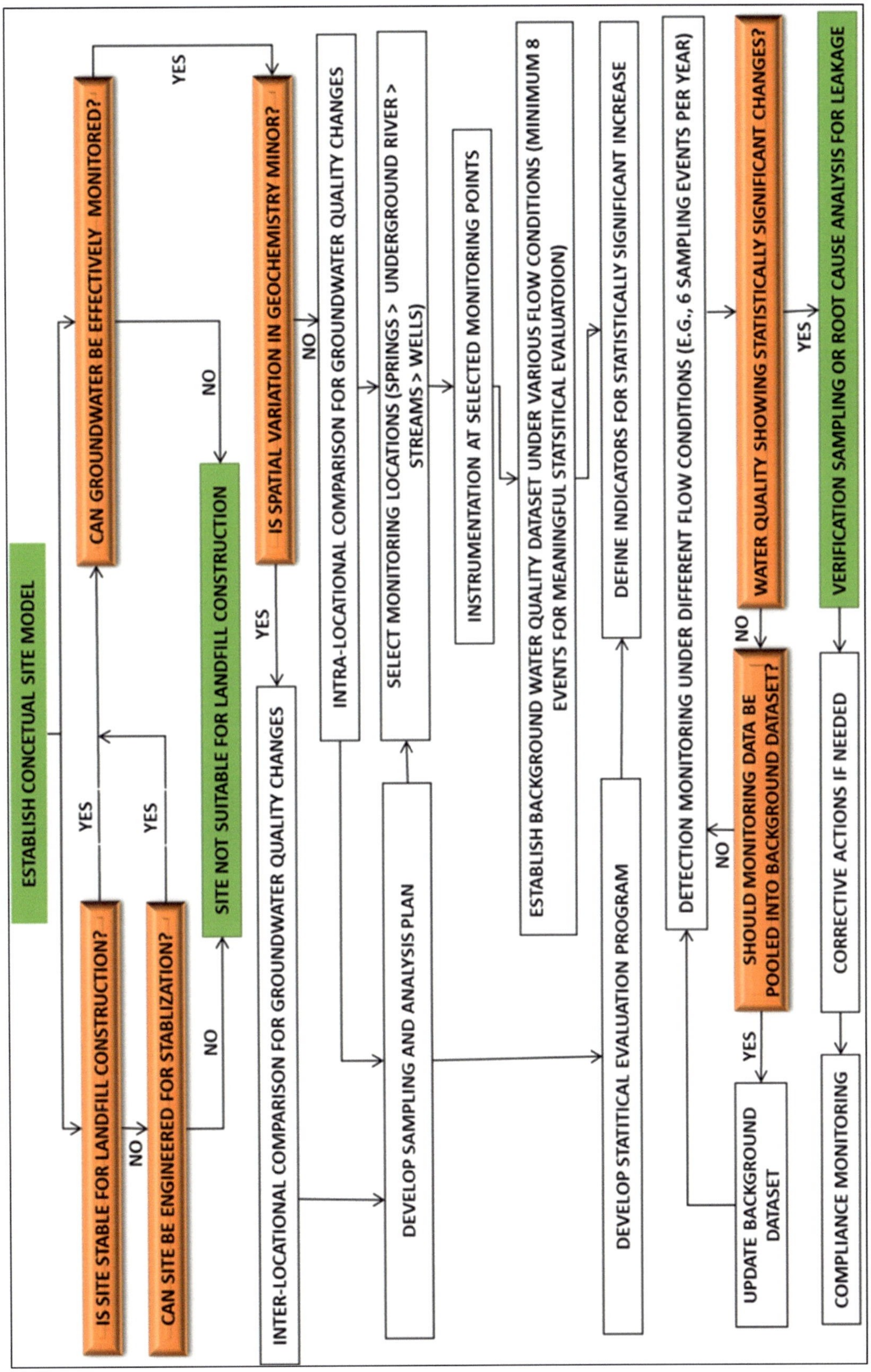

Fig. 7.2 General technical approach of groundwater detection monitoring in karst terranes

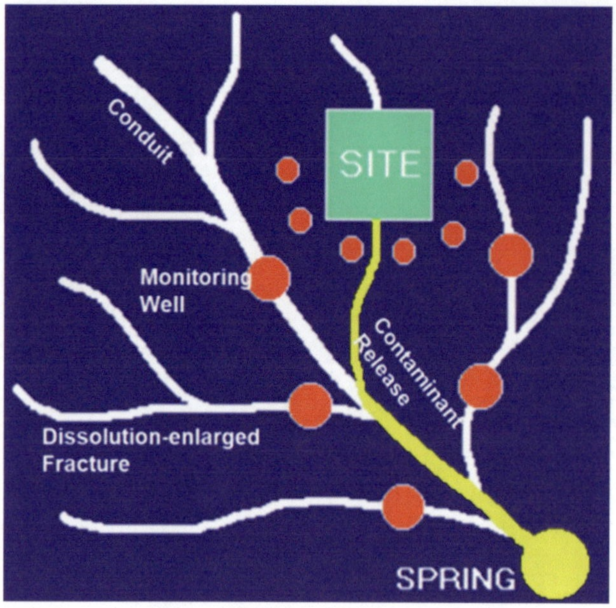

Fig. 7.3 Schematic diagram showing contaminant migration routes (red circles—monitoring wells; yellow line—likely migration pathway of contaminant released from the landfill site)

7.1.3 Intra-locational Versus Inter-locational Comparison

A commonly used statistical comparison of water-quality data involves inter-locational comparison of data, which compares water-quality data between non-impacted upgradient locations and downgradient monitoring locations (typically monitoring wells) that could potentially be impacted by facility releases. An inter-well comparison also requires hydrogeologic conditions that are contiguous and uniform across a site and wells must be screened within the same intervals. With this method, however, statistical tests are oftentimes unable to distinguish between statistically significant differences in water quality caused by a release from the facility versus natural water-quality variations within the aquifer. Consequently, in cases where wide variation in water quality can occur (such as within a karst setting where a storm pulse can cause large variations in spring water quality), the intra-location comparison method is the preferred statistical approach.

The intra-locational method is fundamentally different from the inter-locational analysis methods in that the intra-locational analysis compares each compliance monitoring point (well, spring, etc.) against a background composed of its own historical data. The primary advantage of this is that there are no spatial variability issues since only data from a single location is compared against itself. Due to the absence of spatial variability, the uncertainty in measured concentrations is decreased making intra-location comparisons more sensitive to real releases (thereby reducing false-negatives), while false-positive results due to spatial variability are eliminated.

However, according to Gibbons and Coleman (2001), intra-locational comparisons are only appropriate when pre-disposal data are available, and it can be demonstrated that the facility has not previously impacted a given monitoring location. Therefore, in an ideal situation, all background data for any given site would be collected prior to the placement of any waste material. For many projects, however, this was not possible due to time constraints, instrument-related delays, weather delays, and difficulties obtaining storm-pulse sampling data early in the monitoring program. The background levels of groundwater constituents during a variety of aquifer flow conditions can be generally established prior to significant use of the landfill if the contaminant breakthrough time through the clay liner of the landfill via natural migration is on the order of several years. In addition, outliner identification and trend analysis should be conducted to ensure lack of evidence for positive trends in the constituents of concern. Under these circumstances, it can be assumed that no releases from the landfill have occurred prior to the final background sampling event, and the background data can be collected while the landfill is in use.

7.1.4 Monitoring Time and Frequency

Applying conventional sampling strategies to rather non-conventional karst settings is not effective. Groundwater monitoring in karst settings often requires high-frequency sampling in conjunction with aquifer recharge events (i.e., rainfall or snowmelt) that cause storm-induced pulses or increased discharge at springs, cave streams, or wells that have been appropriately dye traced. This type of sampling allows characterization of any changes in natural groundwater chemistry that occur in response to the increased flow of water into and through the karst aquifer. According to USEPA Guidance (1989), sampling of storm-pulse events in karst settings must also include sampling before, during, and after precipitation to ensure that the overall variability for any given event can be captured. With regards to variability, it should be noted that a well-designed karst monitoring program must also include sampling events conducted during different times of the year, under variable flow conditions, i.e., base-flow, low-flow, and high-flow, and both with and without precipitation.

It is without question that long-term, high-frequency sampling of storm-induced pulses from karst springs is a difficult and challenging task. As stressed by Quinlan (1990), reliable monitoring of groundwater in karst terranes can be done but is not cheap or easy. For long-term, high-frequency sampling to be effective in a karst setting, a non-conventional approach and an extraordinary amount of effort must be invested. A site-specific sampling and analysis plan is an effective approach to start the detection monitoring efforts.

7.1.5 Phased Approaches to Monitoring Program

An effective monitoring program in karst terranes typically consists of four sequential phases, background sampling, detection monitoring, compliance monitoring, and corrective action monitoring. During the background phase, background levels of groundwater constituents are established prior to operation of the landfill. Following this starts the detection monitoring in which multiple rounds of monitoring samples are collected and analyzed annually. Analytical results of the compliance monitoring will then be statistically pooled and compared to background concentrations to determine whether a statistically significant increase over background levels for a given target constituent is occurring. Depending on the number and specific compounds exhibiting statistically significant increases, the compliance monitoring program starts to ascertain whether the constituents released to the environment are exceeding acceptable concentration levels or groundwater protection standards. Once an exceedance of the groundwater protection standards has been detected, corrective action could be required, and such a correction action initiates the correction action monitoring. A generalized flow chart illustrating the major steps of the phased-approach process is included in Fig. 7.4.

The phased approach can also be reflected in spatial monitoring locations. Based on landfill use planning, some monitoring locations may start at the early stage of compliance monitoring, whereas other monitoring locations may start being monitored once certain cells of the landfill are used. In the case study presented in this chapter, the initial background samples are collected in both springs QRS and QNS. The compliance monitoring is initiated at QRS only. Initiation of the compliance monitoring for QNS is

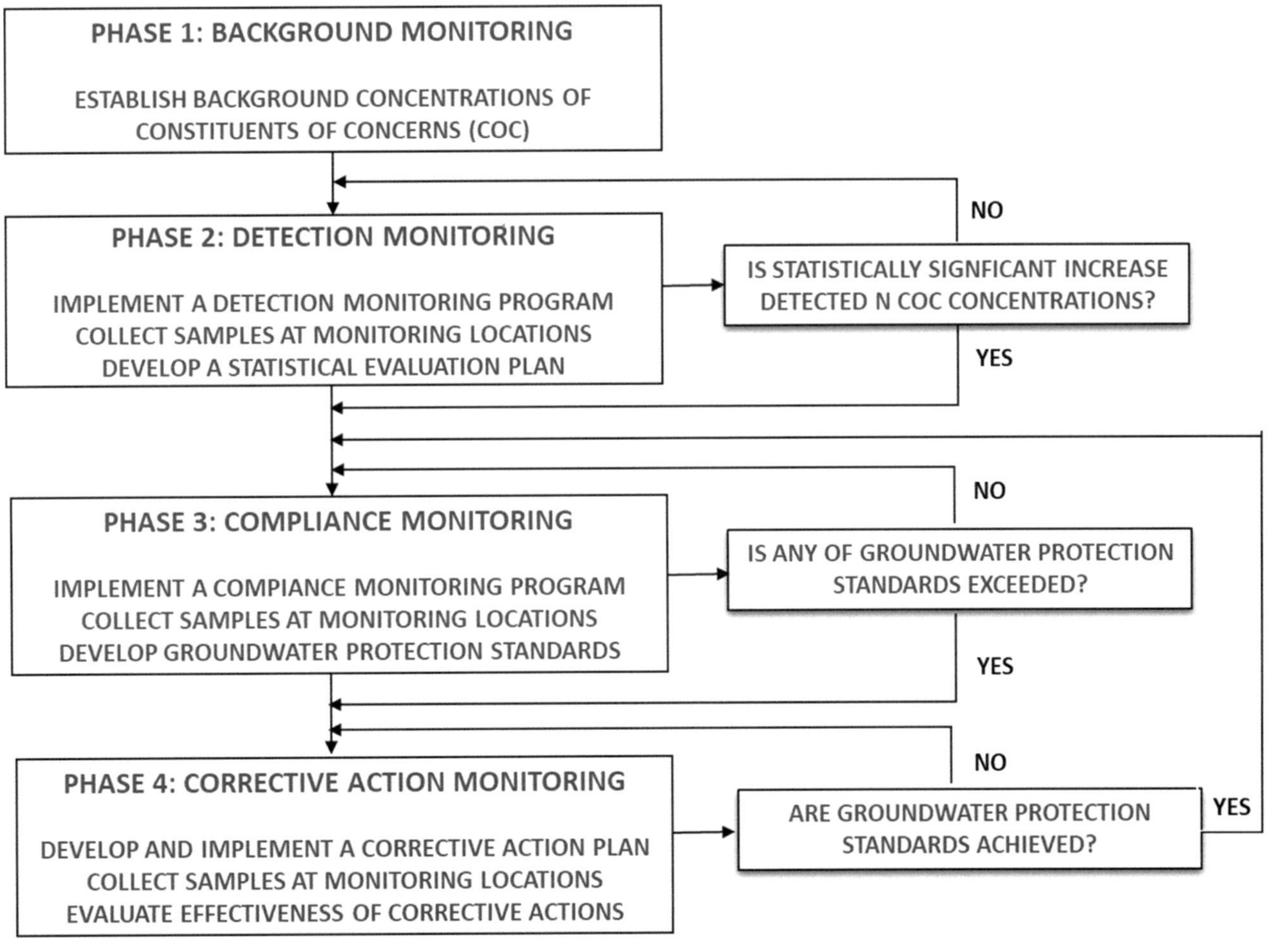

Fig. 7.4 Phased approach in groundwater monitoring in karst terranes

contingent upon use of the last cell of the landfill. While the compliance monitoring is conducted at QRS, the background collection phase continues at QNS, unless the last cell is utilized. Therefore, different chemical parameter lists need to be used for sampling at QRS and QNS.

7.2 Geologic Setting of the Landfill

Figure 7.5 shows the location of the landfill, which was constructed within the eastern slope of a small stream valley, which is entrenched into the Mitchell Plain of southern Indiana. The Mitchell Plain is a 2900-km^2 limestone plateau characterized by an abundance of underground drainage and a lack of streams on the land surface (Powell 1973; McConnell and Horn 1972). Physiographically, it is a subunit of the Highland Rim Section of the Interior Low Plateaus Province (Fenneman 1938). It extends southward into Kentucky as the Pennyroyal Plateau; to the north, it is covered with a thick blanket of Pleistocene till (Palmer and Palmer 1975). A sinkhole plain occupies the westernmost 30% of the Mitchell Plain (Palmer and Palmer 1975).

Detailed geotechnical and geophysical investigations were conducted to evaluate risk of sinkhole occurrence, and the risk of structural collapse was factored into the landfill design (Zhou et al. 1999, 2000). Clayey overburden, up to 12 m thick overlies the St. Louis and Salem limestone bedrock of the Mississippian Period. Seven sinkholes were identified in the vicinity of the study site, and they are labeled as A through G in Fig. 7.5. Several small springs, labeled as QRS, QNS, and QBS, discharge groundwater to a local stream RLB along the base of the valley slope just west of the site. QNS is located at the head of a short, steep valley, whereas QRS and QBS emerge from small openings in the valley floor. GrS is a submerged spring on the RLB riverbed and was discovered through the trace tests, as discussed below.

Permitting the waste disposal facility required extensive hydrogeological and geophysical investigations, ecological evaluation, sinkhole risk assessment, and an effective groundwater monitoring program. Development of the groundwater monitoring program is discussed in this paper. First, one needs to demonstrate that the groundwater system at the study site can be monitored. The monitoring locations must be effectively connected to the groundwater system and not impacted by any other disposal facilities. Secondly, the sampling and analysis plans should be designed to reflect the unique characteristics of the monitoring locations. Characterization of the natural variations in water quality requires sampling efforts under different flow conditions. Thirdly, evaluation of the potential impact of waste disposal units on the groundwater system requires an effective statistical evaluation plan. Due to the heterogeneity of karst aquifers, intra-locational comparison is generally preferred to inter-locational comparison. The intra-locational statistical testing compares data collected prior to operation of the disposal facility versus data collected during operation of the disposal facility at the same sampling point. Because the comparison is made at the same sampling point, concentration differences between monitoring locations due to natural spatial factors do not affect the intra-locational tests. Concentration changes of the chemicals of concern over time cause an intra-locational test to be statistically significant and to show a change in groundwater quality (Interstate Technology & Regulatory Council 2013). The out-of-compliance triggers and quality-control procedures should be developed based on site-specific conditions.

7.3 Dye Tracing to Select Springs for Early Detection Monitoring

A groundwater tracing study consisting of seven tracer tests was conducted to aid in the design of the karst groundwater monitoring program for the proposed landfil facility. Fluorescein and rhodamine WT dyes were introduced into five sinkholes. At sinkholes D and F, tracing tests were repeated under a different flow condition. On-site fluorometric analyses and concentration-dependent sampling were utilized at springs and the stream for dye monitoring. The table inserted in Fig. 7.5 summarizes the dye tracing results. Two examples of dye trace breakthrough curve are shown in Fig. 7.6.

Rhodamine WT inserted into Sinkhole C was detected at RLB-1 and QNS. The discharge from QNS entered RLB downstream of RLB-1, and the dye was not detected at QRS. Therefore, dye detected at RLB-1 must have emerged from an unidentified groundwater discharge feature in the bedrock bottom of RLB. One gallon of Rhodamine WT dye was injected in Sinkhole C on a later date revealed the submerged spring GrS, which occurred at the intersection of bedding plan and a joint. Figure 7.7 shows the location of GrS where Rhodamine WT dye is being discharged.

The dye tracing results indicate that groundwater monitoring at the site can be conducted most appropriately at three natural discharge springs, QRS, QNS, and GrS. GrS is a submerged spring in the stream bed and not readily amenable for monitoring without a significant modification of the stream. Subsequently, springs QRS and QNS were established as the permanent downgradient monitoring stations. Figure 7.8 shows the original discharge points of these two springs.

Based on discussions with the relevant regulatory agency, the monitoring program consisted of the following:

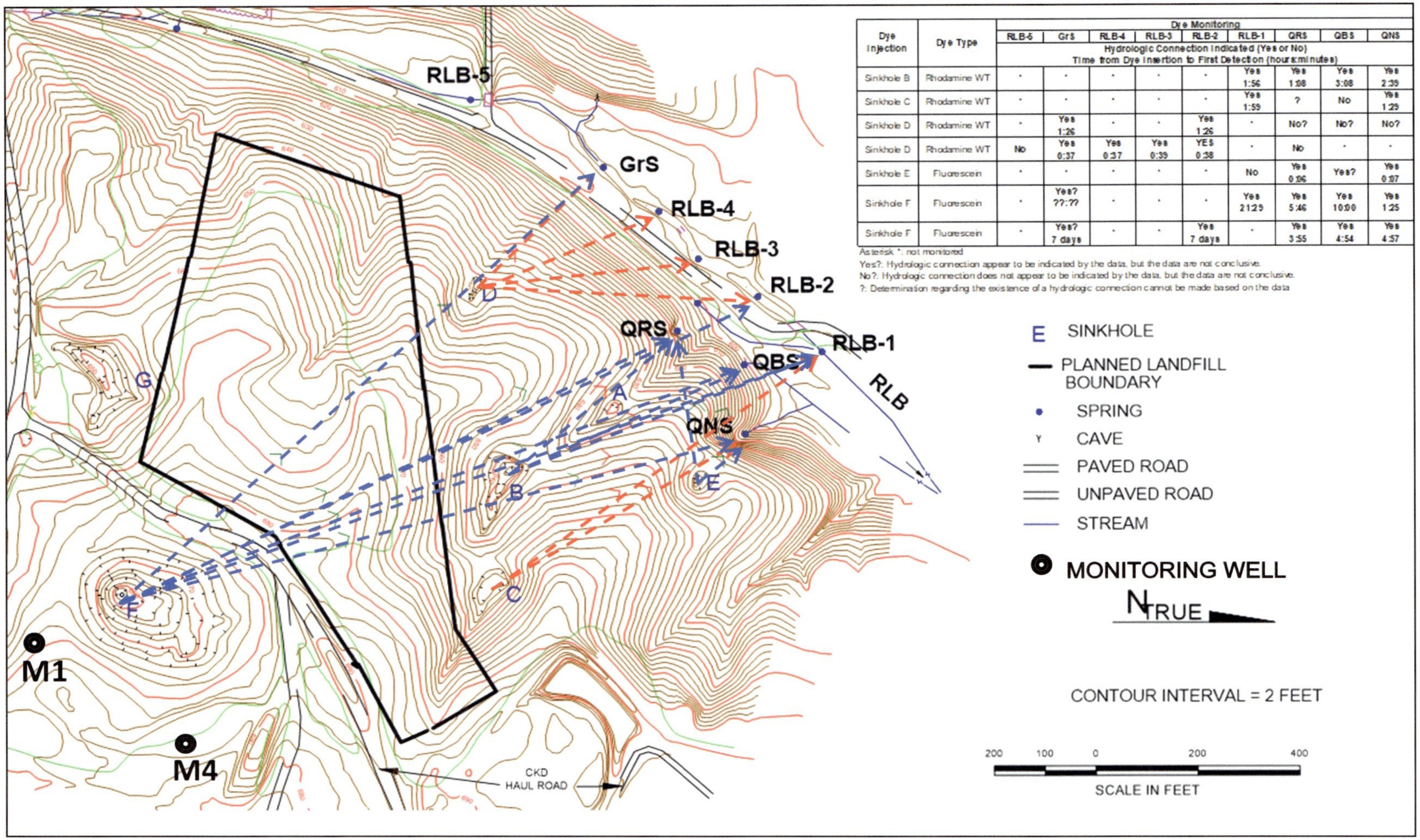

Dye Injection	Dye Type	Dye Monitoring								
		RLB-5	GrS	RLB-4	RLB-3	RLB-2	RLB-1	QRS	QBS	QNS
		Hydrologic Connection Indicated (Yes or No) Time from Dye Insertion to First Detection (hours:minutes)								
Sinkhole B	Rhodamine WT	*	*	*	*	*	Yes 1:56	Yes 1:08	Yes 3:08	Yes 2:39
Sinkhole C	Rhodamine WT	*	*	*	*	*	Yes 1:59	?	No	Yes 1:29
Sinkhole D	Rhodamine WT	*	Yes 1:26	*	*	Yes 1:26	*	No?	No?	No?
Sinkhole D	Rhodamine WT	No	Yes 0:37	Yes 0:37	Yes 0:39	YES 0:38	*	No	*	*
Sinkhole E	Fluorescein	*	*	*	*	*	No	Yes 0:06	Yes?	Yes 0:07
Sinkhole F	Fluorescein	*	Yes? ??:??	*	*	*	Yes 21:29	Yes 5:46	Yes 10:00	Yes 1:25
Sinkhole F	Fluorescein	*	Yes? 7 days	*	*	Yes 7 days	*	Yes 3:55	Yes 4:54	Yes 4:57

Asterisk *: not monitored
Yes?: Hydrologic connection appear to be indicated by the data, but the data are not conclusive.
No?: Hydrologic connection does not appear to be indicated by the data, but the data are not conclusive.
?: Determination regarding the existence of a hydrologic connection cannot be made based on the data

Fig. 7.5 Site location of the landfill and hydraulic connections based on tracer tests

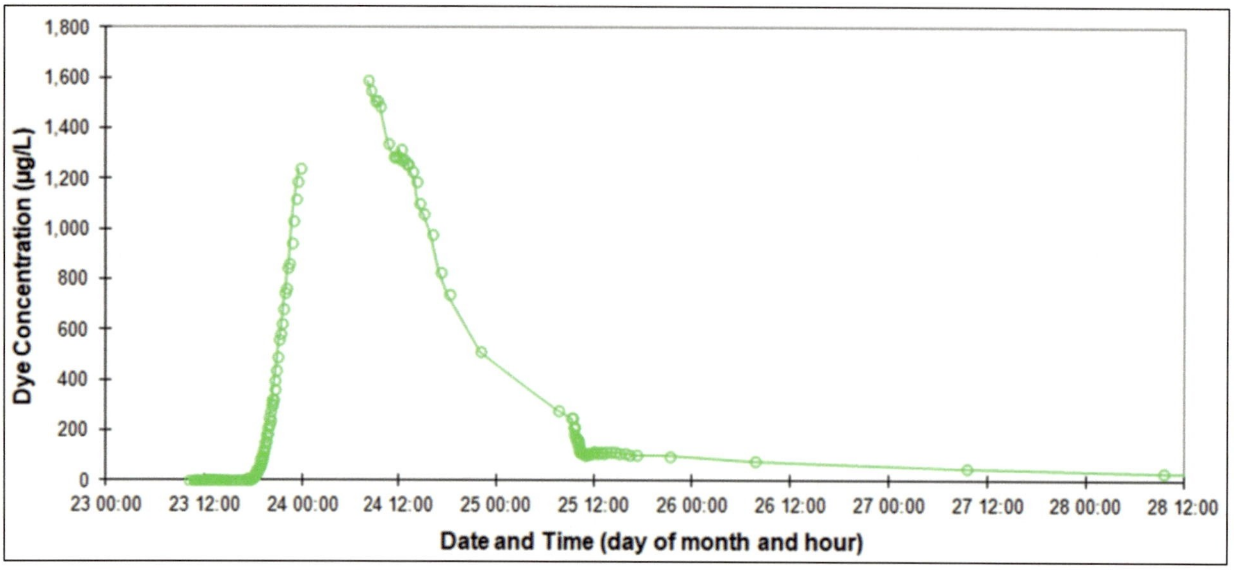

Fluorescein breakthrough curve at QRS in response to dye injection at sinkhole E

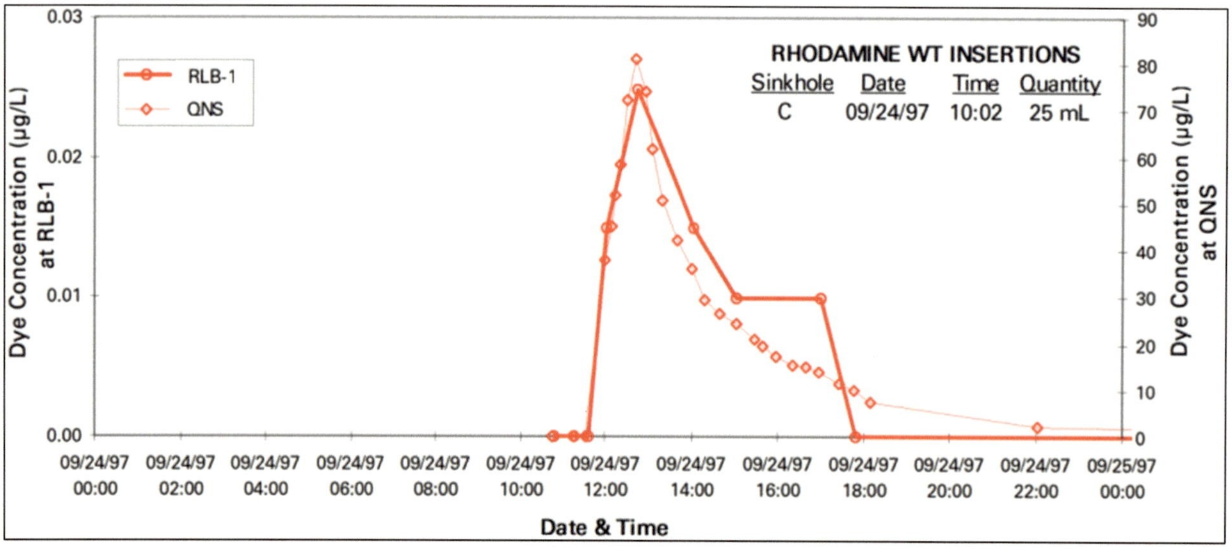

Rhodamine WT breakthrough curve at QNS and RLB-1 in response to dye injection at sinkhole C

Fig. 7.6 Example dye breakthrough curves from tracer tests

- Spring Discharge: Discharge is continuously monitored and recorded (every 15 min) at QRS and QNS. Typical flows for QRS/QNS under low-flow conditions are on the order of five to ten gallons-per-minute (gpm), and range as high as approximately 60–90 gpm under high-flow conditions, with occasional higher peaks.
- Spring Water Quality: Water quality is continuously monitored and recorded (every 15 min) at QRS and QNS. Monitored parameters include temperature, pH, conductivity, turbidity, oxidation–reduction potential (ORP), and dissolved oxygen (DO). This information is currently recorded using permanently installed dataloggers and is measured by various water-quality sensors installed in stilling wells housed inside permanently installed instrument shelters constructed over the spring outlets.
- Spring Chemical Sampling: Water samples for laboratory analysis are collected predominantly during storm-induced recharge events. During these events, samples are continuously collected at intervals ranging from approximately 20 min to four hours using dedicated

Fig. 7.7 Discovery of a submerged spring with Rhodamine WT dye injected into a sinkhole

automatic water samplers installed within the QRS and QNS instrument shelters. It is recommended that a total of six groundwater sampling events be completed annually at these two springs.

- Aquifer Water Levels: Groundwater levels are continuously measured and recorded (every 15 min) within monitoring wells M1 and M4 (Fig. 7.5), which are installed up topographic and hydraulic gradients from the landfill and springs.

7.4 Instrumentation of Springs

Both QRS and QNS discharge water into irregular open channels and eventually flow into the local creek, RLB. Measuring the flow of either spring required construction of a structure to constrict the water flow in its open channel. Figure 7.9 shows the structure design that allowed discharge gauging, water-quality monitoring, and water sampling at QNS. A similar instrumentation structure was constructed at QRS. Every effort was made to minimize the modification of the springs. However, some improvements were necessary to install the instruments and calculate the discharge.

- Both flow measuring devices were to be installed with free fall, or no submergence whenever possible. This required enlargement and regrading of the outfall channel with 0.5–1-foot elevation difference between the weir and the outfall channel.
- Installation of these weirs included an approach section of 5–10 feet straight channel after the spring pools. The weir floor must be level. This required grading the channel section where the weir was to be installed.
- The weirs were designed to be freestanding and required no external support to maintain their dimensional integrity during operation.
- The PVC pipe connecting the weirs should be grouted into the open channel. Grouting the PVC in place lessens the chance of wall deflection and secures the weirs.
- A rock apron was placed beneath the flume outfall to protect against erosion.

The materials used for constructing the quasi-monitoring well and the installation of the well should not measurably alter the physical and chemical quality of the spring water. The packing gravel in the upstream should be grading in size toward the well. The inside packing material can be 2-inch stone and the outside may consist of large rubbles (> 3 in.). These large-diameter packing materials extend the spring orifice to the monitoring well.

A quasi-monitoring well of 6-inch in diameter was installed to house the multi-parameter instruments. Two feet of the well casing were perforated, where the probes were located. The diameter of the slot openings should be in the range of 0.4–0.5 in. The packing materials were quartz cobbles or sandstone fragments. The impermeable liner on top of the packing materials prevented rainwater infiltration. The surface is sloped 15° outward to avoid water ponding.

Karst springs have been known for having high-frequency, high-amplitude changes in water chemistry, especially in response to recharge events. The following parameters were monitored in real time at designated frequencies and stored in dataloggers for regular retrievals:

Fig. 7.8 Original discharge points of QRS and QNS

QRS spring under natural conditions

QNS spring under natural conditions

- Discharge
- pH
- DO
- ORP
- Temperature
- Turbidity
- Specific conductance (SC).

Figure 7.10 shows the spring modification processes and installed sampling devices for the spring monitoring. Figure 7.10a shows the wing wall and pipe construction to keep integrity of the spring, while a still well was installed to facilitate housing of the pressure transducer and water-quality probes (Fig. 7.10b). Figure 7.10c is the installed flume for discharge measurements, whereas Fig. 7.10d illustrates the water-quality monitoring devices and data logging system. The instrumentation is clearly site-specific. The monitorability of spring depends to a large extent on whether its discharge and water quality parameters can be measured. In this case study, for example, QrS was underwater and its monitorability requires a significant effort to modify the discharge point to install the instruments and a regulatory permit for such a modification on the surface creek. Because springs connect groundwater and surface water, it is common that springs are submerged in rivers, reservoirs, lakes, and oceans. The evidence that a

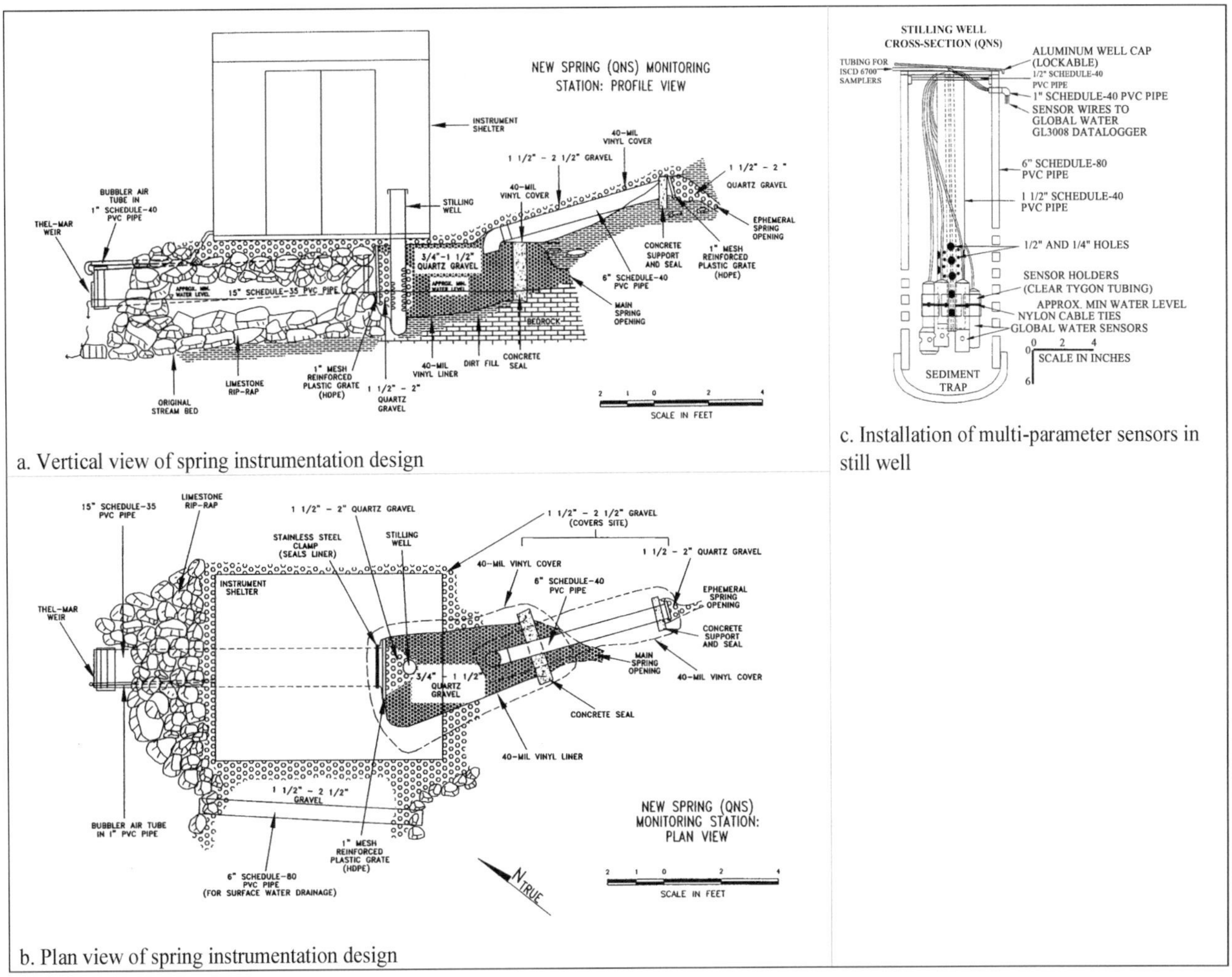

a. Vertical view of spring instrumentation design

b. Plan view of spring instrumentation design

c. Installation of multi-parameter sensors in still well

Fig. 7.9 Structural design for instrumentation at selected springs

spring is hydraulically connected to the study site does not automatically mean that the spring can be used for compliance monitoring.

7.5 Background Sampling and Analysis Results

Nine background sampling events were conducted at the two springs to characterize the background data population. Table 7.1 presents the specific types of background sampling events, approximate precipitation amounts for each event sampling event dates and duration, and analysis results of 11 parameters that are subject to statistical evaluation.

Six of these background sampling events were high-frequency, storm-pulse-type events in response to a variety of aquifer, flow, and precipitation conditions, while three were 24-h base-flow events with no precipitation. Following each background sampling event, samples were judgmentally selected for laboratory analysis based on the spring discharge and field water-quality data (pH, DO, ORP, temperature, turbidity, and SC) collected during each event. Such a sampling approach characterized changes in natural ground water chemistry that occurred under different flow conditions. Table 7.1 summarizes the natural variations of 11 water-quality parameters during the background sampling efforts at Spring QRS. Only data at QRS is presented, whereas the data at QNS are similar. Figure 7.11 shows two examples of specific conductance change in response to recharge events in two background sampling events. Concentrations of measured constituents varied with spring discharge and might be impacted by the antecedent soil moisture conditions, flow conditions, and other factors.

Clearly, these sampling approaches were non-conventional. To perform high-frequency, storm-induced sampling, accurate selection of the appropriate sampling events based on weather forecasts was important. This can be a major difficulty for any karst sampling program. Individual storm-induced sampling events that have been started may need to

a. Wing wall and pipe construction at spring discharge point

b. Still well installation with liner, grating, and gravel pack

c. Installed device for discharge measurement

d. Installed instruments for water-quality parameter monitoring and sampling

Fig. 7.10 Instrumentation at spring QNS

be discontinued because forecasted precipitation does not materialize, or a storm event may not produce a significant hydrographic or chemical response. Discontinuing a sampling event can cause significant unwanted expenses for clients, especially when significant mobilization times exist. Due to the complicated nature of the sampling, the fact that sampling could occur at any time of the day or night, and to facilitate smoother and more efficient handling of sampling-related activities, a plan for monitoring the weather conditions and communication between all parties was developed (Lounsbury et al. 2003). When a weather condition was deemed to be favorable for sampling, the flow conditions at the springs and in the aquifer were immediately evaluated from the field data (groundwater level, discharge, and field water-quality parameters) to guide the sampling efforts.

7.6 Criteria for High-Flow and Low-Flow Conditions

The landfill permit requires that six groundwater sampling events be conducted under variable flow and aquifer conditions annually with a few events associated with storm-induced pulses during high-flow conditions. Determination

Table 7.1 Background sampling events

Date and time	Discharge (gpm)	Alkalinity, Bicarb ($CaCO_3$)	Alkalinity, Carb ($CaCO_3$)	Alkalinity, Total ($CaCO_3$)	Chloride	Solids, dissolved	Sulfate	Calcium	Magnesium	Potassium	Selenium	Sodium
Event 1—Low-flow (Precipitation = 3.0")												
11/6/2000 16:50	16.43	240 J	<1	240		860						
11/8/2000 23:00	16.43	280 J	<1	280	58 J	1000	150	150	16	130 J	<0.005	30 J
11/8/2000 23:40	17.78	250 J	<1	250	60 J	770	140	140	16	120 J	<0.005	27 J
11/9/2000 2:00	47.42	240 J	<1	240	39 J	680	150	120	16	77 J	0.006	18 J
11/9/2000 3:20	46.00	210 J	<1	210	20 J	440	88	110	14	50 J	<0.005	12 J
11/9/2000 3:40	42.44	210 J	<1	210	22 J	450	89	110	15	52 J	<0.005	13 J
11/9/2000 6:20	40.70	260 J	<1	260	44 J	690	180	140	18	94 J	0.006	23 J
11/9/2000 8:00	40.70	270 J	<1	270	37 J	600	64	120	15	85 J	<0.005	20 J
11/9/2000 14:40	38.97	270 J	<1	270	37 J	640	120	130	15	86 J	<0.005	21 J
11/9/2000 15:20	58.28	280 J	<1	280	37 J	610	99	120	15	78 J	<0.005	19 J
11/9/2000 15:40	63.63	280 J	<1	280	36 J	280	100	110	13	66 J	<0.005	15 J
11/9/2000 15:40	69.40	440 J	<1	440	22 J	480	92	100	12	51 J	<0.005	11 J
11/9/2000 16:00	77.40	210 J	<1	210	21 J	480	7.4	110	13	55 J	0.006	11 J
11/9/2000 16:40	81.67	220 J	<1	220	19 J	460	82	110	13	49 J	<0.005	10 J
11/9/2000 17:20	79.51	250 J	<1	250	27 J	530	110	120	15	66 J	<0.005	15 J
11/9/2000 18:00	73.21	240 J	<1	240	18 J	480	110	120	14	59 J	<0.005	12 J
11/9/2000 18:40	60.22	240 J	<1	240	17 J	470	59	110	12	50 J	<0.005	9.7 J
11/9/2000 22:40	54.44	260 J	<1	260	25 J	530	99	120	14	62 J	<0.005	14 J
11/10/2000 2:00	47.42	260 J	<1	260	23 J	540	65	120	15	61	<0.005	14
11/10/2000 7:40	35.62	260 J	<1	260	23 J	550	110	120	15	64	<0.005	15
11/10/2000 7:40	23.27	260 J	<1	260	24 J	560	120	120	15	63	<0.005	15
11/13/2000 16:55	23.27	260 J	<1	260	38 J	640	160	140 J	16 J	83 J	<0.005	21
11/13/2000 16:55	23.27	550 J	<1	550	38 J	680	180	140 J	15 J	79 J	<0.005	21
Event 2—Low-flow (Base) (Precipitation = 0.0")												
1/10/2001 12:15	24.05	180	<1.0	180	38	570	200	100	12	82	<0.005	16
1/10/2001 19:15	25.00	200	<1.0	200	40	590	220	100	12	80	<0.005	16
1/11/2001 3:15	23.27	190	<1.0	190	40	580	210	100	12	82	<0.005	16
1/11/2001 11:15	23.27	210	<1.0	210	41	580	210	110	12	84	<0.005	16
1/11/2001 11:15	23.27	200	<1.0	200	41	610	210	110 i	12 i	85 i	<0.005	17
Event 3—Low-flow (Precipitation = 1.2")												
1/29/2001 18:03	17.78	250	<1	250	42	550	200 J	120	15	70	0.017 J	17
1/29/2001 21:43	19.11	240	<1	240	40	580	200 J	120	16	68	0.019 J	16
1/30/2001 3:43	27.79	220	<1	220	40	550	190 J	120	15	63	0.05 J	16
1/30/2001 4:43	29.22	230	<1	230	42	580	200 J	120	16	66	0.034 J	16
1/30/2001 6:03	30.80	230	<1	230	56	700	250 J	140	18	88	0.07 J	20
1/30/2001 7:03	30.80	220	<1	220	60	730	170 J	140	18	88	0.04 J	20
1/30/2001 8:23	32.37	250	<1	250	60	740	240 J	130	18	90	0.037 J	20
1/30/2001 10:03	32.37	240	<1	240	55	650	220 J	130	17	86	0.033 J	19
1/30/2001 11:43	44.22	250	<1	250	50	620	140 J	130	16	75	0.031 J	17
1/30/2001 12:03	44.22	230	<1	230	41	560	100 J	110	15	63	0.023 J	14
1/30/2001 12:23	42.44	220	<1	220	37	510	92 J	110	15	58	0.026 J	13
1/30/2001 13:23	42.44	240	<1	240	38	540	70 J	110	15	57	0.038 J	13
1/30/2001 15:03	50.70	250	<1	250	37	520	99 J	120	15	57	0.033 J	13
1/30/2001 15:03	54.44	240	<1	240	30	490	130 J	110	14	51	0.024 J	11
1/30/2001 17:43	56.34	240	<1	240	28	500	77 J	100	12	42	0.034 J	9.3
1/31/2001 0:43	61.93	210	<1	210	20	410	68 J	96	11	35	0.033 J	7.4
1/31/2001 3:43	61.93	250	<1	250	19	410	<50 J	99	12	35	0.04 J	7.6
1/31/2001 8:43	61.93	210	<1	210	19	410	60 J	99	12	35	0.038 J	7.5
1/31/2001 8:43	52.55	220	<1	220	19	430	<50 J	96	11	34	0.06 J	7.5
2/1/2001 8:23	47.42	240	<1	240	26	490	130 J	110 i	13	44	0.08 J	9.5
2/1/2001 16:43	35.62	130	<1	130	27	540	180 J	110	14	50	0.057 J	11
2/3/2001 12:43	30.80	230	<1	230	36	640	160 J	120	15	67	0.064 J	15
2/4/2001 16:43	29.22	220	<1	220	35	710	280 J	110	14	62	0.06 J	14
2/5/2001 13:03	29.22	220	<1	220	36	693	92 J	110 iq	14 iq	63 i	0.059 J	14
2/5/2001 13:03	29.22	210	<1	210	38	630	250 J	110	14	61	0.064 J	14

(continued)

Table 7.1 (continued)

Date and time	Discharge (gpm)	Alkalinity, Bicarb ($CaCO_3$)	Alkalinity, Carb ($CaCO_3$)	Alkalinity, Total ($CaCO_3$)	Chloride	Solids, dissolved	Sulfate	Calcium	Magnesium	Potassium	Selenium	Sodium
Event 4—High-flow (Precipitation = 1.2")												
12/12/2001 12:30	27.79	260	<1.0	260	21.0	480	130	111	11.9	45.7	<0.0050	10.20
12/12/2001 16:10	27.79	260	<1.0	260	22.0	490	110	108	11.6	44.1	<0.0050	9.80
12/12/2001 19:10	29.22	250	<1.0	250	21.0	490	110	110	11.8	43.8	<0.0050	9.80
12/12/2001 19:10	37.30	240	<1.0	240	21.0	490	110	110	11.9	44.1	<0.0050	9.8
12/12/2001 19:50	35.62	250	<1.0	250	22.0	480	210	104	11.7	41.1	<0.0050	9.10
12/12/2001 20:30	42.44	250	<1.0	250	17.0	430	94	103	12.7	41.2	<0.0050	9.00
12/13/2001 2:30	46.00	250	<1.0	250	21.0	460	110	110	13.6	42.6	<0.0050	10.10
12/13/2001 7:30	47.42	260	<1.0	260	15.0	430	94	104	12.5	36.2	<0.0050	8.11
12/13/2001 11:10	47.42	260	<1.0	260	15.0	440	85	104	12.7	36.4	<0.0050	8.07
12/13/2001 19:10	47.42	260	<1.0	260	16.0	470	95	104	12.6	37.1	<0.0050	8.27
12/14/2001 6:10	58.28	250	<1.0	250	17.0	480	94	105	12.9	38.0	<0.0050	8.53
12/14/2001 7:10	69.40	250	<1.0	250	15.0	460	96	103	12.7	34.7	<0.0050	7.72
12/14/2001 9:10	73.21	230	<1.0	230	12.0	430	63	98.5	11.7	29.0	<0.0050	6.40
12/14/2001 10:10	73.21	240	<1.0	240	12.0	430	95	98	11.2	28.7	<0.0050	6.14
12/14/2001 11:10	65.65	230	<1.0	230	18.0	380	84	96.6	9.9	24.9	<0.0050	5.20
12/14/2001 15:10	61.93	240	<1.0	240	8.4	430	85	96.4	10.9	24.8	<0.0050	5.34
12/14/2001 21:10	60.22	250	<1.0	250	9.1	450	87	100	12.3	26.5	<0.0050	5.98
12/15/2001 6:10	54.44	260	<1.0	260	12.0	430	99	102	12.7	30.3	<0.0050	6.99
12/15/2001 14:10	69.40	260	<1.0	260	14.0	430	86	105	13.1	33.2	<0.0050	7.75
12/16/2001 1:10	113.51	270	<1.0	270	16.0	440	90	105	13.1	35.2	<0.0050	8.26
12/16/2001 9:10	113.51	260	<1.0	260	18.0	450	100	108	13.5	38.1	<0.0050	8.92
Event 5—High-flow (Precipitation = 1.0")												
1/23/2002 7:28	11.96	260	<1	260	43	680	170	140	19.2	60.0	<0.025	17.6
1/23/2002 19:28	11.96	260	<1	260	46	680	120	138	18.8	57.2	<0.025	17.0
1/23/2002 23:28	11.96	260	<1	260	38	680	87	141	19.5	58.6	<0.025	17.3
1/24/2002 0:48	13.52	260	<1	260	54	630	120	134	18.3	55.0	<0.025	16.1
1/24/2002 2:28	19.11	260	<1	260	35	610	120	137	18.9	53.2	<0.025	16.0
1/24/2002 3:08	35.62	240	<1	240	26	520	98	118	17.0	38.6	<0.025	10.8
1/24/2002 4:08	35.62	190	<1	190	13	250	93	90.6	12.4	21.2	<0.025	5.6
1/24/2002 5:08	33.99	200	<1	200	14	420	100	100	14.0	22.2	<0.025	6.2
1/24/2002 5:28	42.44	230	<1	230	18	480	92	116	15.7	28.2	<0.025	8.3
1/24/2002 7:28	50.70	210	<1	210	18	460	100	101	12.3	29.2	<0.025	7.2
1/24/2002 7:28	52.55	210	<1	210	21	450	120	105	12.6	30.0	<0.025	7.4
1/24/2002 8:48	48.85	210	<1	210	20	490	110	104	12.8	33.8	<0.025	8.6
1/24/2002 10:08	46.00	210	<1	210	16	440	55	108	12.8	30.2	<0.025	7.4
1/24/2002 12:28	40.70	220	<1	220	12	400	63	101	11.9	24.6	<0.025	5.8
1/24/2002 15:08	33.99	220	<1	220	11	390	73	98	11.5	22.0	<0.025	5.2
1/24/2002 21:12	23.27	230	<1	230	15	430	93	102	12.1	22.2	<0.025	5.6
1/25/2002 3:12	19.11	230	<1	230	13	410	72	103	12.4	23.4	<0.025	6.1
1/25/2002 8:12	35.62	230	<1	230	15	420	66	108	13.0	26.8	<0.025	7.0
1/26/2002 8:12	35.62	250	<1	250	22	490	84	114	14.4	31.4	<0.025	9.2
1/28/2002 11:12	33.99	260	<1	260	29	520	140	126	16.0	40.4	<0.025	11.6
1/27/2002 8:12	33.99	250	<1	250	26	520	150	120	15.0	37.0	<0.025	11.0
Event 6—Low-flow (Precipitation = 2.0")												
9/19/2002 13:40	1.51	310	<1	310	93	1,000	320	180	28.9	101	<0.025	29.5
9/19/2002 20:40	1.51	330	<1	330	90	970	310	171	27.2	95.7	<0.025	27.6
9/20/2002 9:40	2.56	310	<1	310	91	920	310	176	27.7	97.2	<0.025	28.6
9/20/2002 13:40	2.33	330	<1	330	92	940	320	178	28.2	98.7	<0.025	28.7
9/20/2002 14:40	4.22	280	<1	280	96	950	330	159	24.5	84.5	<0.025	25.7
9/20/2002 18:40	4.22	300	<1	300	94	970	320	177	28.3	97.7	<0.025	29
9/20/2002 23:40	8.82	310	<1	310	91	940	300	182	28.7	101	<0.025	29.2
9/21/2002 4:40	8.82	300	<1	300	66	750	220	149	23.1	80	<0.025	21.5
9/21/2002 9:40	8.82	300	<1	300	52	650	170	136	20.6	65.7	<0.025	18.2
9/21/2002 14:40	7.66	230	<1	230	47	630	160	124	19	56.4	<0.025	15
9/21/2002 20:40	7.66	240	<1	240	43	600	150	119	18	51.9	<0.025	14
9/22/2002 5:40	6.42	250	<1	250	43	570	150	126	18.8	52.6	<0.025	14.6
9/22/2002 9:40	5.18	250	<1	250	44	610	160	114	17.4	48.7	<0.025	13.9

(continued)

Table 7.1 (continued)

Date and time	Discharge (gpm)	Alkalinity, Bicarb ($CaCO_3$)	Alkalinity, Carb ($CaCO_3$)	Alkalinity, Total ($CaCO_3$)	Chloride	Solids, dissolved	Sulfate	Calcium	Magnesium	Potassium	Selenium	Sodium
9/22/2002 12:40	5.18	230	<1	230	45	610	170	122	19.1	53.9	<0.025	15.2
9/23/2002 0:40	4.70	270	<1	270	51	680	190	128	20	55.6	<0.025	16.9
9/24/2002 9:40	4.70	270	<1	270	55	710	190	126	20.3	59.2	<0.025	18.2
9/25/2002 7:45	4.22	270	<1	270	55	670	190	127	20	58.7	<0.025	18.1
9/25/2002 7:45	4.22	290	<1	290	55	680	190	129	20.5	60	<0.025	18.5
Event 7—Low-flow (Base) (Precipitation = 0.0")												
10/8/2002 11:03	3.39	290	<1	290	56	660	190	150	20	53	<0.010	17
10/8/2002 11:07	3.39	210	<1	210	57	670	200	122	20.6	57.4	<0.010	18.6
10/8/2002 15:03	3.39	270	<1	270	57	640	190	124	21.5	59.6	<0.010	19.4
10/8/2002 19:03	3.39	220	<1	220	57	660	170	118	20.1	55.1	<0.010	17.9
10/8/2002 23:03	3.39	250	<1	250	58	660	200	134	22.2	61.4	<0.010	19.9
10/9/2002 3:03	3.39	260	<1	260	58	670	190	132	22.3	62.2	<0.010	20.1
10/9/2002 7:03	3.39	250	<1	250	59	640	190	103	17.7	62.7	<0.010	16
10/9/2002 10:07	3.39	270	<1	270	58	640	200	126	21.6	60.3	<0.010	19.6
Event 8—High-flow (Base) (Precipitation = 0.0")												
4/3/2003 12:03	10.97	200	<1.0	200	38	560	180	114	13.9	82.2	<0.005	16.8
4/3/2003 12:08	10.97	220	<1.0	220	38	570	180	113	13.6	81.3	<0.005	16
4/3/2003 16:03	10.97	200	<1.0	200	38	570	180	111	13.8	80.9	<0.005	16.3
4/3/2003 20:03	10.97	210	<1.0	210	39	580	180	117	14.3	85.9	<0.005	17.1
4/4/2003 0:03	10.97	190	<1.0	190	39	580	180	119	14.6	87.8	<0.005	17.4
4/4/2003 5:03	10.97	210	<1.0	210	39	590	180	121	14.9	89	<0.005	17.6
4/4/2003 10:03	10.97	210	<1.0	210	40	590	180	108	13.3	78.9	<0.005	15.8
Event 9—High-flow (Precipitation = 0.8")												
5/17/2003 0:33	17.78	230	<1.0	230	38	560	180	115	13.9	65.3	<0.0050	14.5
5/17/2003 9:03	17.78	230	<1.0	230	39	580	190	112	13.4	63.4	<0.0050	13.9
5/17/2003 19:03	19.11	240	<1.0	240	37	570	170	116	14.2	64.7	<0.0050	14.3
5/17/2003 23:03	20.44	230	<1.0	230	35	560	170	112	13.1	54.7	<0.0050	12.8
5/18/2003 3:03	21.86	230	<1.0	230	33	560	170	109	13.6	59.1	<0.0050	13.2
5/18/2003 10:03	20.44	230	<1.0	230	32	530	170	108	13.2	54.9	<0.0050	12.2
5/19/2003 8:03	19.11	230	<1.0	230	35	560	180	113	13.5	61.8	<0.0050	13.3
5/19/2003 8:03	16.43	210	<1.0	210	34	560	180	113	13.3	60.7	<0.0050	13.0
5/20/2003 2:03	15.08	240	<1.0	240	36	580	190	111	12.9	63.4	<0.0050	13.3

of specific criteria that can be used to distinguish between high- and low-flow conditions is necessary for guiding the site-specific groundwater monitoring program.

The terminology of high- and low-flow conditions has been used for a long time and quite frequently in surface water hydrology. However, its use seems to be more based on common sense than on a scientific analysis. The high-flow has been used as synonymous to wet season and the low-flow has been used as synonymous to dry season (Duke 2001). In some cases, the high-flow means the time when it rains, and the low-flow means the time when it does not rain when a collection of water samples is concerned. In other cases, the flow condition is determined by the interaction between groundwater and surface water. The high-flow occurs when the surface water recharges the groundwater, and the low-flow occurs when the groundwater recharges the surface water (Bernd et al. 1998).

The terminology of high- and low-flow conditions is also used in karst hydrology (American Society for Testing and Materials [ASTM] 1995). Because conduits/caves tend to develop at multiple levels in karst aquifers, groundwater flow, and quality may vary depending on the flow conditions. The relatively quick response of karst aquifers to recharge events also necessitates the study of groundwater flow and quality under different conditions. It is apparent that the high- or low-flow bears different meanings from that used in surface hydrology. However, similar to that in surface hydrology, there is no consensus definition regarding the high- or low-flow condition in karst literature either. There are no established criteria that can be followed to help define the flow conditions in karst aquifers. Considering the wide range of hydrogeologic conditions in karst, a universal criterion for the definition of the flow conditions may not be reasonable and the criterion has to be tailored to represent the site-specific circumstances.

QRS and QRS are two monitoring points at the landfill site. The discharge from both springs changes very rapidly in response to rain events and the response is closely related to the antecedent conditions and the rainfall intensity. In addition, the discharge at the springs is affected by many

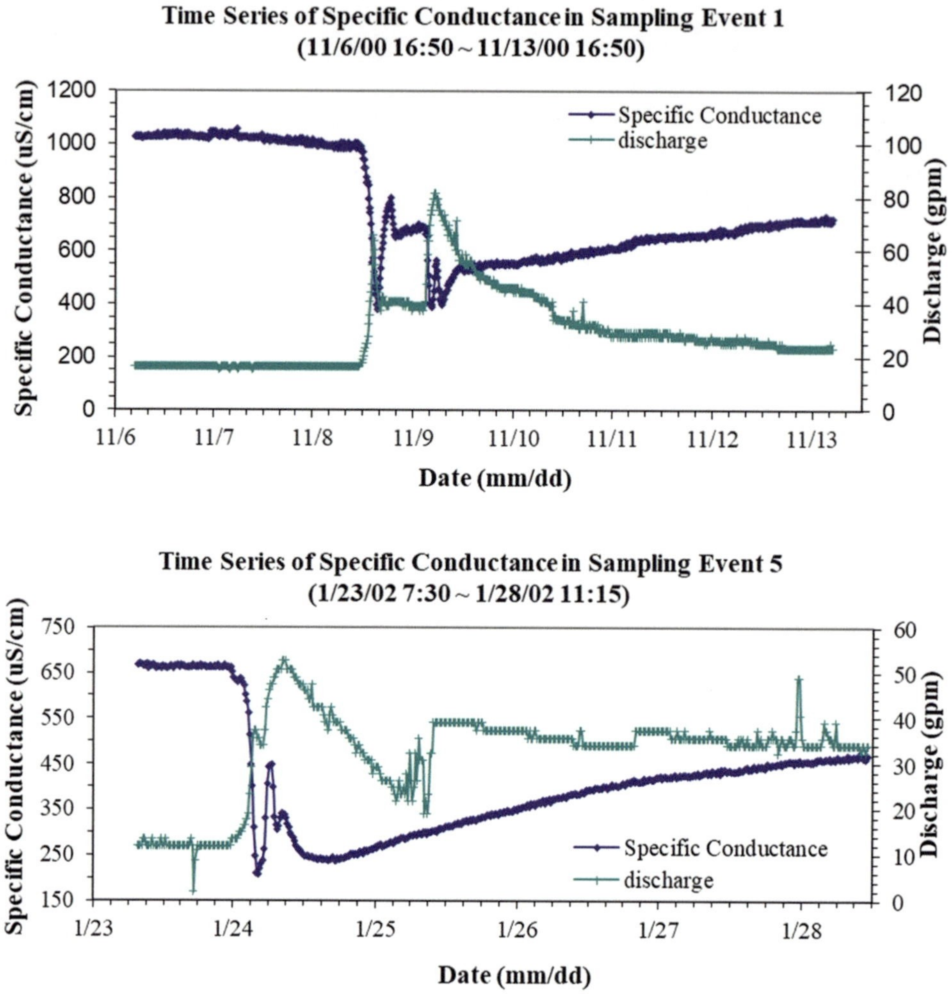

Fig. 7.11 Examples of water-quality variations in response to recharge events

other factors including hydraulic gradient, fracture/conduit size, and flow-through area. As a result, the discharge is too variable and is affected by too many factors to make it a useful indicator of the flow conditions. The same discharge at either QRS or QNS at different times may not necessarily represent the same aquifer conditions. The water level data in the aquifer may be more suitable to indicate the flow conditions.

7.6.1 Groundwater Level in Karst Aquifers

Using water level data to define the flow conditions at the site is not straightforward either. No single aspect of karst aquifers has received more discussion than the existence of the karst water level. There was an argument in Europe in the nineteenth century about whether a consistent water level exists in karst (White 1988). One school regarded cave streams as underground rivers that were not connected and thus, did not flow at common levels. The other school argued that the karst aquifer was an integrated groundwater body with a well-defined water table. Much of the confusion over the water table, however, arises from two sources—the highly dynamic response of the conduit system to storm events and the low-permeability of the matrix block in the absence of solution openings. Karst aquifers are often conceived and modeled as storage reservoirs penetrated by trunk conduits (Smart 1999). In response to storm events, conduits permit exceptionally rapid transfer of water and chemical constituents, while the storage reservoirs (fractures/matrix blocks), which contain the majority of the water in the karst aquifer, slowly adjust to autogenic recharge from sinkholes and back flooding from the primary conduits. During recession periods, the head loss in conduits is much lower than that in surrounding fractures or matrix blocks and the water stored in the matrix gradually drains into the conduits to sustain the spring flow. If time permits equilibrium between the water flow in the conduit and that in the matrix, a unified water level is obtainable. However, the aquifer is a dynamic system. When the aquifer is receiving water from rain events, the water levels in the matrix and in the spring are dynamic and they are different. Using water level data to define the critical flow condition should take into account the water levels in both the conduits and the storage reservoir.

7.6.2 Historical Average Aquifer Groundwater Levels at Monitoring Wells

The drilling of monitoring wells M1 and M4 encountered small fractures, bedding planes, and small voids. The water levels within these two wells can be generally considered to represent the average aquifer conditions in the storage reservoir (matrix). To evaluate aquifer conditions, the monthly average water level data from monitoring wells M1 and M4 from December 1999 to November 2001 were correlated with monthly precipitation data using multivariate linear regression analysis. The correlation coefficient R was used to represent the goodness of the correlation between these two factors. The maximum R was obtained when the water levels at either M1 or M4 were correlated to fifteen months' precipitation. The following are the regression equations developed for the two monitoring wells:

Monitoring well	Regression equation	Correlation coefficient
M1	$H_i = 624.2 - 0.07P_i - 0.19 * P_{i-1} + 0.15P_{i-2} + 0.28P_{i-3} - 0.07P_{i-4} - 0.12P_{i-5} + 0.2P_{i-6} + 0.13P_{i-7} - 0.43P_{i-8} - 0.08P_{i-9} + 0.04P_{i-10} + 0.08P_{i-11} - 0.01P_{i-12} + 0.12P_{i-13} + 0.32P_{i-14}$	0.94
M4	$H_i = 623.97 - 0.02P_i - 0.09 * P_{i-1} + 0.08P_{i-2} + 0.1P_{i-3} + 0.08P_{i-4} + 0.11P_{i-5} + 0.12P_{i-6} + 0.08P_{i-7} - 0.14P_{i-8} + 0.12P_{i-9} - 0.03P_{i-10} - 0.01P_{i-11} + 0.13P_{i-12} + 0.14P_{i-13} + 0.45P_{i-14}$	0.93

where

H monthly average water level elevation in feet;
P monthly precipitation in inches;
i index for current month when the water level is calculated;
$i - 1$ one month before; and
$i - 2$ two months before.

These regression equations indicate that the impact of precipitation on the water levels at M1 and M4 is cumulative and has a memory length of approximately 15 months. The water level is affected not just by the current month but by the last 15 months of precipitation. This conclusion is based on the data available at M1 and M4 and it is subject to change as more data is available. Figure 7.12 compares the measured water levels and the predicted water levels from January 2000 to November 2001. The maximum error between the measured and predicted water levels is 0.3 feet.

The regression equations were used to retrospectively predict the water levels at the locations of M1 and M4 using the precipitation data from 1947 to 2001, as shown in Fig. 7.13. The historical average water level elevation for M1 is 626.9 feet and the historical average water level elevation at the location of M4 is 628.3 feet.

To evaluate the impact of climate change on the water level prediction, the weather data were divided into five 10-year segments (1952–1961; 1962–1971; 1972–1981; 1982–1991; and 1992–2000). The calculated results and using only the most recent 10 years precipitation data yielded similar predicted water level values for both M1 and M4.

With these calculated average water levels for M1 and M4, we estimate (based on all current information available) that the aquifer is in high-flow conditions when the water levels measured in M1 and M4 exceed 626.9 and 628.3 feet above mean sea level, respectively. Alternatively, the aquifer is in low-flow condition when the water levels in these two monitoring wells are below these values.

7.6.3 Discharge from Upper-Level Springs

The groundwater tracing and spring discharge monitoring provide tools for diagnosis of the primary conduits. At the study site, springs are present at various elevations because of the gentle bedding plane dip toward the local creek, RLB. GrS is located on the riverbed of RLB, which has an elevation of 595 feet above mean sea level and spring QSP has the highest elevation of 612 feet. QRS, QNS, and QBS occur at approximately the same elevation of 602 feet. When the groundwater level is below 602 feet elevation in the vicinity of springs, flow only occurs at GrS. As the groundwater level exceeds 612 feet, deeper conduits fill and the groundwater "spills" over into conduits at higher elevations and eventually discharges at QSP. This phenomenon indicates that the groundwater flows to different springs under high-flow conditions than it does during low-flow conditions and that the flow system configuration may change under different aquifer conditions. This flow configuration change has the potential to lead to chemistry change at the monitored springs (Quinlan and Ewers 1989).

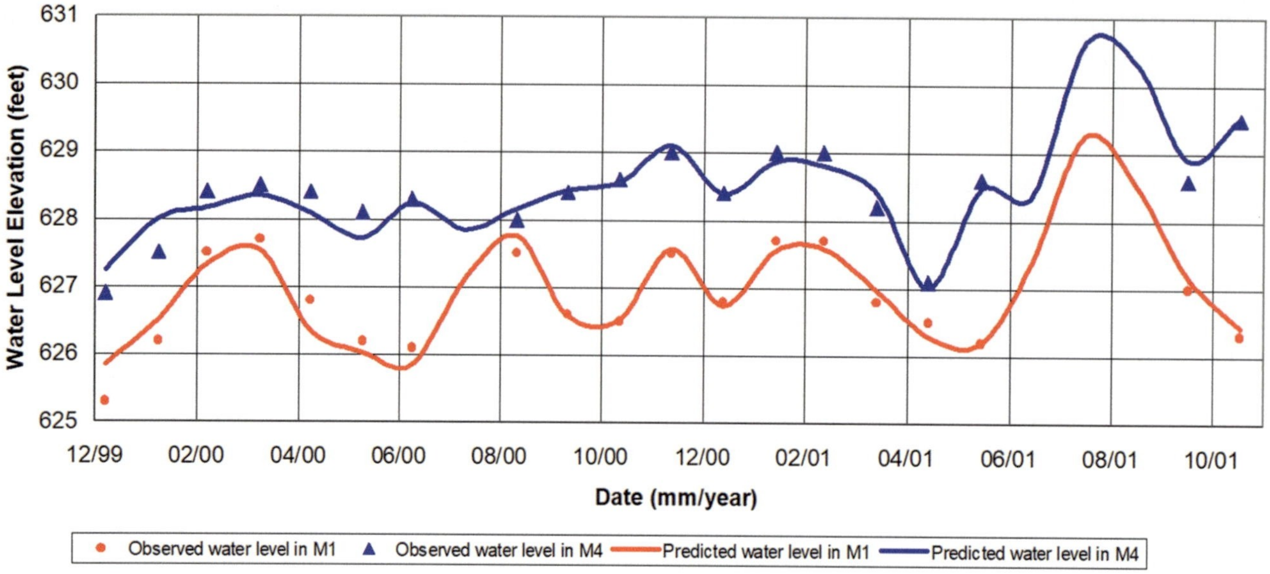

Fig. 7.12 Comparison of measured and predicted water levels at monitoring wells M1 and M4

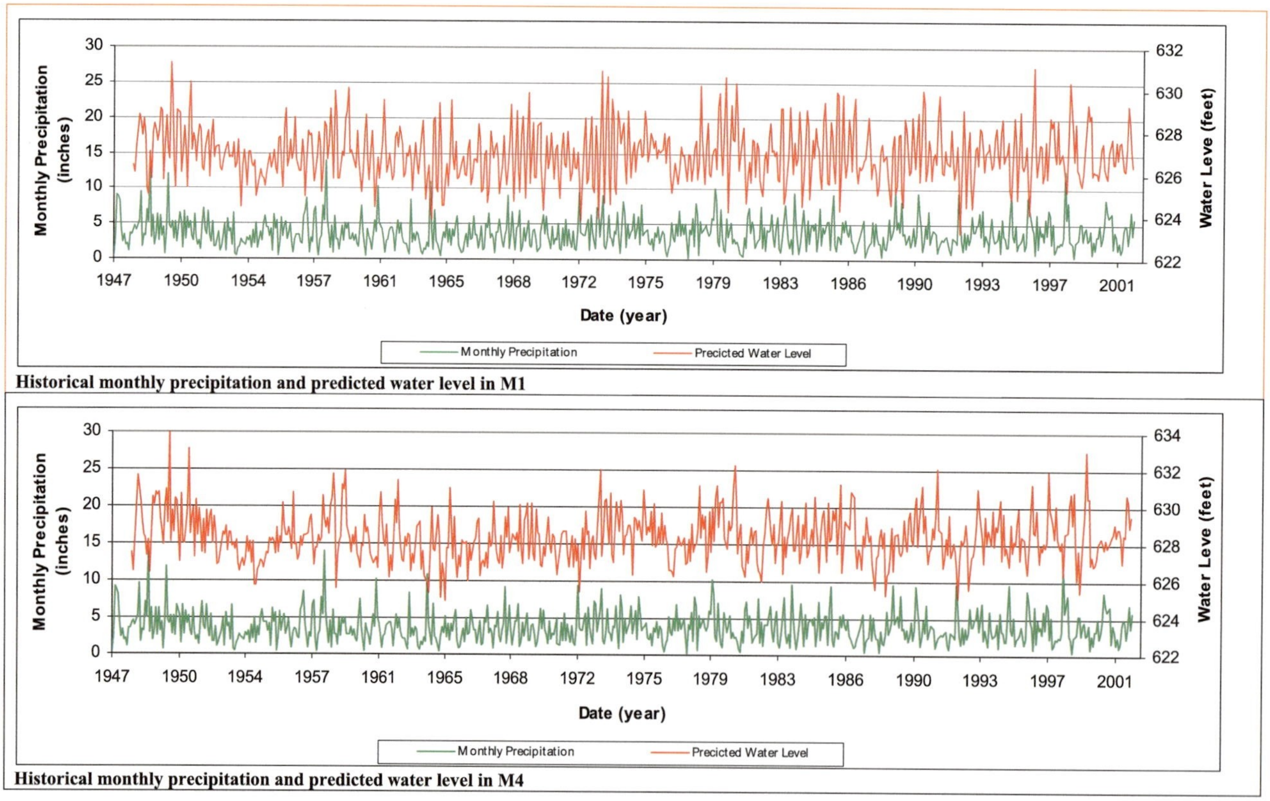

Fig. 7.13 Predicted water levels in M1 and M4 based on historical precipitation data

As a result of direct-point recharge through the sinkholes at the site and vicinity, and from groundwater flow through discrete fractures or conduits, the spring flow at QRS and QNS is very responsive to recharge events and these springs flow continuously throughout the year. However, other upper-level springs (such as QSP only flow intermittently during the year when the uppermost portions of the aquifer become filled. Based on visual observations during the first two high-flow background sampling events, QSP appears to flow at times that are representative of high-flow. The upper-level springs in general could be used as alternative indicators of high-flow aquifer conditions.

Based on the data available, the calculated average historical water level for monitoring well M1, which in general is more responsive to precipitation and in sync with the karst springs, appears to correlate well with the discharge at the upper-level springs and seems to be the most reliable indicator of high-flow conditions at the site.

7.7 Censored Data

Environmental monitoring data often contains a wide variety of results for any given constituent. This variability is oftentimes greater for data collected from karst settings where levels of groundwater constituents are constantly changing in response to recharging into the aquifer. Laboratory reports usually contain results that are quantifiable at a certain level (i.e., a compound is positively detected at some known concentration). However, in other cases, the results of a given target compound are not known and are reported as a non-detect value (i.e., less than some certain value such as a detection or reporting limit). In these cases, data are said to be censored. Censored data is quite common in environmental monitoring data.

According to Gibbons and Coleman (2001), the incorporation of non-detects (or censored data) into estimates of summary statistics and corresponding hypothesis testing and interval estimates is one of the most difficult problems in the analysis of environmental monitoring data. Furthermore, there is considerable disagreement regarding appropriate methods for the handling of censored data in statistics. However, the procedures outlined below and developed specifically for this project will continue to be utilized for the handling of all future censored data. Note that much of the guidance reviewed for determining appropriate techniques and methods for handling censored data for this project is adopted from Gibbons and Coleman (2001).

A total of 11 detection monitoring parameters are subject to statistical analysis. However, three of these parameters contained censored data within the nine background data sets. These three constituents include sulfate, selenium, and carbonate alkalinity, which contained 1%, 80%, and 100% censoring of data, respectively.

Review of Gibbons and Coleman (2001) indicates that the specific amount (or percentage) of censoring for any given parameter is a key factor in determining how the censored data for that parameter should be handled. More specifically, parameters with up to 15–20% censoring need to be handled differently from those parameters with 50% or more censoring. For these reasons, and as outlined below, the censored data for Sulfate (1% censoring) will be handled differently from the censored data for Selenium and Carbonate Alkalinity (both with > 50% censoring). The specifics of handling low and high levels of censoring are outlined below.

7.7.1 Low Levels of Censoring

A commonly used method, for handling low levels of censored data is to use a simple substitution method where one-half (½) of the detection limit (DL) is used in place of the censored value. This substitution method is generally accepted and used by many regulatory agencies including the USEPA and other state regulators. Although this frequently used procedure generally yields reasonable results in most cases when censoring is < 20%, as Gibbons and Coleman (2001) point out, this is the wrong procedure and instead, the quantitation limit (QL) should be used as the censoring mechanism instead of the DL. The reasons are that values above the DL and below the QL are detected but not quantifiable, and that use of the DL produces data with a wide variety of uncertainty and also violates the assumption of homoscedasticity (Gibbons and Coleman 2001). For this project, a simple substitution method will be used whereby the full QL will be substituted for censored values when censoring is < 20%.

Many laboratory reports list what is referred to as a reporting limit (RL), but do not directly list or contain the QL information in the summary reports. For this reason, specific QL information may need to be obtained directly from the laboratory.

7.7.2 High Levels of Censoring

Gibbons and Coleman (2001) list several statistical methods for handling censored data when censoring levels are > 20%. These methods include Maximum likelihood estimators, restricted maximum likelihood estimators, linear estimators, alternative linear estimators, delta distributions, regression methods, and substitution of expected values of order statistics. These methods are typically employed for censoring levels between 20 and 50%. However, because these methods make assumptions about data distribution that cannot be validly done for this project due to Outliers, none of these methods can be validly used for this project. Furthermore, with censoring levels > 50%, the methods listed above for handling censored data will not work well and alternative handling methods need to be utilized (Gibbons and Coleman 2001). For these reasons, the alternative strategies, as developed and presented below, must be used consistently for this project when censoring levels are > 20% for any given sampling event.

Carbonate alkalinity and selenium are the two detection monitoring parameters which contain high levels of censoring. These two compounds contained censoring of approximately 100% and 80%, respectively, in all data for the nine background sampling events. Although the carbonate alkalinity results for all background samples are less than the RL (i.e., < 1.0 mg per liter—mg/L), selenium was detected above the RL in each sample from event E3. Selenium was generally not detected in the other eight background sampling events. In addition, three different RLs were reported for selenium in the background samples for sampling events E1, E2, E4, E8, and E9 (0.005 mg/L), events E5 and E6 (0.025 mg/L); and event E7 (0.01 mg/L). Multiple RL values at different levels make it difficult to compute FWCs for each sampling event for the censored parameters.

Correlation analysis of data for each storm-pulse sampling event indicates that discharge at QRS is the main factor that affects concentrations of the water-quality parameters. Although such correlation could not be established between carbonate alkalinity and selenium because of censored data, it is reasonable to assume that their concentrations are also related to the discharge at the spring. Because both carbonate alkalinity and selenium can be considered to be trace elements when they are below their RLs, it can also be assumed that the total mass of these elements remains constant in the karst aquifer at any given time. It is also assumed that the censored elements are completely mixed within the aquifer. Under these assumptions (note: these assumptions are made specifically for carbonate alkalinity and selenium and are not applicable to other water-quality parameters), the loading rate, the product of discharge at the spring, and the concentration of the censored parameter remain constant.
Therefore:

$$Q_i C_i = Q_j C_j \tag{7.1}$$

where

Q is the spring discharge, C is the concentration for either carbonate alkalinity or selenium, i and j are the time indexes when water samples were collected and $i \neq j$. It should be noted that C is a value below the RL.

When a compound is present at some unquantified value below the RL in the laboratory analysis, the RL can be considered as the maximum concentration, as used in many environmental applications (Gibbons 1994; Helsel and Hirsch 2002). At the study site, it is assumed that this maximum concentration corresponds to the smallest discharge at the spring within a sampling event. Then the concentration at any sampling point can be calculated by

$$C_i = \frac{Q_{\min} \mathrm{RL}}{Q_i} \tag{7.2}$$

where

RL is the reporting limit for the sampling event; $Q_{\min}$ is the lowest flow rate at the spring among the samples with concentration values less than the RL.

Equation (7.2) was used to calculate values for each censored sample for carbonate alkalinity and selenium, as presented in Table 7.2. These values are then used to calculate the FWCs for carbonate alkalinity and selenium. It is important to note that these values are calculated based on the implicit assumptions discussed above.

These procedures for handling higher levels of censoring (i.e., > 20%) in carbonate alkalinity and selenium data will continue to be used consistently for the life of the project. Note that the above procedure is not required when non-censored data (i.e., quantifiable values above the RL) are reported in the laboratory analytical reports. New procedures (and assumptions) may be required in the future if higher levels of censoring (i.e., > 20%) become present in data for other detection monitoring compounds.

7.8 Outliers

An outlier can be loosely defined as a measurement that is considerably higher or lower than most of the measurement population. In environmental data, outliers can be quite common, but can also be extremely problematic. Outliers can arise from a variety of sources including but not limited to transcription errors, inconsistent sampling procedures, instrument failure, calibration or measurement errors, underestimation of spatial or temporal variability, and other factors.

One of the biggest problems with outliers is that in many cases, specific explanations (or causes) are not available, which leads to questions of whether outliers are truly valid data points. Review of outlying data sometimes indicates that outliers have arisen from clearly obvious deterministic mistakes (i.e., calculation, recording, gross measurement errors, etc.). In these cases, rejection or correction of the data are the only remedies (Barnett and Lewis 1998). However, in other cases, the specific cause of an outlier will never be known, and making decisions about whether we use the data or not, accommodate it (through statistical testing), or reject it outright can be difficult. To make things even more problematic for this project, valid measurements of either extreme high or low values (i.e., apparent outliers) are common in hydrological data for karst aquifers.

In some cases, outliers can be accommodated through statistical methods designed to draw valid inferences regarding the population from which a random sample has been obtained, without being seriously distorted by the presence of outliers. A variety of statistical methods exist. It comes down to a matter of either using the outliers in

Table 7.2 Censored data and their calculated values

Sample ID	Date and time	Spring discharge (gpm)	Alkalinity, Carb. (Reported) (mg/L)	Alkalinity, Carb. (Calculated) (mg/L)	Selenium (Reported) (mg/L)	Selenium (Calculated) (mg/L)
Event 1—Low-flow						
QRS-1-15	11/6/2000 16:50	16.43	<1	1.00	NA	NA
QRS-5-22	11/8/2000 23:00	16.43	<1	1.00	<0.005	0.005
QRS-5-24	11/8/2000 23:40	17.78	<1	0.92	<0.005	0.0046
QRS-6-7	11/9/2000 2:00	47.42	<1	0.35	0.006	0.006
QRS-6-11	11/9/2000 3:20	46.00	<1	0.36	<0.005	0.0018
QRS-6-12	11/9/2000 3:40	42.44	<1	0.39	<0.005	0.0019
QRS-6-20	11/9/2000 6:20	40.70	<1	0.40	0.006	0.006
QRS-7-1	11/9/2000 8:00	40.70	<1	0.40	<0.005	0.0020
QRS-7-21	11/9/2000 14:40	38.97	<1	0.42	<0.005	0.0021
QRS-7-23	11/9/2000 15:20	58.28	<1	0.28	<0.005	0.0014
QRS-7-24	11/9/2000 15:40	63.63	<1	0.26	<0.005	0.0013
QRS-8-1	11/9/2000 16:00	69.40	<1	0.24	0.006	0.006
QRS-8-3	11/9/2000 16:40	77.40	<1	0.21	<0.005	0.0011
QRS-8-5	11/9/2000 17:20	81.67	<1	0.20	<0.005	0.0010
QRS-8-7	11/9/2000 18:00	79.51	<1	0.21	<0.005	0.0010
QRS-8-9	11/9/2000 18:40	73.21	<1	0.22	<0.005	0.0011
QRS-8-21	11/9/2000 22:40	60.22	<1	0.27	<0.005	0.0014
QRS-9-7	11/10/2000 2:00	54.44	<1	0.30	<0.005	0.0015
QRS-9-24	11/10/2000 7:40	47.42	<1	0.35	<0.005	0.0017
QRS-10-13	11/10/2000 22:00	35.62	<1	0.46	NA	NA
QRS-10-26	11/13/2000 16:55	23.27	<1	0.71	<0.005	0.0035
Event 2—Low-flow (Base)						
QRS-1-1	1/10/2001 12:15	24.05	<1.0	0.97	<0.005	0.0048
QRS-1-8	1/10/2001 19:15	25.00	<1.0	0.93	<0.005	0.0047
QRS-1-16	1/11/2001 3:15	23.27	<1.0	1.00	<0.005	0.0050
QRS-1-24	1/11/2001 11:15	23.27	<1.0	1.00	<0.005	0.0050
Event 3—Low-flow						
QRS-1-4	1/29/2001 18:03	17.78	<1	1.00	0.017 J	0.017 J
QRS-1-15	1/29/2001 21:43	19.11	<1	0.93	0.019 J	0.019 J
QRS-2-9	1/30/2001 3:43	27.79	<1	0.64	0.05 J	0.05 J
QRS-2-12	1/30/2001 4:43	29.22	<1	0.61	0.034 J	0.034 J
QRS-2-16	1/30/2001 6:03	30.80	<1	0.58	0.07 J	0.07 J
QRS-2-19	1/30/2001 7:03	30.80	<1	0.58	0.04 J	0.04 J
QRS-2-23	1/30/2001 8:23	32.37	<1	0.55	0.037 J	0.037 J
QRS-3-4	1/30/2001 10:03	32.37	<1	0.55	0.033 J	0.033 J
QRS-3-9	1/30/2001 11:43	44.22	<1	0.40	0.031 J	0.031 J
QRS-3-10	1/30/2001 12:03	44.22	<1	0.40	0.023 J	0.023 J
QRS-3-11	1/30/2001 12:23	42.44	<1	0.42	0.026 J	0.026 J
QRS-3-14	1/30/2001 13:23	42.44	<1	0.42	0.038 J	0.038 J
QRS-3-19	1/30/2001 15:03	50.70	<1	0.35	0.033 J	0.033 J
QRS-4-1	1/30/2001 15:03	54.44	<1	0.33	0.024 J	0.024 J
QRS-4-3	1/30/2001 17:43	56.34	<1	0.32	0.034 J	0.034 J
QRS-4-24	1/31/2001 0:43	61.93	<1	0.29	0.033 J	0.033 J
QRS-5-9	1/31/2001 3:43	61.93	<1	0.29	0.04 J	0.04 J
QRS-5-24	1/31/2001 8:43	61.93	<1	0.29	0.038 J	0.038 J
QRS-8-23	2/1/2001 8:23	52.55	<1	0.34	0.08 J	0.08 J
QRS-9-24	2/1/2001 16:43	47.42	<1	0.37	0.057 J	0.057 J
QRS-10-11	2/3/2001 12:43	35.62	<1	0.50	0.064 J	0.064 J
QRS-10-18	2/4/2001 16:43	30.80	<1	0.58	0.06 J	0.06 J
QRS-10-23	2/5/2001 13:03	29.22	<1	0.61	0.059 J	0.059 J

(continued)

Table 7.2 (continued)

Sample ID	Date and time	Spring discharge (gpm)	Alkalinity, Carb. (Reported) (mg/L)	Alkalinity, Carb. (Calculated) (mg/L)	Selenium (Reported) (mg/L)	Selenium (Calculated) (mg/L)
Event 4—High-flow						
QRS-1-2	12/12/2001 12:30	27.79	<1.0	1.00	<0.0050	0.0050
QRS-1-13	12/12/2001 16:10	27.79	<1.0	1.00	<0.0050	0.0050
QRS-1-22	12/12/2001 19:10	29.22	<1.0	0.95	<0.0050	0.0048
QRS-1-24	12/12/2001 19:50	37.30	<1.0	0.75	<0.0050	0.0037
QRS-2-2	12/12/2001 20:30	35.62	<1.0	0.78	<0.0050	0.0039
QRS-2-20	12/13/2001 2:30	42.44	<1.0	0.65	<0.0050	0.0033
QRS-3-11	12/13/2001 7:30	46.00	<1.0	0.60	<0.0050	0.0030
QRS-3-22	12/13/2001 11:10	47.42	<1.0	0.59	<0.0050	0.0029
QRS-4-8	12/13/2001 19:10	47.42	<1.0	0.59	<0.0050	0.0029
QRS-4-19	12/14/2001 6:10	47.42	<1.0	0.59	<0.0050	0.0029
QRS-4-20	12/14/2001 7:10	58.28	<1.0	0.48	<0.0050	0.0024
QRS-4-22	12/14/2001 9:10	69.40	<1.0	0.40	<0.0050	0.0020
QRS-4-23	12/14/2001 10:10	73.21	<1.0	0.38	<0.0050	0.0019
QRS-4-24	12/14/2001 11:10	73.21	<1.0	0.38	<0.0050	0.0019
QRS-5-4	12/14/2001 15:10	65.65	<1.0	0.42	<0.0050	0.0021
QRS-5-10	12/14/2001 21:10	61.93	<1.0	0.45	<0.0050	0.0022
QRS-5-19	12/15/2001 6:10	60.22	<1.0	0.46	<0.0050	0.0023
QRS-6-3	12/15/2001 14:10	54.44	<1.0	0.51	<0.0050	0.0026
QRS-6-14	12/16/2001 1:10	69.40	<1.0	0.40	<0.0050	0.0020
QRS-6-22	12/16/2001 9:10	113.51	<1.0	0.24	<0.0050	0.0012
Event 5—High-flow						
QRS-1-23	1/23/2002 7:28	11.96	<1.0	1.00	<0.025	0.0250
QRS-2-7	1/23/2002 19:28	11.96	<1.0	1.00	<0.025	0.0250
QRS-2-19	1/23/2002 23:28	11.96	<1.0	1.00	<0.025	0.0250
QRS-3-3	1/24/2002 0:48	13.52	<1.0	0.88	<0.025	0.0221
QRS-3-8	1/24/2002 2:28	19.11	<1.0	0.63	<0.025	0.0156
QRS-3-10	1/24/2002 3:08	35.62	<1.0	0.34	<0.025	0.0084
QRS-3-13	1/24/2002 4:08	35.62	<1.0	0.34	<0.025	0.0084
QRS-3-16	1/24/2002 5:08	33.99	<1.0	0.35	<0.025	0.0088
QRS-3-17	1/24/2002 5:28	42.44	<1.0	0.28	<0.025	0.0070
QRS-3-23	1/24/2002 7:28	50.70	<1.0	0.24	<0.025	0.0059
QRS-4-3	1/24/2002 8:48	52.55	<1.0	0.23	<0.025	0.0057
QRS-4-7	1/24/2002 10:08	48.85	<1.0	0.24	<0.025	0.0061
QRS-4-14	1/24/2002 12:28	46.00	<1.0	0.26	<0.025	0.0065
QRS-4-22	1/24/2002 15:08	40.70	<1.0	0.29	<0.025	0.0073
QRS-5-6	1/24/2002 21:12	33.99	<1.0	0.35	<0.025	0.0088
QRS-5-12	1/25/2002 3:12	23.27	<1.0	0.51	<0.025	0.0128
QRS-5-17	1/25/2002 8:12	19.11	<1.0	0.63	<0.025	0.0156
QRS-6-17	1/26/2002 8:12	35.62	<1.0	0.34	<0.025	0.0084
QRS-8-20	1/28/2002 11:12	35.62	<1.0	0.34	<0.025	0.0084
QRS-7-17	1/27/2002 8:12	33.99	<1.0	0.35	<0.025	0.0088
Event 6—Low-flow						
QRS-1-1	9/19/2002 13:40	1.51	<1.0	1.00	<0.025	0.0250
QRS-1-8	9/19/2002 20:40	1.51	<1.0	1.00	<0.025	0.0250
QRS-1-21	9/20/2002 9:40	2.56	<1.0	0.59	<0.025	0.0147
QRS-2-1	9/20/2002 13:40	2.33	<1.0	0.65	<0.025	0.0162
QRS-2-2	9/20/2002 14:40	4.22	<1.0	0.36	<0.025	0.0089
QRS-2-6	9/20/2002 18:40	4.22	<1.0	0.36	<0.025	0.0089
QRS-2-11	9/20/2002 23:40	8.82	<1.0	0.17	<0.025	0.0043
QRS-2-16	9/21/2002 4:40	8.82	<1.0	0.17	<0.025	0.0043
QRS-2-21	9/21/2002 9:40	8.82	<1.0	0.17	<0.025	0.0043

(continued)

Table 7.2 (continued)

Sample ID	Date and time	Spring discharge (gpm)	Alkalinity, Carb. (Reported) (mg/L)	Alkalinity, Carb. (Calculated) (mg/L)	Selenium (Reported) (mg/L)	Selenium (Calculated) (mg/L)
QRS-3-2	9/21/2002 14:40	7.66	<1.0	0.20	<0.025	0.0049
QRS-3-6	9/21/2002 18:40	7.66	NA	NA	<0.025	0.0049
QRS-3-8	9/21/2002 20:40	6.42	<1.0	0.23	<0.025	0.0059
QRS-3-17	9/22/2002 5:40	5.18	<1.0	0.29	<0.025	0.0073
QRS-4-1	9/22/2002 9:40	5.18	<1.0	0.29	<0.025	0.0073
QRS-4-2	9/22/2002 12:40	4.70	<1.0	0.32	<0.025	0.0080
QRS-4-6	9/23/2002 0:40	4.70	<1.0	0.32	<0.025	0.0080
QRS-4-11	9/23/2002 15:40	4.22	<1.0	0.36	<0.025	0.0089
QRS-4-17	9/24/2002 9:40	4.22	<1.0	0.36	<0.025	0.0089
QRS-5-1	9/25/2002 7:45	4.22	<1.0	0.36	<0.025	0.0089
Event 7—Low-flow (Base)						
QRS-1-1	10/8/2002 11:03	3.39	<1.0	1.00	<0.010	0.01
QRS-1-5	10/8/2002 15:03	3.39	<1.0	1.00	<0.010	0.01
QRS-1-9	10/8/2002 19:03	3.39	<1.0	1.00	<0.010	0.01
QRS-1-13	10/8/2002 23:03	3.39	<1.0	1.00	<0.010	0.01
QRS-1-17	10/9/2002 3:03	3.39	<1.0	1.00	<0.010	0.01
QRS-1-21	10/9/2002 7:03	3.39	<1.0	1.00	<0.010	0.01
QRS-1-24	10/9/2002 10:07	3.39	<1.0	1.00	<0.010	0.01
Event 8—High-flow (Base)						
QRS-1-2	4/3/2003 12:03	10.97	<1.0	1.00	<0.005	0.005
QRS-1-6	4/3/2003 16:03	10.97	<1.0	1.00	<0.005	0.005
QRS-1-10	4/3/2003 20:03	10.97	<1.0	1.00	<0.005	0.005
QRS-1-14	4/4/2003 0:03	10.97	<1.0	1.00	<0.005	0.005
QRS-1-19	4/4/2003 5:03	10.97	<1.0	1.00	<0.005	0.005
QRS-1-24	4/4/2003 10:03	10.97	<1.0	1.00	<0.005	0.005
Event 9—High-flow						
QRS-1-8	5/17/2003 0:33	17.78	<1.0	0.85	<0.0050	0.0042
QRS-2-1	5/17/2003 9:03	17.78	<1.0	0.85	<0.0050	0.0042
QRS-2-11	5/17/2003 19:03	19.11	<1.0	0.79	<0.0050	0.0039
QRS-2-15	5/17/2003 23:03	20.44	<1.0	0.74	<0.0050	0.0037
QRS-2-19	5/18/2003 3:03	21.86	<1.0	0.69	<0.0050	0.0034
QRS-3-2	5/18/2003 10:03	20.44	<1.0	0.74	<0.0050	0.0037
QRS-3-8	5/18/2003 16:03	20.44	NA	NA	<0.0050	0.0037
QRS-3-16	5/19/2003 0:03	19.11	NA	NA	<0.0050	0.0039
QRS-3-24	5/19/2003 8:03	16.43	<1.0	0.92	<0.0050	0.0046
QRS-4-5	5/20/2003 2:03	15.08	<1.0	1.00	<0.0050	0.0050

J-estimated

statistical evaluation, which could have an impact on the outcoming answers if the outliers are not real data points, or rejecting the outlying data points if there is appropriate justification to do so. Rejecting outliers that represent real data also has major implications for the statistical evaluation, so extreme caution needs to be exercised in the general handling and treatment of outliers. If outliers are not handled in an appropriate manner, unwanted (and potentially unnoticed) bias in statistical calculations can occur, which could result in false-positive and/or false-negative detections. An overall outlier handling process for this project has been developed and is presented in Fig. 7.14.

As stated in various statistical references, the hole in the ozone layer went undetected for many years because the outlying data points associated with it were so extreme that they were initially ignored altogether. As illustrated in this example, elimination of outliers is considered a dangerous and unwarranted practice for the interpretation of any data (and especially water-quality data from karst aquifers), unless substantial objective evidence is present to demonstrate that the outliers are not representative of the population under study. The procedures outlined here for outlier handling and treatment aim to prevent this type of inappropriate elimination of data.

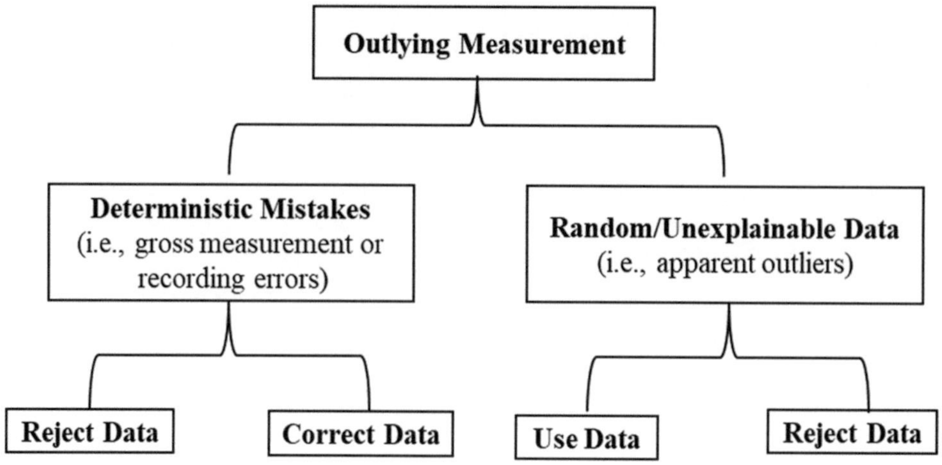

Fig. 7.14 Overall approaches to outlier treatments

As outlined in Barnett and Lewis (1998), a variety of statistical methods have been developed specifically for the testing and handling of outliers through accommodation. However, for these statistical methods to be valid, distributional assumptions must be made regarding the data. This is especially problematic for this project because the groundwater data from this site are predominantly collected in response to recharge or precipitation events, during which water quality is constantly changing (sometimes to extreme values at karst sites). For this reason, assumptions regarding a particular distribution (i.e., normal, lognormal, etc.) of the data from this karst groundwater monitoring site cannot be made. This ultimately means that statistical tests cannot validly be used for this project as a means of determining the validity (or otherwise) of outlying data. Use of a test requiring a normal distribution could produce inaccurate results when outliers affect the population structure. In the present context, whether or not a data point is an outlier or should not be judged by how it fits in some assumed statistical distribution.

The specific ways in which outliers have been identified and handled for this project are explained in detail in the following subsections. It is important that these methods be used consistently for all future outlier handling as the project progresses.

7.8.1 Outlier Definitions

Over the years, outliers have been defined in many ways by various investigators. Gilbert (1987) defined outliers as those values that do not conform to the pattern established by other observations. Barnett and Lewis (1998) define outliers as an observation (or subset of observations) in a set of data which appear to be inconsistent with the remainder of that set of data. The term outlier can generally be used collectively for discordant observations and for contaminants. Iglewicz (1983) defined a discordant observation as an observation that appears surprising or discrepant to the investigator. A contaminant is defined as an observation from a different distribution from the rest of the data, which may or may not be noted by the investigator (Barnett and Lewis 1998). One should immediately note that inherent in all the above definitions of outliers, is a certain degree of subjectivity that is common to the field of outlier identification and handling. To borrow from and expand upon Iglewicz's (1983) outlier definition, the authors have defined outliers, for the purposes of this project, as follows:

Discordant observations within individual or multiple sample sets that appear surprising or discrepant to the investigator, who will be knowledgeable in the specific karst hydrogeologic properties and anticipated aquifer responses to storm-pulse sampling events under base-, low-, and high-flow conditions at this site.

7.8.2 Technical Approach to Outlier Handling

As mentioned above, because the groundwater data for this site is from a karst aquifer, there are no valid statistical tests or techniques that can be used to test for (or accommodate) outliers that should potentially be rejected. Because of this, identification and handling of outliers for this project cannot be based on statistical processes but must instead rely on specific technical knowledge of site hydrogeology, the sampling program, and water-quality data.

The outlier detection and handling process for this project has and will continue to include the specific steps as illustrated in Fig. 7.15. These steps should generally be completed in the order listed below:

- Initial Data Screening: A cursory screening for outliers, anomalous data, or other data problems in general must be completed as soon as preliminary data are available

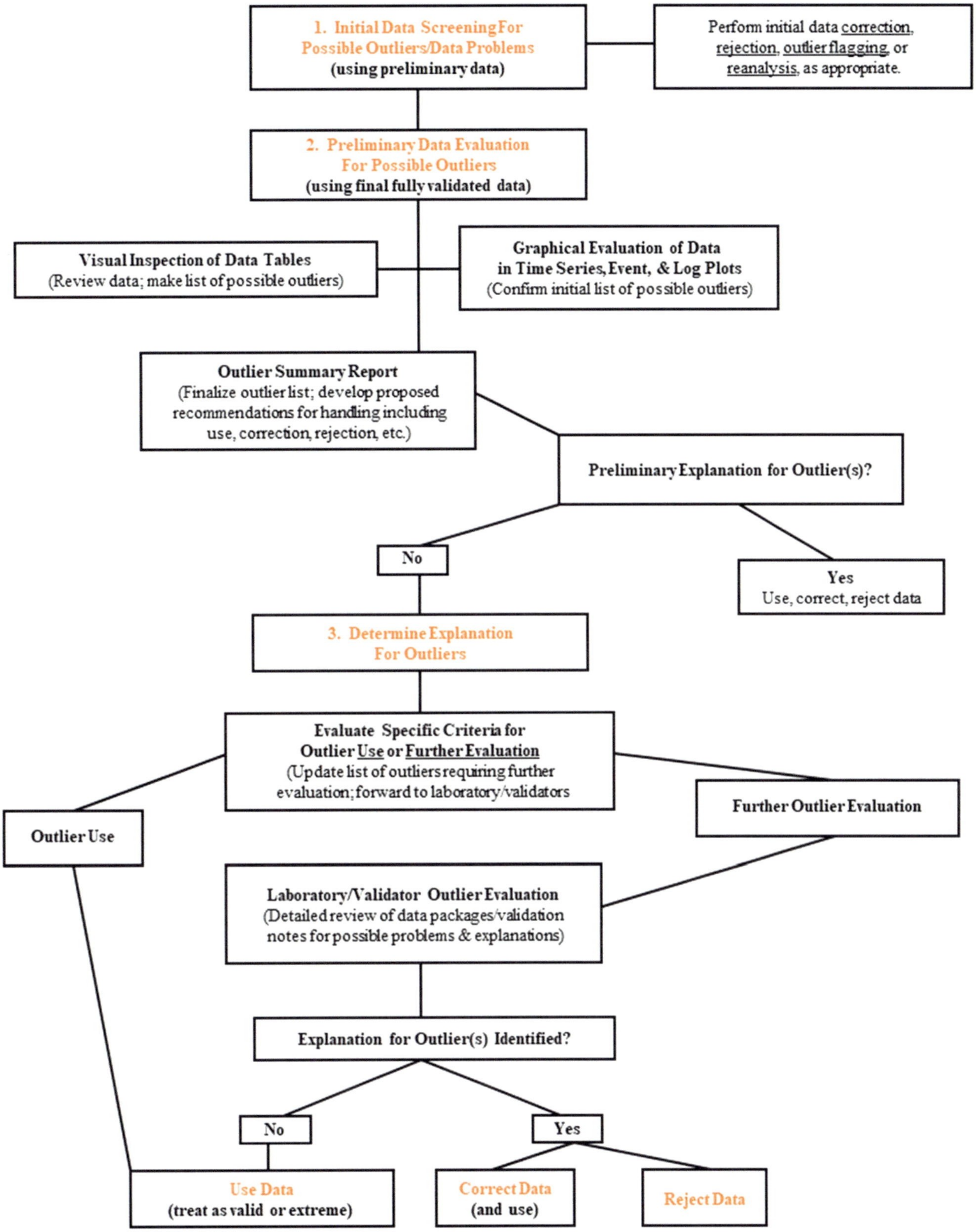

Fig. 7.15 Technical approaches for detection and handling of outliers

from the laboratory. At the first time data for any given event becomes available, karst perspective preliminary reviews should immediately be performed to address potential data problems. These reviews will ensure the highest quality data and will allow addressing of any possible problems with the laboratory data at an early stage. These preliminary data reviews should preferably be completed by someone experienced with karst groundwater monitoring to allow sufficient time for any reanalysis of samples within applicable holding times, should reanalysis become necessary. Depending on the findings of the initial data screening, initial data correction, rejection, outlier flagging, or reanalysis, should be completed, as appropriate. Once final data reports are received and data validation has been completed, a preliminary data evaluation (i.e., more intensive screening of data) for possible outliers will be completed on the fully validated data sets.

- Visual Data Inspection: A visual inspection of the final validated detection monitoring parameter data from all events will be completed to preliminarily identify potential outliers. A list of all suspected outliers identified will be generated.
- Graphical Evaluation: For each individual detection monitoring parameter, time-series plots, log plots (of times series data), and sampling event plots will be created for further evaluation and to confirm previously outlined lists of potential outliers. These plots allow better graphical visualization of the data for identification of possible outliers.
- Outlier Summary Report: A brief report summarizing all potential outliers identified will then be prepared to document all preliminary findings and to allow peer review. Where explanations for the cause or causes of the outliers are identified (or obvious), appropriate actions will be taken to deal with these outliers (i.e., use, correct, or reject the data). Where no explanations for outliers are identified, additional efforts will be completed in an attempt to determine an explanation for the said outlier(s).
- Evaluate Specific Criteria for Outlier Data Use or Further Evaluation: Criteria will then be evaluated to better define the nature of each remaining outlier as a means of determining an appropriate course of action for each outlier.
- External Review: Reviews and/or re-reviews of laboratory data and original data validation information for all remaining outliers (that cannot be explained) will finally be conducted by the laboratory and appropriate data validators to determine explanations for remaining outliers. The qualified and unqualified data points were therefore found to be usable for the purposes of the statistical evaluation.

In the process of completing the above steps, several detection monitoring outliers were identified in the background data that were suspicious or anomalous and/or did not appear to be related to natural aquifer conditions. Additionally, explanations could ultimately not be determined for these outliers. In any such case where outliers exist that may be due to natural aquifer conditions, and explanations for these outliers cannot be identified, there is no choice but to use the data and treat it as valid or extreme data, as was done for several of the outliers identified in the spring background data set.

For this project, all outliers must be appropriately evaluated and handled before use in statistical analysis. The process listed above will continue to be used for all future outlier identification and handling for this project. It is worth repeating that outlier use, correction, or rejection for this project will continue to be based on specific technical knowledge of the site characteristics, the data, and the groundwater sampling program.

7.9 Residence Time

Most statistical methodologies assume that groundwater sampling events are randomly distributed and completely independent. To maintain this independence, groundwater at the springs should only be sampled after water monitored during the previous precipitation event has completely discharged through the springs. Precipitation should not occur within several days prior to a sampling event, and site conditions must be closely evaluated prior to sampling. To help evaluate the degree of independence between each sampling event, the residence time of groundwater and dissolved constituents in the karst aquifer at the site was calculated. This was done by modeling the spring discharge and water-quality responses of the springs to precipitation events, as discussed in more detail below. Most importantly, if any dependence between individual sampling events is found, then the statistical methodology must address this departure from the independence assumption.

7.9.1 Approach for Computing Residence Time

Residence time is effectively the average amount of time a particular substance spends within a designated system (U.S. Geological Survey [USGS] 1997). For this project, residence time is defined as the approximate time that elapses from when water first enters the aquifer (through direct point recharge to sinkholes (precipitation or snowmelt) until it discharges at the spring. The length of residence time is closely associated with the geochemistry of the groundwater and the sensitivity of the karst aquifer to contamination. Shorter travel times may indicate that the infiltrated water has less opportunity to react with the limestone, while longer travel times may indicate that groundwater can reach its chemical saturation before discharging at the spring. The dye-tracing tests at the site indicate that the residence time of water ranges from minutes to hours, depending on the distance between the spring and the sinkhole at which the dye was injected. However, residence times estimated from the previous dye-tracing tests are not representative of the average aquifer conditions because tracing was only completed under one flow condition. Variability in flow condition, hydrodynamic dispersion, incomplete mixing, and diffusion can result in different flow paths and residence times for different substances (Maloszewski and Zuber 1991). Estimation of the residence times at the site is helpful in planning sample collection such that sampling events are independent.

Several techniques can be used to estimate the travel time of groundwater in an aquifer, including the use of tracers, radioactive and stable isotopes, and chlorofluorocarbons (Goode 1996). The underlying connection between the residence-time distribution and transport of an ideal tracer can be exploited to develop the residence times based on the concentration changes at a monitoring point. It should be noted that in discussions of residence times for a tracer, the tracer can be any measurable substance dissolved in the water, or it can be a numerical tracer whose magnitude is related to the concentration of the real-world tracer (Zhou 2000). Danckwerts (1953) describes how the residence-time distribution can be determined experimentally from the outflow concentration (C) of a non-reactive tracer in a chemical reactor. In this context, the mean residence time at a point, determined from temporal integrals of the concentration, is affected by all of the processes accounted for in the solute transport. Based on the concept proposed by Danckwerts (1953), Goode (1996) proposed that the mean residence time in a steady flow domain can be calculated by dividing a concentration-weighted time $\left(\int_0^\infty tC\mathrm{d}t\right)$ by the time interval of the concentration of a tracer $\left(\int_0^\infty C\mathrm{d}t\right)$, which is injected as an impulse at time zero.

The outflow concentration in a chemical reactor is similar to a breakthrough curve of a tracer at the spring. Considering the discrete measurements of the tracer concentration at springs, the average residence time can be computed as (Mull et al. 1988)

$$\bar{t} = \left[\frac{\sum_{i=1}^{N} t_i C_i \Delta t_i}{\sum_{i=1}^{N} C_i \Delta t_i}\right] \tag{7.3}$$

where

C_i Concentration at time t_i;
$\bar{t}$ Mean residence time of tracer;
N Number of measurements.

Equation (7.3) assumes a constant flow rate at the exit or the spring. When the discharge at the spring changes in response to rain events, the following equation, which is modified from an equation in Field (2002), can be used to calculate the mean residence time:

$$\bar{t} = \left[\frac{\sum_{i=1}^{N} t_i Q_i C_i \Delta t_i}{\sum_{i=1}^{N} Q_i C_i \Delta t_i}\right] \tag{7.4}$$

where

Q_i spring discharge at time t_i.

Accordingly, the standard deviation (σ) of the residence time is calculated by

$$\sigma = \left[\frac{\sum_{i=1}^{N} Q_i C_i \left(\bar{t} - t_i\right)^2 \Delta t_i}{\sum_{i=1}^{N} Q_i C_i \Delta t_i}\right]^{1/2} \tag{7.5}$$

Because the average residence time is related to flow conditions and various transport processes, it is very important to select the most representative tracer for its calculation. Experimental dyes and specific dissolved solutes are not good candidates of the tracers because of either limited tests that can be performed or the potential different properties pertaining to the dispersion, adsorption, and diffusion of the dissolved solute. In this study, the electrical conductivity measured at QRS was selected as the representative tracer for residence time calculations because it has been widely accepted that conductivity measured at karst springs is a representative and sensitive parameter to characterize karst aquifers (Quinlan et al. 1991). The water's electrical conductivity is closely related to TDS concentration. During the background monitoring period, in some cases, the conductivity at QRS was as high as 900 microsiemen µS) under base-flow conditions after several days without precipitation. However, the conductivity values in local precipitation ranged from 20 to 36 µS.

This big contrast also favors the selection of conductivity as a tracer. Decreases in conductivity at QRS within several minutes (hours) of moderate rainfall indicate that the spring discharge contains a component of water with generally short residence times. In contrast, an increase in SC after rainfall indicates a contribution of water to the spring discharge with relatively long residence times. The specific correlation between conductivity and various chemical constituents is discussed in the following subsection.

7.9.2 Data Analyses and Correlation

Before residence time could be calculated, two things were considered. The first was determining which field parameters (i.e., discharge, conductivity, temperature, pH, DO, ORP, or turbidity) would be most appropriate to use in the residence-time calculations. The second was how to select an event that would best represent flow conditions and integral chemical concentrations.

The conductivity and discharge were used in the residence-time calculations. This was determined by establishing a correlation between field parameters and the chemistry of groundwater within the karst aquifer. Because the laboratory analytical results can represent water-quality conditions, bicarbonate and total alkalinity, magnesium, potassium, sodium, sulfate, chloride, TDS, and calcium were selected for regression analysis (carbonate alkalinity and selenium were not included due to high levels of censoring or non-detects). Linear regression analyses were then performed on the available QRS data for measured field parameters and laboratory-analyzed parameters listed above for all nine background sampling events. A total of 532 regression graphs were completed as part of these analyses. The R^2 value represents how well the regression model describes the data. An R^2 value near 1.0 indicates that the equation is a good description of the relationship between the independent and dependent variables. After comparison of all R^2 results, conductivity was given first consideration as a representative and sensitive parameter, due to the higher R^2 values obtained.

Sixty-three events were selected following data evaluation and numerous calculations. Finally, the residence times and standard deviations were calculated for each event using the equations presented above. The residence times were calculated by mean travel time of the recharge water. The average residence time value is 2.72 days for all 63 precipitation events. A histogram illustrating all residence times calculated is shown in Fig. 7.16.

Meanwhile, other factors that had the potential to influence residence-time values were also evaluated. These factors included spring-flow conditions at QRS, rainfall conditions, antecedent moisture conditions, and aquifer conditions. The factors of beginning, peak, ending, and total volume discharge, time to reach peak discharge, and hydrograph deflections were grouped into flow conditions. Total rainfall, peak rainfall, and rainfall intensity were evaluated under rainfall conditions. The total number of dry days prior to an event and rainfall 10, 20, and 30 days prior to event were included in the antecedent moisture condition group. The average water levels in monitoring wells M1 and M4, and related high- or low-flow conditions based on data from these monitoring wells, were grouped into aquifer condition. Evaluation results are illustrated in Table 7.3.

7.10 Applicability of Flow-Weighted Concentrations

The challenge of preparing the statistical evaluation plan for QRS is to find a representative value (parameter) to characterize the water quality over each sampling event. Because

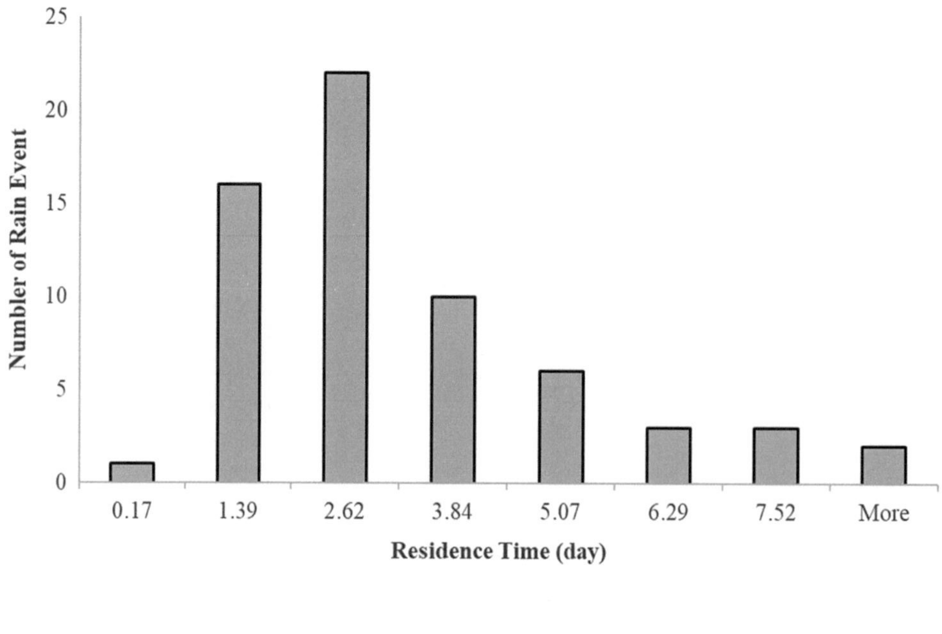

Fig. 7.16 Distribution of calculated residence time

Table 7.3 Factors affecting residence time

Selected event	Factors																	
	Event features				Flow condition at spring QRS					Rainfall conditions		Antecedent moisture conditions				Aquifer condition		
	Duration (days)	Avg. discharge (gpm)	Avg. field condtion (µS/cm)	Residence time (days)	Peak discharge (gpm)	Flow initial (gpm)	Flow end (gpm)	Time to peak discharge (days)	Total volume of discharge (gallons)	Total rainfall (inches)	Rainfall intensity (inch/day)	Dry days prior to event (days)	Rainfall 10 days prior to event (inches)	Rainfall 20 days prior to event (inches)	Rainfall 30 days prior to event (inches)	Avg. water level at M1 (feet)	Avg. water level at M4 (feet)	High/low flow per M1 or M4
1	0.32	87.22	352.38	0.17	91.68	91.68	89.12	0.00	40,508.83			0.00						
2	3.77	50.39	512.30	1.90	113.51	46.00	54.44	3.14	27,1527.88	1.53	0.77	7.00						
3	2.31	36.02	535.80	1.13	81.67	23.27	30.80	0.03	120,066.03	1.50	1.50	5.00						
4	2.05	32.01	549.87	1.01	35.62	23.27	29.22	0.55	94,684.97	0.47	0.24	0.00						
5	9.15	66.80	529.52	3.97	116.57	23.27	26.36	1.09	880,357.40	2.08	0.69	0.00	1.28	2.12	3.76	627.55	628.15	High
6	4.79	26.43	567.77	2.29	33.99	29.22	21.86	2.34	182,410.96	0.23	0.08	0.00	3.01	4.26	5.13	626.80	628.27	Low
7	11.64	25.52	551.78	5.41	77.40	20.44	20.44	0.69	427,288.45	2.48	0.21	0.00	2.13	2.34	2.34	626.53	628.66	Low
8	9.68	29.71	515.78	4.56	47.42	20.44	23.27	0.63	414,241.17	1.23	0.15	4.00	0.96	3.99	3.99	626.70	628.76	Low
9	12.42	65.48	364.18	6.35	151.46	21.86	40.70	0.40	1,170,538.52	2.83	0.26	8.00	0.12	1.34	1.61	627.93	629.26	High
10	2.34	42.39	413.49	1.17	47.42	40.70	40.70	1.83	142,867.66	0.02	0.02	0.00	0.09	2.81	2.93	627.15	628.81	High
11	9.47	19.18	449.07	4.47	23.27	19.11	17.78	2.67	261,379.78	0.30	0.04	1.00	0.02	0.11	0.43	626.62	628.21	Low
12	4.24	47.63	364.32	2.17	81.67	20.44	40.70	2.80	291,185.27	1.12	0.22	1.00	0.42	0.47	0.55	626.61	628.63	Low
13	3.53	51.61	320.02	1.81	88.20	26.36	40.70	0.86	262,899.55	1.09	0.36	2.00	1.21	1.58	1.63	627.23	628.87	High
14	4.44	57.14	316.59	2.54	92.65	27.79	58.28	3.90	365,842.48	0.69	0.17	3.00	1.28	2.51	2.80	628.19	629.13	High
15	0.01	72.07	298.02	1.97	101.85	50.70	44.22	0.08	398,776.04	0.96	0.24	2.00	1.63	2.90	4.14	628.25	629.24	High
16	1.61	38.11	397.70	0.79	40.70	37.30	37.30	0.17	88,420.93	0.39	0.39	1.00	1.46	2.15	3.40	628.32	629.27	High
17	2.09	21.06	491.66	1.02	24.82	20.44	21.86	0.15	63,394.51	0.33	0.33	8.00	0.72	1.79	2.48	627.77	629.01	High
18	2.83	21.51	451.13	1.40	24.82	19.11	19.11	2.81	87,790.53	0.26	0.09	2.00	0.96	1.93	2.72	627.21	628.92	High
19	3.65		593.19	1.78				0.00				9.00						
20	12.25	24.25	392.81	6.05	73.21	5.18	20.72	3.01	428,173.01	7.12	0.51	16.00						
21	11.05	46.57	248.04	6.18	104.70	10.97	29.22	3.20	745,358.03	4.84	0.40	0.00				628.38	628.78	High
22	1.61	33.28	288.37	0.81	44.22	30.80	32.37	0.33	77,397.70	0.60	0.20	0.00	4.18	4.89	4.89	629.10	628.90	High
23	12.50	63.76	212.14	6.52	140.16	27.79	52.55	3.60	1,147,105.01	5.20	0.40	2.00	0.67	5.36	5.38	627.83	629.53	High
24	17.94	31.40	310.00	8.74	86.01	11.96	20.44	8.06	808,186.03	3.01	0.16	3.00				626.53	628.96	High
25	2.65	19.32	445.40	1.29	20.44	17.78	17.78	0.43	73,633.27	2.55	1.28	3.00				626.74	628.32	Low
26	6.39	26.41	347.40	3.16	52.55	17.78	20.44	0.32	242,919.71	1.18	0.39	2.00				627.38	628.85	High
27	4.40	33.12	302.53	2.19	63.63	21.86	23.27	0.20	209,818.30	1.10	0.28	4.00				627.71	629.66	High
28	3.58	53.66	204.21	1.96	83.82	21.86	44.22	0.18	277,218.71	1.98	0.99	2.00				628.23	629.69	High
29	2.43	51.42	207.76	1.21	65.65	46.00	44.22	0.71	179,822.08	0.99	0.50	1.00				628.71	629.51	High
30	4.89	38.67	274.04	2.51	58.28	24.82	29.22	1.07	272,188.94	1.79	0.30	3.00				628.84	629.71	High
31	6.92	33.13	294.15	3.47	71.13	13.52	19.11	1.74	330,183.03	2.39	0.30	2.00				628.20	630.97	High
32	4.50	38.64	279.86	2.41	77.40	19.11	30.80	0.36	250,597.25	2.16	0.43	0.00				629.28	629.22	High
33	3.45	46.48	239.04	1.79	63.63	21.86	38.97	0.58	230,795.55	1.79	1.79	1.00				629.56	629.62	High

(continued)

Table 7.3 (continued)

Selected event	Factors																	
	Event features				Flow condition at spring QRS					Rainfall conditions		Antecedent moisture conditions				Aquifer condition		
	Duration (days)	Avg. discharge (gpm)	Avg. field condtion (µS/cm)	Residence time (days)	Peak dis-charge (gpm)	Flow initial (gpm)	Flow end (gpm)	Time to peak dis-charge (days)	Total volume of discharge (gallons)	Total rainfall (inches)	Rainfall intensity (inch/day)	Dry days prior to event (days)	Rainfall 10 days prior to event (inches)	Rainfall 20 days prior to event (inches)	Rainfall 30 days prior to event (inches)	Avg. water level at M1 (feet)	Avg. water level at M4 (feet)	High/low flow per M1 or M4
34	1.23	42.85	308.75	0.62	44.22	38.97	42.44	0.16	75,877.81	1.96	0.98	2.00				628.53	631.04	High
35	1.67	48.76	187.90	0.83	235.79	21.86	44.22	0.31	117,270.88	2.09	1.05	2.00					630.88	High
36	1.50	47.26	186.18	0.66	110.53	46.00	44.22	0.15	102,110.81	1.69	0.85	0.00					630.98	High
37	6.48	42.37	220.51	3.60	73.21	38.97	37.30	0.08	395,409.32	3.52	0.59	1.00				628.71	630.31	High
38	6.19	7.14	580.44	2.76	9.99	8.82	5.18	0.08	63,656.48	1.00	0.50	1.00				626.75	630.01	High
39	3.90	7.20	563.99	1.98	8.82	7.66	7.66	0.29	40,393.23	0.71	0.18	0.00						
40	3.01	7.11	912.47	1.51	7.66	6.42	7.66	0.19	30,812.19	0.85	0.43	0.00						
41	5.94	4.71	903.57	2.75	7.66	5.18	3.39	0.40	40,261.47	1.32	0.22	0.00	0.00	0.23	0.60	625.08	628.33	Low
42	6.18	5.08	638.39	2.80	8.82	8.82	8.82	0.00	45,107.66	4.48	0.56	3.00	1.82	1.82	1.90	624.61	627.71	Low
43	15.82	5.75	641.67	6.71	15.08	8.82	2.11	0.31	131,042.12	2.71	0.18	5.00	2.42	2.95	2.95	624.60	627.68	Low
44	6.36	3.52	730.05	2.91	5.18	2.56	4.70	0.11	32,225.60	0.79	0.11	7.00	0.10	0.62	5.13	624.43	627.55	Low
45	6.53	10.42	540.34	3.03	16.43	11.96	7.66	0.36	98,029.50	1.56	0.22	0.00	2.21	2.31	2.86	624.63	627.63	Low
46	1.97	10.10	581.26	0.97	10.97	7.66	7.66	0.56	28,659.49	0.50	0.50	0.00	3.11	3.77	4.39	625.13	627.76	Low
47	6.15	14.79	506.89	2.87	27.79	9.99	11.96	0.33	130,987.23	1.53	0.22	0.00	0.75	3.61	4.27			
48	4.68	18.82	434.56	2.19	26.36	20.44	15.08	0.76	126,778.13	1.42	0.28	0.00	5.46	5.48	5.96	625.35		Low
49	4.53	24.43	363.71	1.95	46.00	15.08	13.52	0.70	159,528.94	1.31	0.33	0.00	1.88	6.88	7.30	625.60	628.44	Low
50	9.77	19.60	379.04	4.39	54.44	9.99	8.82	2.15	275,941.86	0.44	0.22	0.00	3.66	6.20	11.20	626.78	628.96	Low
51	5.36	7.17	620.41	2.56	8.82	7.66	6.42	1.50	55,351.78	0.52	0.10	1.00	0.38	0.43	0.43	625.15	628.54	Low
52	5.60	11.60	515.17	2.87	19.11	4.70	11.96	0.98	93,684.85	1.83	0.31	2.00	0.70	1.09	1.77	625.30		Low
53	8.80	27.40	331.08	4.23	63.63	11.96	11.96	1.97	347,489.08	2.36	0.24	0.00	1.81	2.52	2.72	626.60		Low
54	9.34	19.42	320.90	4.44	37.30	11.96	9.99	3.79	261,430.18	0.16	0.03	0.00	2.26	4.06	4.77	627.18		High
55	3.56	15.74	326.03	1.88	21.86	8.82	15.08	1.21	80,807.01	0.64	0.21	5.00	0.05	2.41	4.22	627.36		High
56	3.51	17.60	347.92	1.84	20.44	11.96	17.78	0.33	88,996.81	0.68	0.17	4.00	0.64	0.81	3.75			
57	1.93	16.03	384.33	0.96	17.78	13.52	15.08	0.48	44,505.74	0.61	0.61	2.00	0.68	1.37	3.10	626.86		Low
58	2.56	25.32	324.29	1.22	38.97	13.52	16.43	0.39	93,604.01	0.80	0.40	1.00	1.29	1.93	2.10			
59	2.14	14.05	401.69	1.06	15.08	13.52	13.52	0.46	43,202.85	0.59	0.20	1.00	1.43	2.72	3.36	626.85	629.23	Low
60	2.07	9.30	511.21	1.06	10.97	7.66	8.82	1.11	27,771.99	0.83	0.42	7.00	0.59	2.02	3.31	626.17	629.07	Low
61	4.93	17.65	378.41	2.27	54.44	9.99	11.96	0.17	125,291.76	2.15	0.43	1.00	0.83	2.05	3.64	626.70	629.30	Low
62	3.03	23.99	313.59	1.42	38.97	13.52	15.08	0.25	104,840.71	1.33	0.67	2.00	2.37	3.59	5.00	627.19	629.37	High
63	19.08	31.78	340.29	8.71	420.88	11.96	15.08	5.25	873,657.71	7.52	0.44	1.00	2.18	4.55	5.77	627.85	629.67	High
Avg:	5.54	30.44	419.29	2.72	60.59	20.36	24.54	1.16	241,488.79	1.67	0.41	2.22	1.46	2.67	3.56	627.06	629.07	
Max:	19.08	87.22	912.47	8.74	420.88	91.68	89.12	8.06	1170538.52	7.52	1.79	16.00	5.46	6.88	11.20	629.56	631.04	
Median:	4.44	26.42	379.04	2.19	46.71	18.44	20.44	0.48	131014.67	1.31	0.31	1.00	1.28	2.34	3.36	627.17	628.99	
Min:	0.01	3.52	186.18	0.17	5.18	2.56	2.11	0.00	27771.99	0.02	0.02	0.00	0.00	0.11	0.43	624.43	627.55	

values of measured constituents vary with spring discharge, antecedent soil moisture, flow conditions, and other factors, a technique must be developed to bring all the measurements from each event onto the same 'base level' such that they can be compared. Another challenge is to statistically analyze the dependent measurements within each sampling event, while the sampling events themselves are independent.

The proposed approach to deal with these issues is to calculate FWCs for the 11 detection monitoring constituents and 1 field parameter that is subjected to the statistical analysis for each individual sampling event. The calculation of FWCs for each event is similar to the event mean concentration concept that has been widely used in the evaluation of stormwater-quality data. Information regarding the applicability of FWCs for this project, including literature review findings, calculation procedures, and FWC results are presented in this section.

7.10.1 Literature Review

The applicability of any given statistical procedure depends directly on assumptions about the characteristics of the data being explored. Results from statistical analysis may be meaningless or, even worse, misleading, if the data do not conform to the design assumptions of the statistical methods used. Therefore, careful consideration must be given to which statistical methods are appropriate in terms of the type of data being evaluated. The unique features of water-quality data at QRS affect the suitability of interpretation made by the application of many classical statistical techniques. Sampling at QRS is closely related to the hydrograph and change in field water-quality parameters. One-time grab samples are of little use for estimating water quality, flow rate, and mass loadings because one instantaneous value is not representative of the average conditions at the spring. The variability of the concentration and mass changes also cannot be described by one-time grab samples.

For water-quality evaluation at QRS, the total spring flow load is more important than the individual concentration of specific parameters. This is mainly due to the relatively short residence time of infiltrated rainwater within the aquifer; in most cases, the spring flow and water quality respond quickly to storm events. The effect of the spring water quality on RLB (the receiving water) is rather a response to the total loads than a variable concentration within each specific event. For water-quality studies, a technique is needed to compare constituent loads at QRS between different storms. However, total load cannot be used for comparison because large storm events will generally yield more mass of a particular constituent than small storm events. For comparison of loads between storms with varying characteristics, another method of comparison is needed.

The detailed analysis of the factors within individual sampling events strongly indicates that discharge at QRS is the main factor that affects the water-quality concentrations. Therefore, the average concentration of each water-quality parameter should be adjusted as appropriate with the discharge. Accordingly, this statistical analysis focuses on evaluating FWCs of individual constituents for each individual sampling event. FWCs are calculated as a ratio of mass of a water-quality parameter contained in the sampling event and the associated total volume of flow. To estimate FWCs per sampling event, two methods of sampling are generally used; note that the second method is being utilized for this project:

- Volumes of spring water are sampled and mixed together to obtain a flow-weighted composite sample; for the flow-weighted method, each sample is mixed with a volume of water directly proportional to the flow at the time of sampling. This has been widely used in stormwater sampling.
- Sequential discrete samples are taken at specified intervals and each sample concentration is determined. The event load is determined by multiplying each concentration and flow-weighted volume at the time of collection.

FWC are useful in estimating the typical concentration of a constituent that has been adjusted for spring flow. FWCs are neither a parametric average nor a statistical mean based on any implied probability distribution, rather it is an operational definition used to characterize a time- and flow-integrated water quality at the spring. The flow-weighted average is less prone to bias being introduced by a non-uniform sampling frequency than the arithmetic average when using sampling plans such as those recommended by Quinlan and Alexander (1987). Currens et al. (1987) analyzed 566 water samples collected at Pleasant Grove Spring to determine the minimum sample set needed to calculate a statistically significant annual mean. He found that flow-weighted means were typically less than arithmetic means, but greater than geometric means in the data set he analyzed. His findings indicate that flow-weighted averages are superior to either geometric or arithmetic averages. Unfortunately, statistical tests to compare the equivalence of flow-weighted averages have not been developed.

FWCs have been widely used as the primary statistic for evaluations of stormwater-quality data and as the stormwater pollutant loading factor in analysis of pollutant loadings to receiving waters (Cooke et al. 2000). FWC is often referred to as event mean concentration (EMC) in stormwater-quality evaluations (USEPA 1992). Although EMC was the terminology used in the original references, in this document, the term FWC is used to maintain consistency.

The study from the USEPA Nation Wide Urban Runoff Program (USEPA 1992) showed that FWCs are weakly correlated to runoff volumes (thus the storm size), and they are independent of rainfall intensity and duration, geographic location, size and slope of sewer system, and runoff volumes. A further analysis was performed to determine whether the observed FWCs conformed to a particular probability distribution (Driscoll et al. 1990) and it was found that the FWC values from an individual stormwater site generally follow a lognormal distribution. The lognormal distribution of FWCs is confirmed by other studies, using a somewhat smaller pollutant database in combined sewer overflows (Driscoll 1986a). In a study dealing with urban stormwater quality in Texas (USGS 1997), FWCs were calculated for 12 water-quality parameters and were best represented by lognormal distributions. There seems to be an increasing body of evidence in stormwater monitoring indicating that, in a wide variety of water-quality situations, pollutant concentrations can be adequately represented by a lognormal distribution (Driscoll 1986b; Pratt et al. 1987). An examination of a possible correlation between FWCs and rainfall volume was also performed (Irish et al. 1998) and a weak correlation pattern between them was found.

These studies and their results contribute substantially to the wide application of FWCs in stormwater-runoff investigations. In stormwater monitoring, there is a natural and appropriate bias that emphasizes resource allocation to moderate and large storm events. Because of differences in local meteorological conditions, the degree of site imperviousness, and other factors, there are appreciable differences in the average sizes of the storms in their database (USEPA 1992; Driscoll et al. 1990). Since little linear correlation was found between storm size and FWCs, such biases and differences are not expected to influence site comparisons to any appreciable extent (USEPA 1992).

7.10.2 Applicability of FWCs to Groundwater Monitoring at Spring QRS

Although FWCs do not appear to have been used previously in regulatory-driven groundwater monitoring programs for statistical evaluation of water-quality data at karst springs, they have been accepted and widely used in stormwater runoff evaluation. The strong similarities between stormwater runoff and spring flow in karst aquifers, as listed in Table 7.4, further justify the adoption of FWCs to water-quality evaluation at QRS.

7.10.3 Calculation of FWCs at Spring QRS

For the groundwater monitoring program under development here, sequential discrete water samples are collected across the hydrograph curve for each storm-pulse sampling event. For base-flow sampling (during periods of no precipitation), samples are collected at constant intervals over a period of approximately 24 h. For each sampling event, one discharge hydrograph (discharge versus time) and one chemograph (concentration versus time) are obtained for each water-quality parameter at QRS. From the discharge hydrograph and chemograph, the load graph (loading rate versus time) can be calculated. The loading rate is defined as the product of concentration and discharge at the time any individual sample is collected. The FWC values are then determined by calculating the cumulative mass of a constituent (loadgraph) and dividing it by the total volume of water (area under the hydrograph). A single FWC value for each constituent for each sampling event is obtained using this process. The schematic diagrams in Fig. 7.17 illustrate the general concept of the calculation process.

The equations used to calculate FWCs are as follows:

$$\text{FWC} = M/V \tag{7.6}$$

where M is the cumulative mass of a water-quality parameter and V is the total volume of water discharged.

The cumulative mass is calculated from the loadgraph by

$$M = \sum_{i=1}^{n-1} (L_i + L_{i+1})(t_{i+1} - t_i)/2 = \sum_{i=1}^{n-1} (Q_i C_i + Q_{i+1} C_{i+1})(t_{i+1} - t_i)/2 \tag{7.7}$$

The total volume of water is calculated from the discharge hydrograph by

$$V = \sum_{j=1}^{m-1} \left(Q_j + Q_{j+1}\right)\left(t_{j+1} - t_j\right)/2 \tag{7.8}$$

where

- M cumulative mass of water-quality parameter
- L_i Loading rate at the ith sample
- V total volume of water discharged
- Q flow rate at QRS
- C concentration of a water-quality parameter
- t time
- i ith water sample
- j jth flow measurement
- n number of samples collected
- m number of flow measurements taken.

FWCs calculated from the above equations represent the average concentrations in the receiving waters when the waters are completely mixed, according to the basic principle of mass conservation (balance). This is another

Table 7.4 Stormwater runoff versus karstic spring flow comparison

Category	Stormwater runoff	Groundwater flow at karst springs
Sequential discrete sampling	One-time grab sampling is of little use for estimating water quality, flow rate, and mass loadings because one instantaneous value is not representative of the average conditions of stormwater runoff. Furthermore, the variability of the concentration and mass changes with time cannot be described	One-time grab sampling is of little use for estimating the water quality, flow rate, and mass loadings because one instantaneous value is not representative of the average conditions at the spring. Furthermore, the variability of the concentration and mass changes cannot be described
Response to runoff	Water quality changes quickly with stormwater runoff with little lag time between the peak concentration and the maximum flow rate	Generally, water quality changes with an increase in discharge at the springs. Although the lag time may vary, depending on the characteristics of the karst aquifer, at QRS, the average lag time between the peak concentration and the maximum discharge is 30 min or less
Non-repeatability	Stormwater runoff measurements are observational, not experimental because of rainfall intensity and antecedent moisture conditions. More specifically, statistical regularity cannot be demonstrated for the water-quality data by repeating a controlled experiment because recreating the many natural and anthropogenic influences that affect each measurement is impossible. Varying rainfall patterns result in runoff flows, constituent concentrations, and constituent loads that vary considerably within and between storm events. Different antecedent conditions, and different patterns of precipitation intensity make each storm a unique event	Measurements at QRS are observational, not experimental because of the continuous flow of groundwater, antecedent conditions, rainfall intensity, and constant reaction with the surrounding environment. Recreating the many natural and anthropogenic influences that affect each measurement is impossible. This is especially true for the karst aquifer under study. The water quality and flow have a dynamic response to rainfall events. Ancillary and metadata (information about a given data set, including explanatory and data-quality information) pertinent to the statistical characteristics of the sampled population are investigated using the data from nine background sampling events. Because statistical regularity cannot be demonstrated through controlled experiments at the site, ancillary data are important to quantify possibly confounding variables that may preclude meaningful interpretation of data. Any modifications on the land use upstream of the landfill site should be documented
A lower limit of zero	Negative values are not possible for the stormwater quality data. For example, negative values of precipitation, flow, or concentration cannot occur do not have meaning	Negative values are not possible for the water-quality data at QRS. For example, negative values of precipitation, flow or concentration cannot occur do not have meaning
Censored data	Limits in methods for sample collection and analysis cause data to be reported as either above or (more typically) below one or more reporting limits, which produces a censored population of data. The effect of censored data can be especially problematic for interpretation of water-quality data. Laboratory detection limits change with time and can be dramatically different from laboratory to laboratory and may even be different from method to method within a laboratory Detection-limit artifacts also affect statistical properties of individual data sets. When a data set contains values reported as less than one or more detection limits, an overestimation of central-tendency measures and an underestimation of dispersion measures will be caused by truncation of the lower tail of the true population. Because the relative uncertainty in the accuracy and precision of individual values tend to increase as reported concentrations approach the detection limit, the percent error expected for measurements near detection limits is much higher than for values well within the measurement range of the method of analysis (Granato et al. 1998)	Data for karst springs can have similar problems as listed in the left column Also, at QRS, the censored data is significant in two of the detection monitoring water-quality parameters subject to statistical analysis, including selenium and carbonate alkalinity. Procedures have been developed to handle both low and high levels of censoring

(continued)

Table 7.4 (continued)

Category	Stormwater runoff	Groundwater flow at karst springs
Meaningful outliers	Valid measurements that are considerably higher or lower than most of the measured population are common among stormwater runoff data sets. The presence of meaningful high-end outliers (actual but extreme values) contributes to the positive skew and is a factor producing non-standard distributions. High-end outliers represent times when, for example, regulatory criteria may be exceeded, and the health of the local ecosystems may be affected. Those outliers, however, produce a host of potential problems for interpretation of data sets assembled for If an outlier is discovered to have a strong influence on the slope of a regression line (the slope or the correlation coefficient changes significantly when the point is omitted), then one must determine whether the outlier represents extreme values for a single process or if a secondary process is characterized by the outlier. Measurement and documentation of explanatory variables such as precipitation or flow; real-time measures of water-quality characteristics such as specific conductance, pH, temperature, and turbidity, use of ratios between constituents of interest, and results from a comprehensive QA/QC program can be used to identify and explain outliers in terms of the potential effect of real physicochemical processes as opposed to the effect of sampling artifacts	Data at karst spring has the same problems as discussed in the left column. The outlier issue is obvious for the data at QRS. Because of the complexity of sampling and analysis at the spring, strict QA/QC measures are taken to ensure the meaningful outliers are not excluded from the statistical analyses Certain assumptions and conditions need to be considered in terms of the decision to include or exclude individual data points or even entire data sets, among which are the following: It is often assumed that outliers are not meaningful. This assumption may be misleading unless documented QA/QC information indicates problems with sample collection, processing, and analysis that would justify elimination of suspected data An extreme outlier may be meaningful and may represent the effects of a process that should be considered in a synthesis designed to characterize the normal water -quality at QRS, but this process may obscure the effects of other process-related variables In general, elimination of outliers is considered a dangerous and unwarranted practice for the interpretation of water-quality data at the site, unless one has substantial objective evidence demonstrating that the outliers are not representative of the population under study
Positive skewness	Data sets that are not symmetrical around mean or median values are typical for stormwater runoff data sets because the combined effects of a lower bound of zero, censoring, and meaningful outliers tend to produce data sets in which the right tail of the distribution is extended, and the left tail truncated	Data sets that are not symmetrical around mean or median values are typical for spring water-quality data sets because the combined effects of a lower bound of zero, censoring, and meaningful outliers tend to produce data sets in which the right tail of the distribution is extended, and the left tail truncated
Autocorrelation and independence	An event is said to be independent of another event when the occurrence of one does not affect the occurrence of another. Natural and anthropogenic effects tend to cause conditions in which consecutive measurements are correlated. The natural and anthropogenic processes controlling stormwater quality and the methods for sampling, processing, and analysis often cause problems with autocorrelation (also referred to as serial correlation or correlation—the dependence of residuals in a time sequence because data reflect the effects of preceding conditions). Time-series effects may also occur between subsequent sampling events. For example, Irish et al. (1998) indicate that the duration, the volume of runoff, and the intensity of the runoff are significant causal variables in a regression model developed for highway runoff in Texas. Autocorrelation can be important because it affects the optimization of regression coefficients, affects estimates of population variance (invalidating results of hypothesis tests), and produces confidence and prediction intervals that are too narrow for the real population being sampled	Spring peak flow rates separated by a long period of time may be independent, but two peaks close to one another may not be independent. This is true when the recession limb of the first hydrograph at a spring becomes part of the rising limb of the next hydrograph. Natural and anthropogenic effects tend to cause conditions in which consecutive measurements are correlated. Autocorrelation can be a problem within the data sets at QRS if not planned properly because many of the variables used for analysis are pairs of data in a time series
Interdependence	Changes in one characteristic of interest (such as rainfall intensity) cause changes in other characteristics (such as measured flows and concentrations)	Changes in one characteristic of interest (such as rainfall intensity) cause changes in other characteristics (such as measured flows and concentrations)
Temporal variation	Measured water-quality characteristics vary cyclically at time scales of hours, days, seasons, years, and even decades because of both natural and anthropogenic influence. Temporal variation may also increase variability in data and affect the comparability of data between sites	Temporal variation is a characteristic of many karst springs. Seasonality is an obvious factor that may affect a population of water samples at QRS. By using flow-weighted concentrations, the effects of seasonality have been minimized
Decision making	Knowledge of within-event constituent concentration fluctuations provided little useful information to the decision-maker regarding control measures that might be required	Knowledge of within-event constituent concentration fluctuations provided little useful information to the decision-maker regarding control measures that might be required

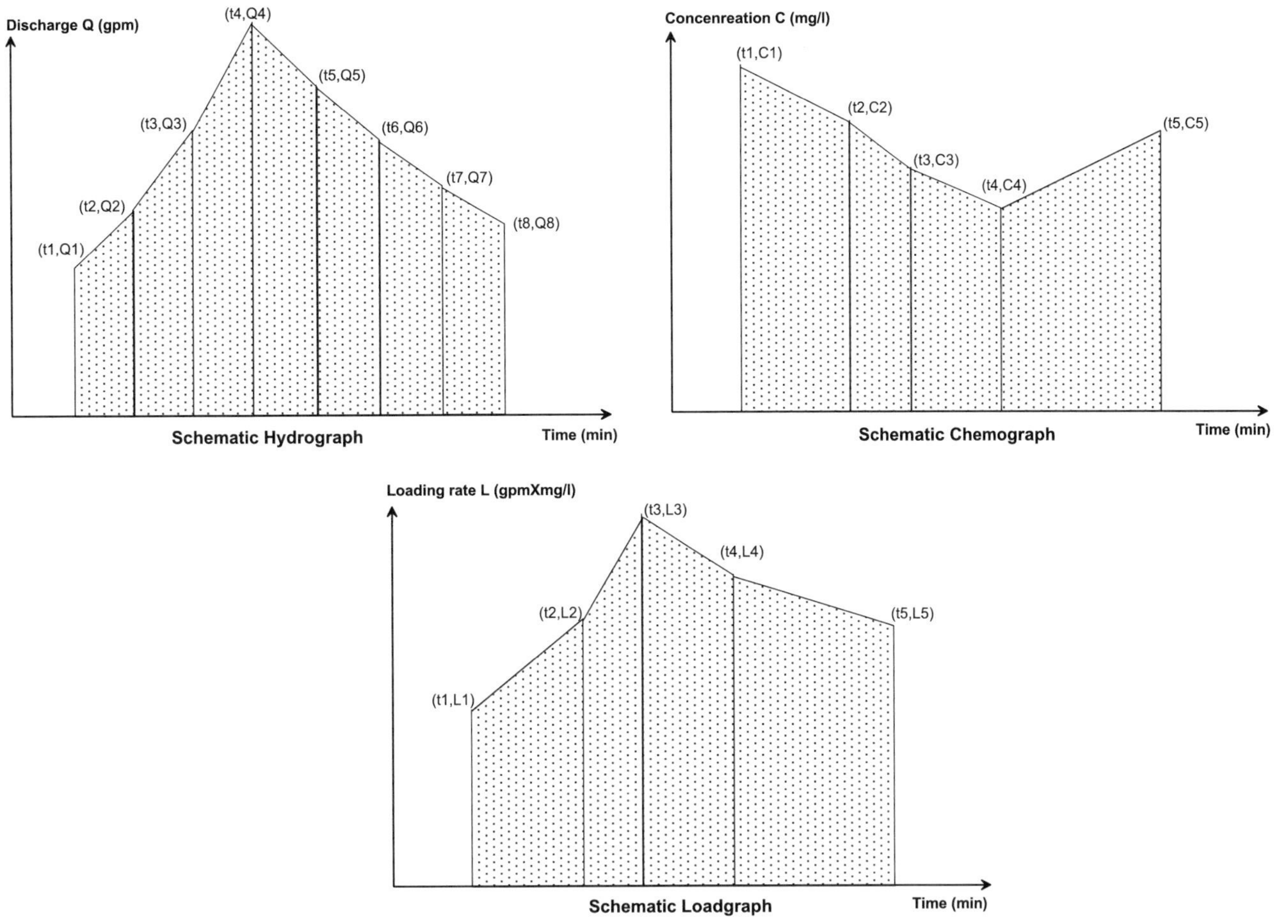

Fig. 7.17 Schematic diagrams for calculation of FWCs

important reason why the FWC is preferred in this study for each sampling event. In the above schematic example, the arithmetic mean of the concentrations can be calculated by

$$\text{Arithmetic Mean} = \sum_{i=1}^{n} C_i/n \tag{7.9}$$

The geometric mean of the concentrations can be calculated by

$$\text{Geometric Mean} = \sqrt[n]{C_1 C_2 \ldots C_n} \tag{7.10}$$

Using either arithmetic or geometric mean as the average concentration in the receiving waters cannot guarantee mass balance. However, the FWC will equal the arithmetic mean when the spring flow rate remains unchanged throughout the entire sampling event, as may be observed during baseflow conditions. The geometric mean is always smaller than the arithmetic mean unless all numbers in a data set are identical (Parkhurst 1998), whereas the FWC values are often between the arithmetic mean and the geometric mean for the background data set, in which both flow and concentration vary with time.

7.10.4 FWCs for Background Sampling

Calculated FWC values for the 11 detection monitoring parameters for the nine background sampling events completed are illustrated in Table 7.5, whereas Table 7.6 presents calculated FWCs for SC for 62 selected rain events during the background monitoring period.

7.10.5 Field Parameters Affecting FWCs

In this study, it is important to determine whether or not the calculated FWCs are independent and random variables; otherwise, key assumptions for specific statistical methods could be violated. This section presents an evaluation of natural factors that may influence the characteristics of the calculated FWCs at QRS.

Table 7.5 Calculated FWCs of 11 water-quality parameters for nine sampling events (unit in mg/L)

Sampling event	Alkalinity, Bicarbonate ($CaCO_3$)	Alkalinity, Carbonate ($CaCO_3$)	Alkalinity, Total ($CaCO_3$)	Magnesium	Potassium	Selenium	Sodium	Sulfate	Chloride	Solids, Dissolved	Calcium
1	258	0.6	258	20.9	71.8	0.003	22.6	131.1	53.6	772	143
2	195	1.0	195	18.5	81.7	0.005	22.2	211.7	65.7	727	121
3	212	0.4	212	20.2	53.5	0.05	17.9	149.1	55.5	698	130
4	256	0.5	256	23.0	34.0	0.003	17.6	93.0	56.1	676	134
5	242	0.4	242	20.4	33.5	0.010	15.1	107.4	46.1	620	132
6	274	0.3	274	22.7	65.8	0.008	20.4	205.5	64.5	748	141
7	255	1.0	255	22.2	59.5	0.010	19.9	188.9	62.9	684	129
8	203	1.0	203	21.9	85.0	0.005	24.1	180.0	68.9	749	138
9	232	0.8	232	22.5	60.6	0.004	21.9	177.8	71.0	761	138
FWC mean or mean log	236	– 0.5	236	21.3	60.6	– 5.0	20.2	160.5	60.5	715	134
Standard deviation	28	0.4	28	1.5	18.3	0.9	2.9	42.6	8.1	49	7
Distribution	*N*	Log-*N*	*N*	*N*	*N*	Log-*N*	*N*	*N*	*N*	*N*	*N*

N—Normality not rejected; Log-*N*—Lognormality not rejected.

Table 7.6 Calculated FWCs of specific conductance

Event	FWC of S.C (μS/cm)	Event	FWC of S.C (μS/cm)	Event	FWC of S.C (μS/cm)
1	431.61	22	265.57	43	943.80
2	622.01	23	381.99	44	687.15
3	658.35	24	586.65	45	747.90
4	676.92	25	431.25	46	640.98
5	619.29	26	377.12	47	557.81
6	702.52	27	254.08	48	441.12
7	676.66	28	271.77	49	429.95
8	648.86	29	344.68	50	821.47
9	443.58	30	346.43	51	659.06
10	550.20	31	344.17	52	385.94
11	602.05	32	307.83	53	407.96
12	461.06	33	405.66	54	422.50
13	389.74	34	217.23	55	461.27
14	421.77	35	202.88	56	509.65
15	394.34	36	278.47	57	415.65
16	541.30	37	743.30	58	530.21
17	662.92	38	724.77	59	667.69
18	606.88	39	1164.63	60	438.83
19	452.44	40	1131.74	61	392.77
20	274.93	41	799.04	62	372.11
21	371.16	42	769.34		

7.10.5.1 Natural Factors

Based on site characteristics and all available data, the following natural factors were evaluated:

- **Sampling Event Duration**: The duration of each sampling event is defined as the time span from the first sample collected (typically prior to the onset of precipitation) to the last sample collected. This factor is only related to storm-pulse sampling events. As recommended by Quinlan et al. (1991), storm-pulse samples are collected across the corresponding hydrograph. The first and last samples for laboratory analysis are selected based on review of the discharge hydrograph and the SC chemograph for QRS.
- **Average Spring Water Temperature**: For each sampling event, spring water temperature at QRS is monitored every 15 min. The average water temperature is the arithmetic mean of the measured temperatures over an entire sampling period. In some cases, temperature may be an important factor that affects the average water quality.
- **Residence Time**: Residence time of groundwater within the karst aquifer is very closely related to the characteristics of the aquifer. These characteristics determine the degree of mixing of the newly infiltrated water and the existing water within the aquifer, and the strength of the chemical reaction for dissolving the minerals in the rock and soil.
- **Flow Conditions**: Discharge at QRS is monitored every 15 min. During the background monitoring period, flow varied from less than 2 gpm to over 300 gpm. QRS generally responds to recharge from rainfall; however, the antecedent conditions, evapotranspiration, rainfall intensity, seasonal conditions, and other factors affect the extent to which the spring responds. For example, a storm event with approximately 1.0 in. of precipitation occurred in the early summer of 2002 (just after the aquifer transitioned to low-flow conditions) which yielded no hydrographic or water-quality response at QRS. Because of the variations in flow conditions, it is very difficult to use any single parameter to represent

the flow conditions at QRS. In this study, the following five parameters are used to define the flow conditions at QRS:

(a) Discharge at the beginning of the sampling event
(b) Peak discharge over the sampling event
(c) Discharge at the end of the sampling event
(d) Total volume of water discharged during the sampling event
(e) Time elapsed to reach peak flow
(f) Time elapsed for increased discharge following onset of precipitation.

- **Precipitation Conditions for Site and Adjacent Areas**: Because QRS may drain a wide area, off-site precipitation can also affect flow conditions at QRS. Precipitation data evaluated for this study were obtained from weather monitoring station located immediately adjacent to the site. For each weather station, the following parameters are considered to represent the rainfall condition:

(a) Total rainfall
(b) Peak rainfall intensity
(c) Rainfall intensity.

- **Antecedent Moisture Conditions**: The antecedent moisture condition, as defined for this study, is the soil moisture content preceding a given sampling event. The moisture conditions prior to a rainfall event appear to have an influence on the infiltration rate, the flow rate at QRS, and thus the water quality. According to the U.S. Department of Agriculture, Natural Resources Conservation Service, formerly the Soil Conservation Service (1985), the antecedent moisture condition is defined by total rainfall 5 days immediately preceding a rainfall event. Felton (1994) used this definition to relate peak flow rates at a Kentucky spring to antecedent moisture conditions and several other variables including precipitation, antecedent flow rate, and time to peak flow. Of importance to this work are the findings that antecedent moisture conditions are strongly related to peak flow rate at his study site. In a study of karst spring flow in Illinois, Barfield et al. (2000) developed an antecedent moisture index to represent moisture conditions using the 5-, 10-, and 15-day precipitation totals as well as the rainfall for the storm being evaluated.

  Water in the soil zone is capable of movement but under a negative pressure gradient. During storm events, the soil zone becomes fully saturated and begins releasing water to the underlying aquifer. As the soil desaturates, some of the pores become air-filled, the hydraulic conductivity decreases, and tortuosity increases (Hillel 1980). Recharge to the karst aquifer also occurs directly through sinkholes on and in the vicinity of the site; however, the soil zone still has the potential to regulate enough infiltrating water to have some effects on the saturating of the epikarst zone.

  At the study site, it is likely that antecedent moisture conditions also affect storage in the epikarst zone. A relatively high hydraulic conductivity in the epikarst zone is responsible for rapid drainage of stored water, which partially causes high rates of recession and a flashy flow at QRS. For water-quality evaluation, it is critical to recognize the importance of the storage within the epikarst zone because water in this zone may have different physiochemical properties from that in the saturated zone (Field 1989). The following parameters are used to better understand and quantify the effects of antecedent moisture conditions on FWCs at the site:

(a) Dry days preceding the sampling event
(b) Rainfall total 10 days preceding the sampling event
(c) Rainfall total 20 days preceding the sampling event
(d) Rainfall total 30 days preceding the sampling event.

- **Aquifer Storage Conditions**: As a result of direct point recharge through sinkholes at the site and vicinity and groundwater flow through discrete fractures or conduits, spring flow at QRS is generally responsive to recharge events. It has also been observed that the flow system configuration changes under different aquifer conditions. At the study site, springs are present at various elevations as a result of the gentle bedding plane dip toward the local creek—RLB. Spring GrS occurs on the creek-bottom of RLB, which has an elevation of 595 feet above mean sea level (amsl). The spring QSp has the highest elevation of 612 feet amsl. The springs QRS, QNS, and QBS occur at approximately the same elevation of 602 feet amsl. If the groundwater level were to drop below 602 feet amsl, there is a potential that only GrS would have water flow. However, as the groundwater level exceeds 612 feet amsl, deeper conduits fill and the groundwater spills over to conduits at higher elevations and discharges at QSp. This phenomenon indicates that the groundwater flows to different springs under high-flow conditions than it does during low-flow conditions. This flow configuration change has the potential to lead to chemistry changes at the monitored springs (Quinlan and Ewers 1989). Because the karst aquifer under study is unconfined and springs can occur at any elevations between 595 and 612 feet amsl at the site, aquifer storage conditions can be practically determined by the average water levels within the aquifer. Karst aquifers are generally conceived and modeled as storage reservoirs penetrated by trunk conduits (Smart 1999). Head loss in conduits is much lower than in surrounding fractures or matrix blocks. As a result, conduits permit exceptionally rapid transfer of water and chemical constituents, while the storage reservoirs contain most of the water in the karst aquifer and slowly adjust to autogenic recharge from sinkholes and back flooding from the primary conduits. Previous groundwater tracing and spring-discharge monitoring provide tools for diagnosis of the primary conduits, while the water level measurements in monitoring wells M1 and M4 tend to represent the average aquifer conditions.

Table 7.7 summarizes quantified information regarding the natural factors pertinent to the 9 background sampling events for the 11 detection monitoring parameters.

The degree of the linear association between X and Y is indicated by the square of the correlation coefficient or R^2 value. If the linear regression equation derived from the analysis predicts exactly each Y value, the regression equation will have a coefficient of correlation of 1.0. If the coefficient of correlation is zero, the variables are considered to be uncorrelated or not linearly associated. This does not mean that the variables are statistically independent. Correlation coefficients have only relative meanings mathematically. No absolute values of superiority or inferiority can be associated with coefficients obtained from different data sets (Mann 1987). For example, a correlation coefficient of 0.8 from one data set cannot be compared indiscriminately with a coefficient of 0.7 from another data set and be concluded to be superior. However, for a given data set the correlation coefficient for X versus Y can be compared with a coefficient for Z versus Y, where X and Z are the different natural factors that are discussed above, although the values of the coefficients remain relative. These basic understandings of the linear regression analysis can be used to guide the selection of the best predictors for each individual FWC by using the R^2 values.

In selecting the best FWC predictors for each data set, it is important to understand that the decision as to what defines a substantial correlation is entirely subjective, especially considering the limited number of data points available. A correlation coefficient becomes more attractive as it approaches 1.0. However, to say that it must be greater than a specific value to be considered large is an entirely subjective judgment. This judgment, however, must also be based on an understanding of the hydrogeological conditions of the site. Using multiple parameters to represent one factor further complicates this issue. For example, the factor of precipitation is described by three parameters including total rainfall, peak rainfall intensity, and rainfall intensity. A factor is considered to be linearly associated with the FWC only when the correlation coefficients remain consistently high for all parameters that were initially used to define the factor.

The data analysis performed indicates that the rainfall conditions, flow conditions, residence time, average water temperature, and antecedent moisture conditions are not likely to be the factors that affect any of the calculated FWCs. However, the aquifer storage condition, which is represented by the average water level in M1, was found to be the only natural factor that affects water-quality parameters (SC, magnesium, sodium, chloride, TDS, and calcium). Additionally, for every affected water-quality parameter, levels decreased with an increase in the average water level in M1. Figure 7.18 shows the regression chart for SC as an example. As previously mentioned, the average water level in M1 has been used to determine the threshold for high- and low-flow conditions at the site.

7.10.5.2 Discussions on Flow Conditions

In karst areas, it is generally recognized that groundwater flow and water quality vary with flow conditions because conduits and caves tend to develop at multiple levels. The relatively quick response of the karst aquifer to recharge events also suggests a potential change in water quality under different flow conditions. Karst aquifers like the one under study are often conceived and modeled as storage reservoirs penetrated by trunk conduits (Smart 1999). In response to recharge events, conduits permit exceptionally rapid transfer of water and chemical constituents, while the storage reservoirs (fractures and matrix blocks), which contain the majority of water in the karst aquifer, slowly adjust to autogenic recharge from sinkholes and back flooding from the primary conduits. During recession periods, the head loss in conduits is often much lower than in the surrounding fractures or matrix blocks and the water stored in the matrix gradually drains into the conduits to sustain the spring flow. If time permits, equilibrium between the water in the conduits and that in the matrix occurs, and a unified water level is obtainable. The constant exchanging process also leads toward equilibrium between the chemical constituents within the conduits and matrix.

Because most of the water is stored in the non-conduit portion of the karst aquifer, the groundwater level in M1 is an indication of storage conditions of the aquifer. The unconfined nature of the aquifer makes the relationship between the volume of water stored in the aquifer and the water level in M1 more obvious. Like surface reservoirs, the volume of water in the aquifer is directly proportional to its water level. Because of a dilution effect in the aquifer, the concentrations of the water-quality parameters tend to be larger when the total volume of water stored in the aquifer is smaller (or under low-flow conditions). This relationship between chemical concentration and water level in M1 clearly explains why the average water level in M1 is related to the FWCs for the six mentioned parameters. This dilution effect is also a reflection of the seasonal effects. However, simple binary division into wet and dry seasons (or high-flow and low-flow conditions) over-simplifies site conditions.

Table 7.7 Linear regression results for factors affecting FWCs

Parameter	Factors linearly associated with FWCs	Regression equation	Explanations
Specific conductance	Average water level at M1	$Y = -111.37X + 70350$ $R^2 = 0.62$	Most *R*-square values are less than 0.5. The average water level in M1 has the strongest correlation with its FWCs. The *R*-square values for the beginning and ending discharges are greater than 0.5. *R*-square value for the total volume is only 0.28
Alkalinity, Bicarbonate ($CaCO_3$)	None		There seems to be a strong association with the average temperature with an *R*-square value of 0.75. Because the infiltrated water generally has a different temperature from the existing water in the aquifer, this effect is not conclusive
Alkalinity, Carbonate ($CaCO_3$)	None		The lack of strong linear correlation with various factors may result from the high percentage of censored data
Alk., Total ($CaCO_3$)	None		See the explanation for carbonate alkalinity
magnesium	Average water level at M1	$Y = -2.35X + 1487.9$ $R^2 = 0.82$	The association with the average water level in M1 is the strongest. The second possible factor is the beginning discharge at QRS with an *R*-square value of 0.81. The correlation does not appear to be strong for the ending discharge at QRS and the total volume of water with *R*-square values being 0.41 and 0.16, respectively
Potassium	None		There is no obvious correlation with any of the factors analyzed. The maximum *R*-square value is 0.35 for rainfall intensity
Selenium	None		There seems to be some correlation with peak rainfall and antecedent moisture conditions for total rainfall 20 days preceding a sampling event
Sodium	Average water level at M1	$Y = -2.24X + 1416.8$ $R^2 = 0.65$	The average water level in M1 has the second greatest *R*-square value. The greatest *R*-square is for the ending discharge at QRS. The *R*-square values for the beginning discharge and the total volume of flow are relatively small
Sulfate	None		The flow conditions at QRS seem to have a close association with the FWCs. All of the *R*-square values for the peak discharge at QRS, ending discharge at QRS, and the total volume of water exceeds 0.6. Review of the regression equation for the total volume reveals that the slope is very small (i.e., on the order of ten thousandths) and the correlation for the beginning discharge is relatively insignificant
Chloride	Average water level at M1	$Y = -9.34X + 5892.5$ $R^2 = 0.80$	The *R*-square for the average water level is the greatest for this data set. Other factors that may affect the FWCs include the beginning discharge, ending discharge, peak discharge, and the total volume of water. Considering chloride is closely related to SC, while the R-squares between the flow condition factors and SC are relatively small, we select the average water level in M1 as the most significant factor
Solids, dissolved	Average water level at M1	$Y = -52.47X + 33,468$ $R^2 = 0.80$	The *R*-square value for the average water level in M1 is the greatest. Other factors that may affect the FWCs are beginning and ending discharge. See explanations for SC for discussion as these two sets are similar
Calcium	Average water level at M1	$Y = -6.89X + 4431.9$ $R^2 = 0.63$	The *R*-square value for the average water level in M1 is the second largest. The largest is for the beginning discharge at QRS, with a value of 0.73. The *R*-square values for the ending discharge, peak discharge, and the total volume of water are small, which do not support the beginning discharge as a major factor affecting the FWCs

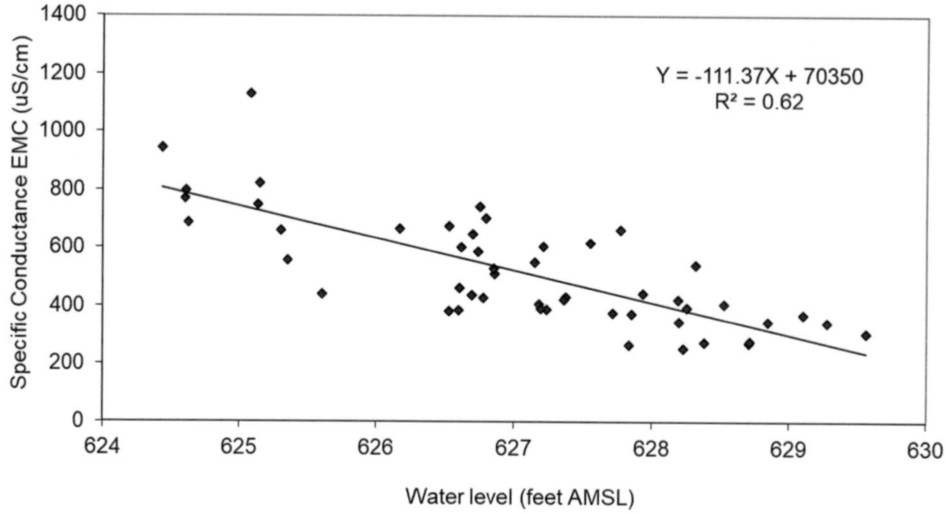

Fig. 7.18 Regression plot of SC and groundwater levels in monitoring well M1

The linear correlation between the average water level in M1 and the FWCs is shown in only 5 of 11 detection monitoring parameters and in SC. It is obvious that other geochemical processes need to be considered in the aquifer, especially for alkalinity (carbonate, bicarbonate, and total), selenium, potassium, and sulfate. The karst aquifer at the site is in the St. Louis and Salem Limestone which is of Mississippian age. As in most limestone aquifers (Drever 1997), water in this aquifer contains Total alkalinity that approximately equals the bicarbonate alkalinity. This is because carbonate alkalinity is only present in trace amounts below a pH of 8.3, and as a result, carbonate alkalinity is below the quantitation limit for all of the samples at the study site (i.e., the data are censored). Because of the high percentage of censoring in the carbonate alkalinity data, the linear correlation analysis is inconclusive. The same applies to selenium, which has approximately 80% censored data.

Karstification is the process of mineral dissolution in a carbonate aquifer. The discussion of the dissolution process is beyond the scope of this study; however, it is very complex and is affected by many factors (Dreybrodt 1988). Gypsum layers are also known to be present at the site, which further complicates the geochemical interactions of various constituents. These complex geochemical processes may cause the effect of the average groundwater level in monitoring well M1 on the FWCs to be less important in some water-quality parameters.

7.10.6 Sampling Methods Affecting FWCs

The sampling procedures and methodologies in the karst system are more complex than those for traditional groundwater sampling programs. For FWCs to be valid statistical variables, they should not be affected by sampling methods. Otherwise, the FWC could be too sensitive to the sampling methods to be used as a valid statistical variable. SC was selected as a representative parameter to test whether the developed sampling methods affect the FWCs. It has been long recognized that SC (or conductivity) is an accurate and sensitive indicator of the characteristics of karst aquifers (Hess and White 1988; Ford and Williams 1989; Quinlan et al. 1991).

7.10.6.1 Effect of Sampling Event Starting and Ending Times on FWCs

Unlike stormwater–runoff sampling, the spring water sampling at QRS always contained a base–flow component. This base-flow component made it difficult to clearly define the starting and ending points for a given sampling event for the purposes of calculating FWCs. The aquifer properties and antecedent moisture conditions further complicated this issue. Similar rainfall events could also result in a totally different response from the spring. As stated by ASTM (1995), at the start of a recharge/discharge event, it is impossible to know how significant it would be. At its middle or end, it would be too late to collect samples that might have characterized its beginning. Accordingly, it was necessary to commence sampling at the start of an event. After the event, the decision to analyze or not to analyze the samples should be based on professional judgment and evaluation of the significance of the event. Sampling should continue until SC is within 10% of pre-event values. Alternatively, if the recharge/discharge event continued for an extended period of time, sampling would be continued until SC values were within 10% during three consecutive 15-min measurements. Therefore, the starting and ending times for the storm-driven sampling events were subjective to evaluation of the measured discharge and SC data. Some inconsistencies were expected in these decisions. For the FWC to be a variable representative of the average water quality for the sampling event, it is critical that the calculated FWCs were not severely impacted by the potential decisions that may slightly affect the moments to start and end the collection of water samples.

The sensitivity of the calculated FWCs for SC to different starting and ending times is shown in Table 7.8. The difference between the FWC for each sampling event and the FWCs for other scenarios ranges from 0 to 13.5%. In general, the difference is proportional to the difference in the time duration within each individual event. However, the effect of each scenario on the FWC depends on each individual sampling event because the time to reach the deflection in SC varies substantially with the event. In events 1 and 6, the time to reach the deflection was over one day, whereas the time was only a few hours for events 3 and 4. The time to reach 90% of recovery in SC also contributed to the phenomenon. The effect of the starting and ending time on FWC was event-specific.

On average, the FWC of SC for the sampling events was not sensitive to the beginning and ending times for the

Table 7.8 Effects of sampling times on FWCs

Event	Category	Beginning time (mm/dd/yy hh:mm)	Ending time (mm/dd/yy hh:mm)	Total data point	SC FWC (μS/cm)
1	Sampling event beginning and ending	11/6/00 17:00	11/13/00 16:40	833	673.21
	One day later beginning and ending	11/7/00 17:00	11/13/00 16:40	785	642.16
	SC change beginning and ending	11/8/00 22:30	11/13/00 16:40	686	600.68
	Sampling beginning and 90% ending	11/6/00 17:00	11/27/00 17:30	2726	747.61
	One day later beginning and 90% ending	11/7/00 17:00	11/27/00 17:30	2678	738.49
	SC change beginning and 90% ending	11/8/00 22:30	11/27/00 17:30	2579	727.88
2	Sampling event beginning and ending	1/10/01 12:14	1/11/01 11:10	139	595.67
3	Sampling event beginning and ending	1/29/01 18:00	2/5/01 13:00	979	498.68
	One day later beginning and ending	1/30/01 18:00	2/5/01 13:00	835	482.98
	SC change beginning and ending	1/29/01 21:10	2/5/01 13:00	960	497.58
	Sampling beginning and 90% ending	1/29/01 18:00	2/3/01 1:10	620	461.79
	One day later beginning and 90% ending	1/30/01 18:00	2/3/01 1:10	475	431.53
	SC change beginning and 90% ending	1/29/01 21:10	2/3/01 1:10	601	459.81
4	Sampling event beginning and ending	12/12/01 12:30	12/16/01 9:00	186	299.63
	One day later beginning and ending	12/13/01 12:30	12/16/01 9:00	138	290.00
	SC change beginning and ending	12/12/01 18:00	12/16/01 9:00	175	297.35
	Sampling beginning and 90% ending	12/12/01 12:30	12/16/01 11:00	190	301.15
	One day later beginning and 90% ending	12/13/01 12:30	12/16/01 11:00	142	292.26
	SC change beginning and 90% ending	12/12/01 18:00	12/16/01 11:00	179	298.99
5	Sampling event beginning and ending	1/23/02 7:30	1/28/02 11:15	496	391.62
	One day later beginning and ending	1/24/02 7:30	1/28/02 11:15	400	375.53
	SC change beginning and ending	1/23/02 23:30	1/28/02 11:15	432	377.23
	Sampling beginning and 90% ending	1/23/02 7:30	1/28/02 23:15	544	404.39
	One day later beginning and 90% ending	1/24/02 7:30	1/28/02 23:15	448	391.07
	SC change beginning and 90% ending	1/23/02 23:30	1/28/02 23:15	480	391.71
6	Sampling event beginning and ending	9/19/02 13:45	9/25/02 7:45	553	838.86
	One day later beginning and ending	9/20/02 13:45	9/25/02 7:45	457	812.34
	SC change beginning and ending	9/20/02 22:15	9/25/02 7:45	423	784.36
	Sampling beginning and 90% ending	9/19/02 13:45	9/27/02 4:30	732	847.77
	One day later beginning and 90% ending	9/20/02 13:45	9/27/02 4:30	636	828.09
	SC change beginning and 90% ending	9/20/02 22:15	9/27/02 4:30	602	807.97
7	Sampling event beginning and ending	10/8/02 11:00	10/9/02 10:00	93	832.48
8	Sampling event beginning and ending	4/3/03 12:00	4/4/03 10:00	89	571.59
9	Sampling event beginning and ending	5/17/03 0:00	5/20/03 2:00	223	499.12
	One day later beginning and ending	5/18/03 0:00	5/20/03 2:00	Water quality data missing	
	SC change beginning and ending	5/17/03 9:45	5/20/03 2:00		
	Sampling beginning and 90% ending	5/17/03 0:00	5/19/03 8:30	227	499.47
	One day later beginning and 90% ending	5/18/03 0:00	5/19/03 8:30	131	473.31
	SC change beginning and 90% ending	5/17/03 9:45	5/19/03 8:30	188	492.71

sampling events. The sampling event was best defined by when the first samples were collected (i.e., when SC first began to change), and when the last sample was collected (or SC had recovered to within 90% of its pre-event value). The data analysis for storm-pulse sampling events indicated that the calculated FWCs for SC were not seriously affected by slight differences in decisions to start and stop collecting water samples as long as the sampling event covers the duration from the deflection of SC to 90% recovery of its pre-event value.

7.10.6.2 Effect of Judgmental Versus Constant-Interval Sample Selection on FWCs

For storm-driven sampling events, the SC chemographs, which vary with changes in sampling frequency, can potentially be affected by a phenomenon called aliasing. Aliasing is a misrepresentation of data signals caused by not collecting samples at frequent enough intervals and is caused by the dynamic nature of karst springs. As indicated by Quinlan et al. (1991), aliasing of water-quality data can be a serious problem because it can inadvertently and seriously bias interpretation. The aliasing effect can only be addressed by frequently collecting water samples; however, there is currently no consensus as to an optimal sampling frequency (Quinlan and Alexander 1987), and the frequency used should be site-specific. Considering the expenses for laboratory analyses of multiple analytical parameters, most studies do not have the budget to collect and analyze water samples as frequently as should be. In addition, a small group of representative samples can often be chosen to characterize the water quality over a given sampling event. This assumption led to the application of judgmental selection of samples for analyses in the background study for this study.

For the recharge-discharge sampling events during the background collection phase, the frequency of field sample collection at QRS depended on the nature of the event and the associated response of the spring with respect to discharge and the six field water-quality parameters: temperature, pH, conductivity, turbidity, ORP, and DO. After all samples were collected, judgmentally selected samples were shipped to the designated laboratory for chemical analyses. The judgmental selection of samples for laboratory analysis was subject to knowledge in karst hydrogeology, experience with the site, and availability of field data.

In theory, judgmental selection of samples can better characterize the water-quality variation over the sampling period than the constant-interval selection of samples. However, the use of judgmental selection should be evaluated by how the site-specific data will be used in the decision-making process. Table 7.9 compares the calculated FWCs for SC using judgmental sample selection with those from constant-interval sample selection. The most frequent data interval is 15 min, and the results are listed in the column Base Line Time.

As shown in Table 7.9, the FWCs calculated by these two methods are consistent for different sampling frequencies within each sampling event. The maximum change caused by different sampling frequencies is less than 3% when compared to FWCs in the Base Line Time. Thus, the judgmental sampling strategy does not seem to be superior to a constant-interval selection of samples. Flow measurements using different intervals also do not appear to affect the FWCs substantially. This analysis suggests that storm-pulse samples for laboratory analysis can be based on constant rate intervals across an entire event. This is true because the FWC concept is being utilized and the spring discharge was the main factor affecting concentrations. It should also be noted that the insensitivity of FWCs to SC is based on the nine background sampling events at QRS (especially the six storm-pulse sampling events), and this assumption should not necessarily be used as a rule for karst sampling.

7.11 Statistical Techniques

The applicability of any given statistical procedure depends directly on assumptions about the characteristics of the data being explored. Results from statistical analysis may be meaningless or, even worse, misleading, if the data do not conform to the design assumptions of the statistical methods used. Therefore, careful consideration must be given to which statistical methods are appropriate in terms of the type of data being evaluated. The unique features of water-quality data at QRS affect the suitability of interpretation made by the application of many classical statistical techniques. The specific statistical methods and procedures are used to determine whether statistically significant increases in chemical parameters are occurring for any given sampling event, with the overall goal of balancing false-negative and false-positive detections.

7.11.1 Data Properties

Statistical methods should be developed with careful consideration of the distribution properties of the data. Many statistical methods assume that measurements are normally distributed, or that they can be mathematically transformed into normal distributions. In this study, FWCs (not individual data points) are the variables used in statistical analyses. However, only when calculated FWCs are independent random variables can their probability distributions be approximated. Probability distributions are mathematical tools to describe the behavior of random variables.

Table 7.9 Comparison between constant interval sample selection and judgmental sample selection

Event	Category	Base line time	1 h	2 h	4 h	8 h	Judgmental
1	Total data point	1009	169	85	43	22	21
	Total volume (gallon)	297,813	298,407	298,776	300,320	301,991	305,381
	SC-EMC (μS/cm)	677	678	678	677	675	677
	FWC (μS/cm)	677.26	680	681	683	684	695
2	Total data point	139	24	13	7	4	4
	Total volume (gallon)	32,440.76	32,283	32,249	32,482	32,206	32,958
	SC-EMC (μS/cm)	595.67	597.19	598.88	596.51	596.78	590.29
	FWC (μS/cm)	595.67	594.29	595.34	597.27	592.48	599.71
3	Total data point	979	164	83	42	22	23
	Total volume (gallon)	408,882	409,540	407,725	406,021	405,094	408,538
	SC-EMC (μS/cm)	498.68	498.89	498.98	498.16	498.80	497.08
	FWC (μS/cm)	498.68	499.69	497.56	494.67	494.18	496.66
4	Total data point	186	94	48	25	13	20
	Total volume (gallon)	299,698	300,685	301,370	307,665	303,130	309,639
	SC-EMC (μS/cm)	299.63	299.70	299.84	299.37	299.38	300.41
	FWC (μS/cm)	299.63	300.69	301.52	307.33	302.81	310.38
5	Total data point	496	125	63	32	17	20
	Total volume (gallon)	240,318	239,852	240,250	244,088	246,102	224,996
	SC-EMC (μS/cm)	382.88	383.61	382.56	382.47	390.60	383.02
	FWC (μS/cm)	382.88	382.86	382.45	388.47	400.00	358.60
6	Total data point	553	139	70	36	19	19
	Total volume (gallon)	39,444	39,397	39,480	39,280	38,960	39,204
	SC-EMC (μS/cm)	838.86	838.82	839.69	838.40	835.98	841.76
	FWC (μS/cm)	838.86	837.82	840.45	834.90	825.71	836.64
7	Total data point	93	24	13	7	4	7
	Total volume (gallon)	4580	4580	4580	4679	4679	4679
	SC-EMC (μS/cm)	832.48	832.93	833.45	833.17	832.58	833.17
	FWC (μS/cm)	832.48	832.93	833.45	851.27	850.66	851.27
8	Total data point	89	23	12	7	4	6
	Total volume (gallon)	14,513	14,542	14,483	14,483	14,483	14,483
	SC-EMC (μS/cm)	571.59	571.65	573.04	572.87	572.85	570.97
	FWC (μS/cm)	571.59	572.81	571.87	571.70	571.69	569.80
9	Total data point	223	57	29	15	8	9
	Total volume (gallon)	63,417	63,654	63,778	63,837	63,356	64,222
	SC-EMC (μS/cm)	499.12	499.55	498.24	499.75	500.17	500.35
	FWC (μS/cm)	499.12	501.42	501.08	503.06	499.69	506.71

Base line time is the data measured interval; 10 min. interval in sampling events 1, 2, and 3; 15 min. in Events 5, 6, 7, 8 and 9; 30 min. in Event 4
Other frequencies are chosen by 1, 2-, 3-, 4-, and 8-h intervals and judgmental time
Judgmental time is the sampling time during the event period in which samples were selected for laboratory analysis
Total data point is the data number used for the EMC calculation depending on base line time, 1, 2, 4, and 8 h, and judgmental time
Total volume is the sum of flow volume in the base time interval in different frequencies
SC-EMC is specific conductance event mean concentration = total load of constituent/total flow volume
FWC is the flow-weighted concentration, total load of constituent was divided by the base flow volume

Among the 11 detection monitoring parameters and one field water-quality parameter, subject to the statistical evaluation, the FWCs for alkalinity (total, carbonate, and bicarbonate), selenium, sulfate, and potassium were considered to be independent variables, as these values were not found to be associated with any of the factors analyzed. However, FWCs for SC, TDS, chloride, sodium, magnesium, and calcium, were found to be related to aquifer storage conditions, or more specifically to the average groundwater level in on-site monitoring well M1. Because of this correlation for these constituents, the FWCs needed to be adjusted to a consistent water level in M1. Otherwise, the FWCs directly calculated from the discrete sample points were not comparable. For example, the FWC for magnesium from a sampling event in drought conditions would be higher than that from a sampling event under rainy conditions due to aquifer conditions. This difference could not be used in evaluation of the integrity of the landfill because it resulted from a natural cause (i.e., a change in natural weather conditions) that was independent of the landfill. Therefore, appropriate adjustment of the calculated FWCs to compensate for the effect of aquifer storage helped minimize both false-positive and false-negative rates.

7.11.1.1 Adjustment of FWCs

Table 7.10 presents the equations used to adjust FWCs for each of the 6 parameters that were found to be related to the average water level at M1. The adjusted FWCs corresponded to an arbitrary water level elevation of 624 feet amsl. Linear regression analysis was conducted between the adjusted FWCs and the average water level in M1, and the R^2 values were found to be extremely small (i.e., less than 2E−10). The simple and linear adjustments had little effect on the correlation with other factors. The R^2 values between the adjusted FWCs and other factors remained little changed.

Table 7.11 illustrates the adjusted and non-adjusted FWCs for magnesium, sodium, chloride, TDS, and calcium, whereas Table 7.12 presents the adjusted and non-adjusted FWCs for SC. Because of a lack of water level data in M1 for some of the selected events, FWCs were adjusted for 50 selected precipitation events. Although the FWCs for alkalinity, potassium, and sulfate did not need any adjustments, their FWCs were considered as adjusted in the following probability and statistical tests.

7.11.1.2 Testing of Non-normality

Several methods for testing non-normality of environmental data have been described (Gibbons 1994), however, no single method is suitable for generic use because of the complexity of data patterns. The primary reason to test FWCs for normal distribution is to determine whether parametric test procedures can be employed. The null hypothesis for all tests of normality is that the data are normally distributed (Helsel and Hirsch 1992). Rejection of this hypothesis indicates that it is doubtful that the data are normally distributed. However, failure to reject the hypothesis does not prove that the data are normally distributed, especially for small sample sizes. It simply says that normality cannot be rejected with the evidence at hand. Because there are only nine FWCs for each detection monitoring analytical parameter (and 50 for SC), the overall population characteristics will be inferred from the limited sample number. A population is defined as the entire collection of potential measurements about which one wants to make a statement.

Four methods are used to search for non-normality of the data sets including histograms; empirical cumulative distribution function plots; probability plots; and the Shapiro–Wilk goodness-of-fit tests. The first three methods provide a visual assessment of the data distribution, and the Shapiro–Wilk test computes a test statistic to determine whether the

Table 7.10 Equations for adjustment of calculated FWCs

Parameter	FWC adjustment equation	R^2 between FWC* and average water level in monitoring well M1
Calcium	$FWC^* = FWC + 6.88(\overline{H} - 624)$	7E−11
Dissolved solids	$FWC^* = FWC + 52.17(\overline{H} - 624)$	2E−10
Chloride	$FWC^* = FWC + 9.3(\overline{H} - 624)$	8E−12
Sodium	$FWC^* = FWC + 2.22(\overline{H} - 624)$	1E−10
Magnesium	$FWC^* = FWC + 2.35(\overline{H} - 624)$	2E−10
Specific conductance (SC)	$FWC^* = FWC + 111.01(\overline{H} - 624)$	2E−10

Table 7.11 Adjusted FWCs of laboratory analytes

Event	Magnesium, original	Magnesium, adjusted	Sodium, original	Sodium, adjusted	Chloride, original	Chloride, adjusted	TDS, original	TDS, adjusted	Calcium, original	Calcium, adjusted
1	15.18	20.87	17.19	22.57	31.07	53.58	645.72	772.06	126.51	143.18
2	12.00	18.52	16.00	22.16	39.86	65.66	581.74	726.50	101.69	120.79
3	13.80	20.17	11.88	17.91	30.31	55.52	556.91	698.33	110.85	129.51
4	12.56	23.00	7.72	17.59	14.79	56.09	443.96	675.72	103.53	134.10
5	14.41	20.35	9.48	15.10	22.62	46.14	487.94	619.92	114.95	132.36
6	21.35	22.69	19.12	20.39	59.18	64.51	717.88	747.77	136.58	140.52
7	20.82	22.16	18.63	19.90	57.63	62.94	653.91	683.71	125.28	129.21
8	14.26	21.86	16.95	24.14	38.84	68.93	579.77	748.61	115.93	138.21
9	13.46	22.52	13.28	21.85	35.18	71.04	559.40	760.61	111.41	137.95

Table 7.12 Adjusted FWCs of field parameter SC

Selected event	Original FWC	Adjusted FWC
1	619.29	1013.19
2	702.52	1013.06
3	676.66	957.42
4	648.86	948.84
5	443.58	879.96
6	550.20	899.76
7	602.05	893.13
8	461.06	750.56
9	389.74	748.78
10	421.77	887.10
11	394.34	866.59
12	541.30	1020.51
13	662.92	1081.21
14	606.88	963.01
15	274.93	761.34
16	371.16	937.64
17	265.57	691.11
18	381.99	662.84
19	586.65	890.82
20	431.25	806.10
21	377.12	789.51
22	254.08	723.65
23	271.77	794.52
24	344.68	882.50
25	346.43	812.21
26	344.17	930.19
27	307.83	925.06
28	405.66	908.18
29	278.47	801.57
30	743.30	1048.58
31	1131.74	1251.63
32	799.04	866.76
33	769.34	835.94
34	943.80	991.83
35	687.15	756.74
36	747.90	873.63
37	557.81	708.04
38	441.12	619.27
39	429.95	738.21
40	821.47	948.90
41	659.06	803.83
42	385.94	674.76
43	407.96	761.13
44	422.50	795.49
45	509.65	827.08
46	530.21	846.45
47	667.69	908.82
48	438.83	738.25
49	392.77	746.96
50	372.11	799.60

normality assumption is unacceptable to a statistically significant extent. Brief discussions regarding each of the four test methods are presented as follows:

- **Histograms**: Histograms summarize the distribution of the data by placing observations into intervals and counting the number of observations in each interval. A histogram is a useful way of displaying the distribution of a data set, but the appearance of a histogram depends on the number of intervals the data are divided into (Helsel and Hirsch 1992). The general guideline is to choose an interval width such that data accuracy does not get lost, but there is no consensus rule for what to choose. When a large number of data points exist and the interval is very fine, a density plot can be created, which can be used to see whether it resembles a normal distribution density plot. However, with only nine FWCs for each water-quality parameter, histograms are not used as a primary tool for judging the normality of the data and are used as a reference tool only.
- **Empirical Cumulative Distribution Function Plots**: An empirical cumulative distribution function (cdf) plot displays the ordered data on the *x*-axis versus the estimated cumulative probabilities on the *y*-axis. Because this kind of plot portrays the quantiles or percentiles (quantiles times 100) of the distribution of the data set, they are also called quantile plots. The *p*th quantile of a population is the number such that a fraction of *p* of the population is less than or equal to this number. As noted by Helsel and Hirsch (1992), the quantile plots do not require the data to be divided into classes, and every data point is displayed in the graph. The quantile plots are used to compare the empirical cdf of the data set for each water-quality parameter with the cdf from a normal distribution. If the data under study are consistent with a normal distribution, the empirical cdf will roughly fit the theoretical cdf, appearing as an approximate straight line on the associated plot.
- **Probability Plots**: A probability plot or quantile–quantile (i.e., *Q*–*Q*) plot is used to graphically compare the adjusted FWC data set for each water-quality parameter to a normal distribution. The assumption is that if the distributions of two populations are the same, then they have the same quantiles (percentiles), so a plot of the quantiles for the first distribution versus the quantiles for the second distribution will fall on a straight 0–1 line with intercept 0 and slope 1. Each *Q*–*Q* plot displays every data point and a 0–1 line. Data points from a normal distribution tend to fall on the line. Probability plots for normality tests are recommended by USEPA (1992).

- **Shapiro–Wilk Goodness-of-fit Tests**: As indicated by Gibbons (1994), from the perspective of testing for non-normality in multiple independent samples, the Shapiro–Wilk test appears to be the most rigorous. This test is also recommended by USEPA (1992) for non-normality tests for water samples less than or equal to 50 in number. If a set of data is normally distributed, the ordered values should be highly correlated with corresponding quantiles taken from a normal distribution. The Shapiro–Wilk test statistic (W) tends to be large when the probability plot of the data indicates a straight line. When the sample size is 9, normality of the data should be rejected if the statistic W is lower than the 5% critical value of 0.829 (USEPA 1992). If the calculated W is greater than or equal to 0.829, it is assumed that the data are sufficiently normally distributed and can be used in further statistical analyses. In cases when the Shapiro–Wilk tests for the adjusted FWCs showed statistically significant evidence of non-normality, the data were transformed using logarithms, and data were then tested for lognormality using the same method. When the sample size is 50 (SC), the 5% critical value is 0.947 for the normality test. The limitations of the Shapiro–Wilk test procedure have been acknowledged as these tests do not really provide a measure of whether the fit is good enough to allow accurate use of normal-based statistics. Table 7.13 presents a summary of the normality test results.

7.11.1.3 Independence Between Sampling Events

Testing for independence between sampling events is an essential task in developing a valid statistical evaluation plan that employs intra-spring statistical analysis. The statistical methodologies being utilized for this project (as outlined in detail in the following subsections) assume that sampling events are unique, randomly distributed, and completely independent. Violation of this assumption would render selected statistical tests invalid unless appropriately compensated for. Ideally, to maintain this independence, groundwater should only be sampled after water monitored during the previous precipitation event, has completely discharged through the springs. In theory, this may be possible, but in reality, it is understood that there will almost always be some small amount of residual water from previous precipitation events within the karst groundwater system.

To help ensure that the nine background events previously conducted were independent of each other and independent of other preceding precipitation events, the following procedures shall follow:

- All sampling events will have no precipitation immediately preceding the event. More specifically, major precipitation should not occur within several days prior to a sampling event. A storm event projected for sampling must also generally have at least two days with no precipitation following it.
- Subsequent sampling events will not be initiated until spring discharge levels and field water-quality parameters have returned to (or near to) pre-storm levels. For most major storm-pulse events, a minimum of approximately five to seven days without precipitation will be sufficient to allow conditions to stabilize.

Although these procedures were in place to help maintain sufficient independence between sampling events, the actual verification of independence between sampling events was still required to ensure the overall validity of the statistical methodologies selected. Any substantial dependence between data sets collected in separate sampling events would have to be recognized and accounted for in the statistical evaluation. The independence testing completed for this project involved the evaluation of water quality data and the calculation of residence times of groundwater and dissolved constituents within the karst aquifer. The calculated average residence time value was found to be 2.72 days. The independence assumption for the statistical tests being performed appears valid, and the data from all nine background events can be reasonably assumed to be independent of all other precipitation events. Therefore, adjustments to remove any dependency were not required.

Because maintaining independence between sampling and other precipitation events is critical to ensuring valid statistical tests, it is recommended that independence between future sampling and precipitation events be verified at a minimum of every 2 years or prior to the pooling of data to background. Verification of independence can be

Table 7.13 Summary of normality tests

Water-quality parameter	Mean or mean-log FWC*	Standard deviation	Normality test result
Specific conductance (SC)	855.65	120.99	Normal
Alkalinity bicarbonate ($CaCO_3$)	236.28	27.47	Normal
Alkalinity carbonate ($CaCO_3$)	– 0.49	0.44	Log-normal
Alkalinity total ($CaCO_3$)	236.28	27.47	Normal
Magnesium	21.35	1.48	Normal
Potassium	60.60	18.31	Normal
Selenium	– 5.00	0.94	Log-normal
Sodium	20.18	2.87	Normal
Sulfate	160.45	42.60	Normal
Chloride	60.49	8.13	Normal
Solids, dissolved	714.80	49.39	Normal
Calcium	133.98	6.90	Normal

completed by calculating and evaluating residence times, evaluating field water-quality and hydrograph data for return to pre-storm levels prior to sampling, determining whether sampling procedures were appropriately followed, plotting FWC data, and evaluating autocorrelation, or by other methods.

7.11.1.4 Testing for Trend

The characteristics of environmental data may vary cyclically at time scales of hours, days, weeks, seasons, years, and even decades due to both natural and anthropogenic influences. Evaluating the nature of possible effects in environmental data caused by temporal variation can be useful in data interpretation and in understanding site hydrogeology. However, for analytical data obtained during the nine background sampling events, it appears that little can be done to evaluate the specific effects of temporal variation because the sampling events were randomly completed over a period. To properly evaluate temporal variation, data must be obtained at some type of regular frequency (such as quarterly, monthly, weekly, daily, and so forth). For these reasons, it does not appear that the analytical data from the nine background sampling events will lend themselves to evaluation of temporal trends. It should also be noted that by using FWCs for this project, any temporal variation in the analytical data (even if present) is likely already compensated for, although indirectly. In summary, it does not appear that temporal variation can be usefully evaluated in the background analytical data given the random nature of sampling conducted in any given year.

Although testing for seasonal variation was not performed for reasons outlined above, general testing for long-term trends of the 12 constituents of concern was conducted using the Mann–Kendall nonparametric test for trend (Kendall 1975; Mann 1945). This specific method, which is recommended for 40 or fewer measurements (Gilbert 1987), was used because it does not make any distributional assumptions regarding the data, and it allows for irregularly spaced sampling events. This testing was performed to evaluate the potential for positive (increasing) trends in data collected during the background collection phase, for the purposes of evaluating overall trends and determining whether the background data were truly background (i.e., not affected by the landfill), since all background sampling was completed after placement of some waste within the landfill.

Using the methods outlined in Gibbons and Coleman (2001) for the Mann–Kendall test, trend testing was performed on background FWCs for the 11 detection monitoring parameters subject to the statistical evaluation (for nine events) and on SC FWCs (for 50 events). The testing was performed using adjusted FWCs versus individual analytical data values for each constituent for each event because the dependency of various other factors needed to be removed to properly evaluate trend. The general mechanics of the procedure first involves ordering of the FWC measurements by date and time (i.e., X_1, X_2, …, X_n) and then computing as follows:

X_1	X_2	X_3	…	X_{n-1}	X_n
	$X_2 - X_1$	$X_3 - X_1$		$X_{n-1} - X_1$	$X_n - X_1$
		$X_3 - X_2$		$X_{n-1} - X_2$	$X_n - X_2$
				$X_{n-1} - X_3$	$X_n - X_3$
				–	–
				–	–
				–	–
				$X_{n-1} - X_{n-2}$	$X_n - X_{n-2}$
					$X_n - X_{n-1}$

Using the above procedure, the Mann–Kendall statistic (S) is then calculated as the overall number of positive differences minus the total number of negative differences. This statistic and the total number of sampling events (n) can then be used to determine an associated probability (Gibbons and Coleman 2001).

As illustrated in Table 7.14, the results of this testing indicate that two constituents (potassium and sodium) showed no trend ($S=0$), SC, selenium, and bicarbonate alkalinity and total alkalinity showed a negative trend, and the remaining constituents of concern showed a positive trend. However, the positive trend was statistically significant only for chloride ($P=0.038$) and approached significance for manganese ($P=0.130$). Although these findings suggest a lack of strong evidence for positive trends (except for chloride), they do not provide any evidence that there is no such trend. A Kendall test procedure was also performed, which in general revealed similar findings to the Mann–Kendall trend testing results.

7.11.2 Determination of Statistical Tests

It is very important to emphasize that on the basis of a statistical analysis alone, it can never be concluded that a waste disposal facility has or has not impacted groundwater (ASTM 1998). A statistically significant increase over background levels indicates that the new measurement from a particular monitoring location is inconsistent with chance expectations based on the background data available. Although carefully designed statistical tests and procedural consistency will help in minimizing false-positives and false-negatives, knowledge in how to apply these specific techniques to a karst groundwater setting is also critical. The following paragraphs outline the details of the specific

Table 7.14 Summary of trend analysis

Test method	Potassium, FWC	Sodium, FWC	SC, FWC	Selenium, FWC	Alkalinity, bicarbonate, FWC	Alkalinity, total, FWC	Chloride, FWC	Magnesium, FWC	Calcium, FWC	Sulfate, FWC	Solids, dissolved, FWC	Alkalinity, carbonate, FWC
Mann-Kendall statistic, S	0	0	**-299**	**-8**	**-2**	**-2**	18	12	4	2	2	1
Probability (no trend)	0.540	0.540	Small #	0.238	0.460	0.460	0.038	0.130	0.381	0.460	0.460	0.500
Kendall Test statistic, z	0	0	**-2.49**	**-1.36**	**-0.10**	**-0.10**	1.77	1.15	0.31	0.10	0.10	0
Probability, p-value (no trend)	1.0	1.0	0.013	0.175	0.917	0.917	0.076	0.251	0.754	0.917	0.917	1.0

Shading indicates no trend (i.e., $S=0$).
Negative S indicates a decreasing trend (bold values).
Positive S indicates an increasing trend (non-bold values).
Probability values indicate the probability of no trend (i.e., $S=0$); lower values can indicate a significant trend (positive or negative).

statistical tests to be used in the future for this groundwater monitoring project.

7.11.2.1 Statistical Quality Control

Like groundwater monitoring programs at other waste disposal facilities, operating the landfill requires statistical methods as the basis for investigating the integrity of the landfill. Therefore, statistical tests will be performed on each of the 12 parameters of concern for QRS that are subject to the statistical evaluation (i.e., SC and 11 detection monitoring parameters). Statistical decision procedures for these tests are based on the properties of data collected during the background monitoring collection period for QRS. Following future statistical tests, appropriate decisions will be made after each compliance-monitoring event as to whether additional investigations or regulatory actions are required. Therefore, it is essential that a realistic and workable model for the measurement data be formulated and used both in the construction and evaluation of decision procedures. Obviously, no statistical measurement model is perfect for a karst groundwater monitoring program. However, the model for a decision procedure should be reasonable and as simple as possible.

During the detection-monitoring phase, statistical tests will be performed to decide whether additional investigation or regulatory action is required following each sampling event. Over the operating life of the landfill facility, a sequence of decisions, rather than just one decision, will be made. The one-decision case is a test of a hypothesis situation in which it makes sense to consider the significance level and power of the test. Where decisions are made sequentially over time, one is dealing with quality-control schemes and should be interested in distributions of run lengths (Starks 1988). An in-control run length is the number of sampling periods from start-up until a decision is made, on the basis of water sample measurements, that additional regulatory action is required when, in fact, there is no leakage from the landfill. An out-of-control run length is the number of sampling periods from the time that a pollutant plume originating from the landfill discharges at QRS until a decision is made that additional regulatory action is required. In comparing decision procedures for the landfill, consideration should be given to the distributions of their run-lengths rather than to their probabilities of Type I and Type II errors on individual applications of the decision rule in each sampling period (Starks 1988). Naturally, one wants to use a quality-control scheme that has, on average, long in-control run lengths and short out-of-control run lengths. From this perspective, it is apparent that the decision procedures pertaining to the groundwater monitoring program at QRS are in truth quality-control schemes.

Quality-control schemes have proved successful in many industrial applications (Lucas 1982, 1985; Lucas and Crosier 1982). Industrial quality-control schemes have been used to monitor effluent from waste treatment plants (Vaughan and Russell 1983; Loftis et al. 1987; Berthouex et al. 1978). Shewhart Control Charts, CUSUM Control Charts, and Combined Shewhart-CUSUM Control Charts are typical graphical and statistical methods of assessing the performance of a system over time and have been widely used to maintain process control (Ryan 2002; Montgomery 1997). Shewhart is the author of the 1931 book entitled Economic Control of Quality of Manufactured Product, in which Shewhart control charts were systematically discussed, whereas CUSUM is an abbreviation for Cumulative Sum. The combined Shewhart-CUSUM control charts are used in statistical tests for this project.

7.11.2.2 Combined Shewhart-CUSUM Control Charts

USEPA (1989) and ASTM (1998) recommend using Combined Shewhart-CUSUM Control Charts for groundwater monitoring in order to detect both large sudden releases and smaller gradual releases. Combined Shewhart-CUSUM control charts are suggested as an alternative to prediction or tolerance limits for intra-locational comparison of monitoring constituent concentrations (USEPA 1989; ASTM 1998). As previously discussed, when justified, intra-locational comparisons are more powerful than their inter-locational counterparts because they eliminate the spatial component of variability. Due to the absence of spatial variability, the uncertainty in measured concentrations is decreased, making intra-locational comparisons more sensitive to real releases, and false-positives due to spatial variability are eliminated (Gibbons 1994).

The combined Shewhart-CUSUM control chart is a mixture of two statistical procedures—Shewhart and CUSUM Control Charts. The Shewhart control scheme is better than the CUSUM scheme in quickly detecting a large ($> 3\sigma$) shift in the mean μ; whereas the CUSUM scheme is usually faster in detecting a small change in μ that persists (Lucas 1982). Because the integrity of the landfill liner is potentially threatened by sudden or catastrophic failure, in addition to more gradual and less severe types of failure, it is essential to account for both large and small changes in the water-quality data at QRS. Combined Shewhart-CUSUM Control Charts take advantage of the good properties of both individual tests and are sensitive to both gradual and rapid releases. Using both Shewhart and CUSUM Control Charts increases the overall false-positive rate for each sampling occasion beyond that obtained using just one of these control charts individually. Using both charts will also reduce the false-negative rate and increase the overall correct-positive rate.

The following properties of the water-quality data at QRS also support the application of the Combined Shewhart-CUSUM Control Charts in the statistical tests for this project:

- Independence of Data: A fundamental assumption for application of control charts is that the observations are independent. Starks (1988) showed that the limits of a control chart can be off by a wide margin when the data are correlated. Berthouex et al. (1978) found that the use of a standard Shewhart or CUSUM chart for certain environmental data was unsatisfactory in that the charts gave too many false alarms because the data are not independent. Linear correlation analyses indicate that 6 of the 12 water-quality parameters of concern are closely associated with storage conditions in the karst aquifer. However, their FWCs have been appropriately adjusted to the average water level measured in the on-site monitoring well M1. As a result of these adjustments, the adjusted FWCs of the 12 water-quality parameters of concern appear to be independent variables.
- Normal Distribution: Another fundamental assumption in the development of control charts is that the underlying distribution of the quality characteristic is normal. Several authors have investigated the effect of departures from normality on control charts. Burr (1967) notes that the usual normal theory control limit constants are very robust to the normality assumption and can be employed unless the population is extremely non-normal. Schilling and Nelson (1976) have also studied the effect and found that in most cases, samples of size 4 or 5 are sufficient to ensure reasonable robustness to the normality assumption. The Monte Carlo simulations conducted by Starks (1988) indicate that the average run-length for the Combined Shewhart-CUSUM Control Charts has a certain degree of sensitivity to the distribution, although in general, there is a considerable amount of distribution-assumption robustness. The distributions of the adjusted FWCs for the 12 water-quality parameters of concern are either normally or log-normally distributed.
- Non-impacted Background Data: The basic premise for using intra-locational comparison is that the background data collected at the monitoring point are representative of the water quality prior to waste placement. Although all background samples were collected after the landfill began to receive waste, evidence indicates that QRS FWC background data have not been impacted by the site. This is supported by the Mann–Kendall trend analysis results on FWC data for the 12 constituents of concern which indicates, with the exception of chloride, the absence of statistically significant upward trends in these constituents.
- Minimum Number of Observations: The adjusted FWCs for SC are calculated from 50 selected events, while the adjusted FWCs for the 11 detection monitoring parameters subject to this statistical evaluation are calculated from nine background sampling events. USEPA (1989)

and ASTM (1998) have suggested a minimum of 8 historical observations to estimate the background mean and standard deviation. In general, it is unwise to perform statistical computations on any less than 8 background samples. Taking any fewer samples will lead to high false-negative rates due to the larger size of the control limits. For example, with four samples and three degrees of freedom, the uncertainty in the true mean and standard deviation (μ and σ), given that sample-based estimates ($\overline{X}$ and S) are enormous, will result in extremely high prediction limits. Conversely, with only a few background measurements, our knowledge of the true sampling variability and the distributional form may be inaccurate leading to a high false-positive rate. The analytical data from the nine background sampling events provides slightly more than the minimum number of historical observations recommended.

- Multivariate and Univariate Control: In the statistical evaluations, the water quality at QRS is defined by 12 water-quality parameters. Some of these parameters are correlated to each other. The total alkalinity and bicarbonate alkalinity have essentially the same concentrations for all of the samples, and thus the same FWCs for all events. The measured SC is closely related to the TDS and to other major ions such as calcium, magnesium, chloride, and sulfate. When the water quality of the spring is monitored by multiple and correlated constituents, there is a need to use multivariate procedure such as Hotelling's T^2 test (Ryan 2002). The problems with multivariate control charts are the likelihood of receiving a false alarm and more importantly, the likelihood of not receiving a signal when the multivariate signal is out of control. When a signal is received on a univariate chart, it is clear how to look for the cause for the water-quality constituent that is being charted. For a multivariate chart, however, a signal could be caused by a mean shift in one or more constituents, a variance shift, or a change in the correlation structure. These uncertainties make it very difficult to determine which factors have triggered the signal and also which constituents are involved.

In addition, the multivariate tests require estimation of the covariance matrix of the vector of measurements. However, the number of sampling events required to obtain the data to estimate the covariance matrix is greater than the dimension of the observation vector. The covariance matrix is unable to be estimated in this investigation because there are only nine sampling events with 12 parameters subject to the statistical evaluation. A simple and commonly used approach to this problem is to use the Bonferroni inequality (Gibbons 1994), in which each water-quality parameter is tested independently, and a separate (univariate) control chart is maintained for each constituent. Alt (1982) demonstrates that the use of Bonferroni's inequality gives a reasonably good approximation to the multivariate control charts when the number of statistical comparisons is less than 50. Although the Bonferroni corrections were not used for this project, their use reduces the frequency of rejecting the null hypothesis, which in turn reduces the false-positive rate and increases the false-negative rate. For these reasons, univariate control charts are used for each parameter of concern for this project.

7.11.2.3 Design of Combined Shewhart-CUSUM Control Charts

Design of a combined control chart requires the determination of the following five parameters:

$\overline{X}$	Estimated mean of the adjusted FWC from the background samples.
S	Estimated standard deviation of the adjusted FWC from the background samples.
H	The value against which the cumulative sum will be compared.
k	A parameter related to the displacement that should be quickly detected.
SCL	The upper Shewhart limits, which is the number of standard deviation units for an immediate release. In this application, the upper control limits are only of interest.

Table 7.15 illustrates the calculated mean ($\overline{X}$) and standard deviation (S) of the adjusted FWC for each water-quality parameter from the background sampling events. $\overline{X}$ and

Table 7.15 Calculated mean and standard deviation of adjusted FWCs

Water-quality parameter	Mean or mean-log FWC*	Standard deviation	Data transformation
Specific conductance (SC)	855.6452	120.9873	–
Alkalinity Bicarbonate ($CaCO_3$)	236.2756	27.4694	–
Alkalinity Carbonate ($CaCO_3$)	– 0.4852	0.4391	Log-transformation
Alkalinity total ($CaCO_3$)	236.2756	27.4694	–
Magnesium	21.3489	1.4763	–
Potassium	60.5978	18.3099	–
Selenium	– 5.0027	0.9433	Log-transformation
Sodium	20.17889	2.8700	–
Sulfate	160.4544	42.5976	–
Chloride	60.49	8.1330	–
Solids, Dissolved	714.8033	49.3857	–
Calcium	133.9811	6.8958	–

S are computed from either nine or 50 sampling events, and they are reasonably precise estimates of their population values μ and σ (Helsel and Hirsch 1992). The means and standard deviations for Carbonate Alkalinity and Selenium are computed as logarithms of the FWCs.

Let the adjusted FWC for any one water-quality parameter during the new sampling event i be x_i. Then the standardized difference z_i is calculated by

$$z_i = \frac{x_i - \overline{X}}{S} \tag{7.11}$$

And the cumulative sum S_i is calculated by

$$S_i = \max\left[0, (z_i - k) + S_{i-1}\right] \tag{7.12}$$

where max[A, B] denotes the maximum of A and B (Gibbons 1999).

As a common practice, $S_0 = 0$ is used, which ensures that only cumulative increases over the background are considered. If a process is in control, such that all future observations come from a normal distribution with a fixed mean μ and standard deviation σ, then the quantity z_i in Eq. (7.11) is approximately distributed as a $N(0, 1)$ random variable and is "bouncing around" 0; so the quantity $z_i - k$ in Eq. (7.12) is "bouncing around" $-k$, and, therefore, the ith upper cumulative sum S will tend to bounce around 0 (Millard and Neerchal 2000).

To illustrate the calculations graphically, the values of S_i (y-axis) and z_i versus t_i (x-axis) will be plotted on a time chart. An out-of-control situation is declared on sampling event i if for the first time, the cumulative increase of one water-quality parameter over its background $S_i \geq h$ or the standardized increase on that measurement occasion $z_i \geq$ SCL.

USEPA (1989) recommends using SCL = 4.5, $k = 1$, and $h = 5$, based on the recommendations of Lucas (1982) and Starks (1988). These values are suggested because they allow a displacement of two standard deviations to be detected quickly (USEPA 1992). For easy application, ASTM (1998) suggested the use of h = SCL = 4.5, which is slightly more robust in detecting leakage and thus reducing the rate of false negatives.

Unlike prediction limits, which provide a fixed confidence level (e.g., 95%) for a given number of future comparisons, control charts do not adjust for the number of future comparisons. The selection of $h = 5$, SCL = 4.5, and $k = 1$ is based on USEPA's own review of the literature and simulation (Lucas 1982; Starks 1988). Since 1.96 standard deviation units correspond to 95% confidence on a normal distribution, there is approximately 95% confidence for this method as well for each comparison.

When the number of background sampling events is over 12, Starks (1988) and USEPA (1992) suggest using $k = 0.75$ and h = SCL = 4. Since 50 FWCs were calculated for SC in the background, $k = 0.75$ and h = SCL = 4 will be used for this parameter. However, for the other 11 parameters subject to this statistical evaluation, there are only nine background sampling events. Therefore, ASTM (1998) recommended values for these 11 parameters will be used (i.e., h = SCL = 4.5, $k = 1$).

It should be noted that these recommendations for values h, k, and SCL were based on analysis of a single statistical comparison. They do not account for adjusting the control chart parameters to account for the number of monitoring points and the number of constituents monitored during each sampling event (Gibbons 1994; Davis 1998). In practice, the statistical comparisons are often more than 1. For this project, 12 statistical comparisons need to be performed at the end of each sampling event. ASTM (1998) also suggests several possible modifications to control charts, one of which includes updating background data (i.e., pooling) to keep statistical power high for each monitoring occasion.

It is also important to note that control charts should be used for variability, as described by Gilbert (1987) in addition to those used for means. This is especially important for those constituents whose FWCs are being log-transformed, because of tendency of logarithms to down-play large values. Although range control charts cannot currently be developed due to insufficient data, they should be developed once data becomes available that defines the overall range of variability.

7.11.2.4 Early Warning of Contaminant Release

Due to complexities in the groundwater sampling program and of the karst aquifer in general, the background monitoring completed over an approximate 2.5-year period may not have characterized the full range of natural conditions. From time to time, inconsistently large or small values can be observed due to sampling, laboratory, transportation, or transcription errors, or even by chance alone. Although the specific procedures attempt to minimize and/or appropriately handle problematic data, being out of control for any given parameter could be caused by factors external to the landfill. Therefore, in some situations, one may opt to demonstrate that an increase is due to causes other than the landfill.

In other cases, additional monitoring phases may be warranted when statistically significant increases are detected. Based on the relative importance of the 12 water-quality parameters of concern, compliance monitoring could also be initiated under either of the two following scenarios:

- Selenium is out of control on the combined control chart.
- Three or more of the other 11 constituents of concern are out of control on the combined control chart.

7.11.2.5 Procedures to Compute Combined Shewhart-CUSUM Control Charts

The following specific procedural steps were used to compute Combined Shewhart-CUSUM Control Charts for each of the 12 constituents of concern for each compliance monitoring event:

(a) Select the three Shewhart-CUSUM parameters h, k, and SCL as follows:
- For SC, use $h=4$, $k=0.75$, SCL$=4$.
- For 11 detection monitoring parameters subject to the statistical evaluation, use $h=4.5$, $k=1$, SCL$=4.5$.
- Note that after the first pooling event, lower control chart variables can be used for the detection monitoring parameters.

(b) Calculate the FWC for each water-quality parameter. For selenium and carbonate alkalinity, use the methods discussed in Censored Data to estimate the FWCs when they have high a percentage of censored data.
(c) Make the necessary FWC adjustments to account for the aquifer storage condition using Equations listed in Table 7.10 for the appropriate parameters.
(d) Denote the new (adjusted) FWC at sampling event I at t_i as x_i.
(e) Compute the standardized value z_i using Eq. (7.11).
(f) Compute the cumulative sum S_i using Eq. (7.12).
(g) Plot the values of z_i, S_i versus t_i on a time chart.
(h) Evaluate charts for statistically significant (i.e., out of control) increases for all 12 constituents of concern. Declare an out-of-control situation for the water-quality parameter at the end of sampling event i if for the first time, $S_i \geq h$ or $z_i \geq$ SCL.
(i) Consider the need to initiate compliance monitoring if selenium is out of control, or if three (3) or more of the remaining 11 constituents of concern are out of control.

As previously discussed, range control charts should also be developed once data becomes available that sufficiently defines the overall range of variability. However, these range control charts cannot be developed until the first 2 years of compliance-monitoring data become available.

7.11.2.6 Verification Resampling

The water quality of karst springs is dynamic. Water quality in karst settings is constantly fluctuating due to a variety of factors including but not limited to antecedent moisture conditions, residence time of storm-pulse waters, seasonal variation, weather patterns, evapotranspiration, frequency and duration of precipitation events, and partial pressure of carbon dioxide. This is in great contrast to porous-medium sites, where conditions are stable at any given location, throughout the course of a given year.

To help control false-positive and false-negative rates, the immediate re-collection of groundwater samples to refute or confirm an unexpected measurement result (also known as verification resampling) has been recommended by many including the USEPA. In an ideal groundwater monitoring situation, when a statistically significant increase in any given constituent is detected, verification resampling can be immediately performed for that constituent to either confirm or refute the initial results. However, in karst groundwater settings, this is not possible because groundwater concentrations are constantly changing in response to aquifer recharge and many other factors, and no single concentration can ever be duplicated for any given storm pulse sampling event. For this reason, verification resampling is not an option for this project.

ASTM (1998) states verification resampling is an integral part of the statistical methodology, and that it allows sequential application of a much smaller prediction limit, thereby minimizing both false-positive and false-negative rates. Because verification resampling cannot be completed for this project (because there is no way to go back and duplicate results of any given event), minimizing both false-positives and false-negatives for this project will be accomplished by using:

- Detailed procedures outlined in sampling and analysis plans.
- Appropriate and consistent handling and treatment of outliers.
- Intra-locational statistical comparison.
- FWCs, and adjustment for aquifer storage condition, where necessary.
- Combined Shewhart-CUSUM Control Charts and appropriate values of h, k, and SCL for various parameters of concern.
- Data pooling techniques, as appropriate.

7.11.2.7 Pooling

Pooling is a method of incorporating newly obtained data into the background data set as a means of increasing the overall size of the background data set. Through the pooling of data, uncertainty in the sample-based mean and standard deviation decreases (as does the size of the prediction limit), thereby minimizing both false-positives and false-negatives (ASTM 1998). Although pooling is not being completed now as part of the statistical evaluation development process, pooling will be completed once an appropriate amount of compliance monitoring data becomes available and it will continue for the life of the monitoring program. Pooling newly obtained data into background data to increase the overall size of the data set is a method used and recommended by the USEPA (1992) and Gibbons (1994). According to ASTM (1998), pooling should be completed every 1–2 years using newly obtained data. Gibbons (1994) recommends that all new data that are tested to be in control be pooled with the initial samples, and $\overline{X}$ and s are recomputed. Note that to make the statistical tests as robust as possible and prevent against concentrations creeping upward (possibly due to the landfill), but not consistently enough to result in out-of-control detections, lower values for control chart parameters will be utilized as sufficient data becomes available.

Pooling should only be performed if data and the monitoring process are shown to be in control (ASTM 1998). Therefore, future compliance monitoring data for this project will only be pooled into the background data set when:

- No evidence of leakage (i.e., no statistically significant increases or out-of-control conditions) is identified using the Combined Shewhart-CUSUM Control chart process for Selenium or three (3) or more of the other constituents of concern.
- No increasing positive trends are identified after performing trend analyses on FWCs using the Mann–Kendall testing procedure.

It is recommended that pooling be completed every 2 years for this project. Pooling of the first 2 years of QRS compliance monitoring data at the site would increase the overall size of the background data sets from 9 to 21 data sets (assuming six compliance monitoring events will be completed annually). A generalized flow chart illustrating the various steps of the pooling process is included in Fig. 7.19.

7.11.3 Investigation of Statistical Power

A robust statistical test procedure is one with large average in-control run lengths and small average out-of-control run lengths. The control procedures developed for this groundwater monitoring program are expected to identify a statistically significant increase as soon as possible if the landfill were in fact to leak but should also give statistically significant increase signals as late as possible when there is no leakage from the landfill. To investigate statistical power, the Combined Shewhart-CUSUM control scheme proposed in this document has been evaluated by Monte Carlo simulations. For these simulations, the following should be noted:

- Up to 20,000 adjusted FWC values of each of the 12 parameters of concern were generated for QRS based on properties as characterized during the background sampling phase. Future values were obtained for each sample by use of the multivariate normal random number generating function in Microsoft® Excel.
- The quality-control procedures were followed.
- The in-control run length is the number of sampling events that have passed before a statistically significant increase is produced when there is no leakage from the landfill (i.e., when the standard unit increase is zero).
- The out-of-control run length is the number of sampling events that have passed before a statistically significant increase is produced when there is leakage from the landfill. The process is simulated by designating constant releases of varying standard deviation unit increases over the background.
- The simulation process was repeated 100 times, and information regarding the empirical distribution of the run lengths was obtained.

The simulation results indicate that the average run length of the developed control charts depends on the standard deviation unit, which reflects the amount of displacement in the concentration means over the background. The average run length decreases rapidly with the increase in the displacement of the concentration means. When the displacement of the mean concentrations is 5 standard deviation units, the control scheme can immediately detect the change at the next sampling event. When the displacement is relatively minor, for example, 0.01 and 0.1, the average run lengths suggest that the control scheme has little chance to detect the change over the life of the groundwater monitoring project. Figure 7.20 illustrates the average run lengths versus the standard deviation units.

USEPA (1992) has established a general guideline for using control charts in statistical approaches, in which an increase of 2 standard deviation units should be detected easily. The Monte Carlo simulations indicate that the average run length for the increase of 2 standard deviation units is 5.4 with the range from 1 to 15, which is very similar to values obtained by Starks (1988) when he suggested the values for h, SCL, and k. The simulation results obtained for this project are also very comparable to those obtained by Lucas (1982) who calculated the average run lengths with the Markov chain approach (Brooke and Evans 1972). Figure 7.21 illustrates the distribution of the run lengths when the increase of the mean concentrations is 2 standard deviation units. Based on 100 realizations in the Monte Carlo simulation, the probability of detecting an increase of 2 standard deviation units at QRS at the next sampling event is only 2%. Assuming six detection monitoring events are completed per year, the probabilities of detecting the same increase are 76% and 99% following the first and second years of compliance monitoring, respectively.

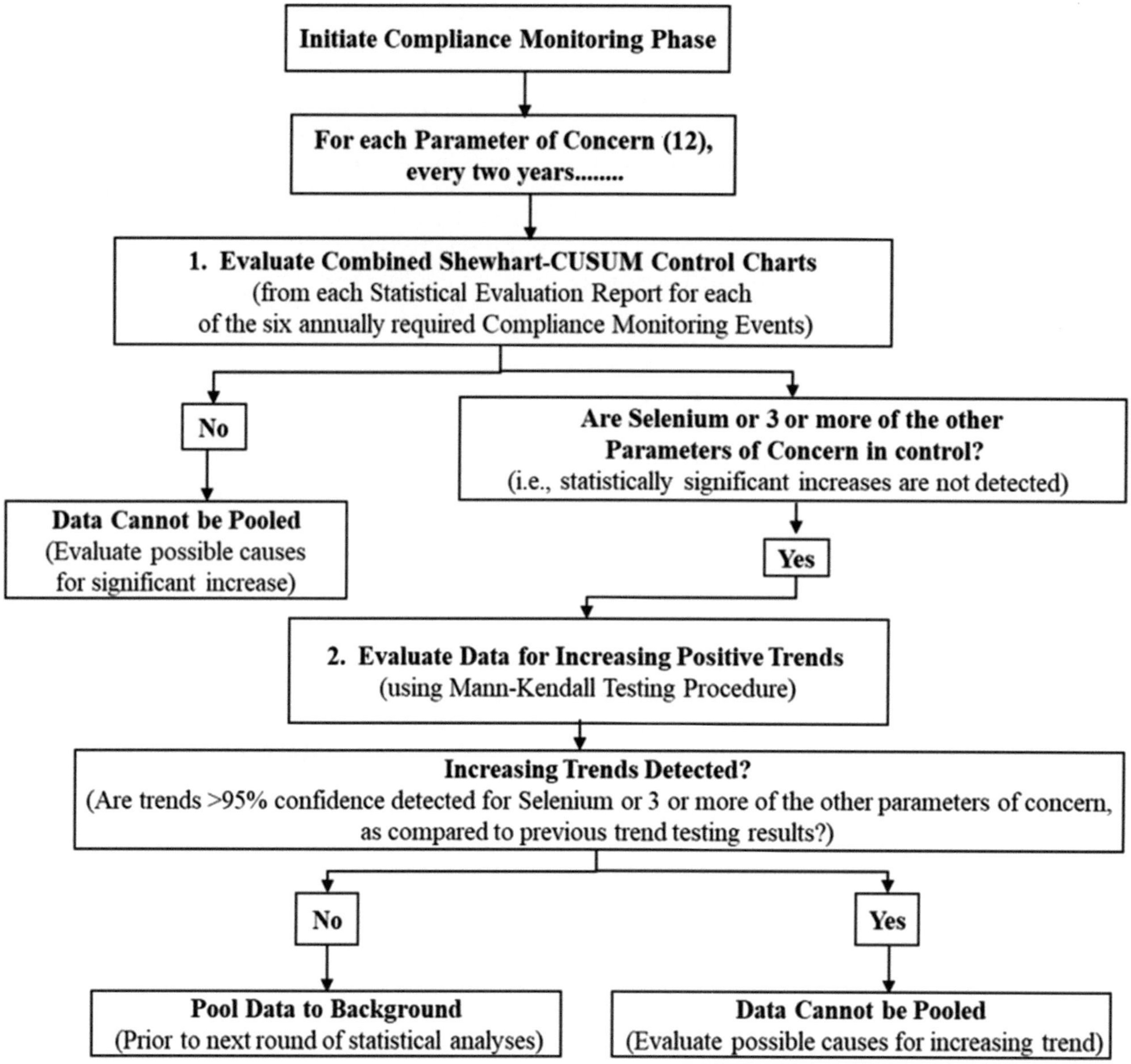

Fig. 7.19 Generalized flow-chart for data pooling process

The simulation results also indicate that selenium is the primary constituent that determines the run lengths when the displacement of the mean concentrations is over 0.5. However, when the displacement for all constituents is minor, the other water-quality parameters play at least an equal role in determining the run length distributions. Of course, if the increases over the background data occur for only the parameters other than selenium, the control charts can also detect them, but with slightly longer run lengths.

With an in-control state (standard deviation unit = 0), the average run length is approximately 4248. If six sampling events are conducted annually, this average run length is equivalent to over 700 years of sampling. On average, the chance to have a false-positive detection at QRS is negligible. Because the minimum in-control run length is 129 and two simulations (out of 100) produced run lengths less than 360, there is a slight chance (~ 2%) that a false-positive detection may occur over 30 years of groundwater monitoring.

We also examined the change in the average run lengths when the control limits for SC changed from the more conservative values ($h = \text{SCL} = 4.5$, $k = 1$) to more aggressive values ($h = \text{SCL} = 4$, $k = 0.75$). Because there are more data points for this field parameter, it is logical to let this parameter carry more weight in the statistical analysis. Table 7.16 compares the average run lengths for the two scenarios

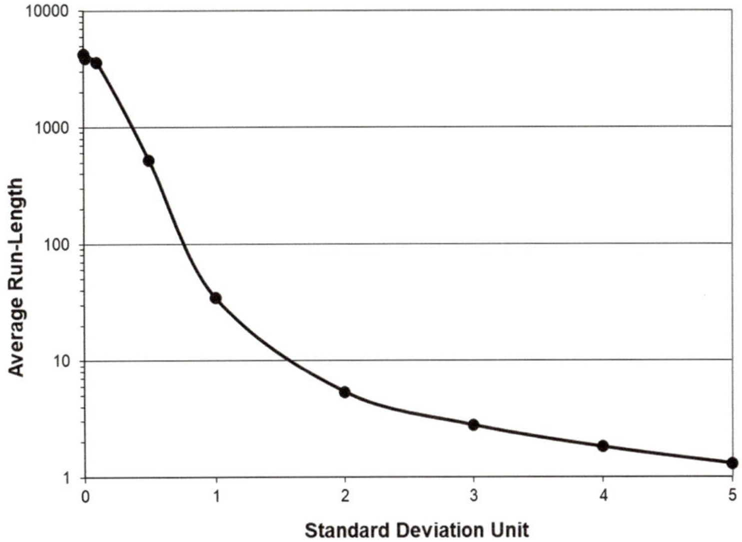

Fig. 7.20 Run length versus standard deviation

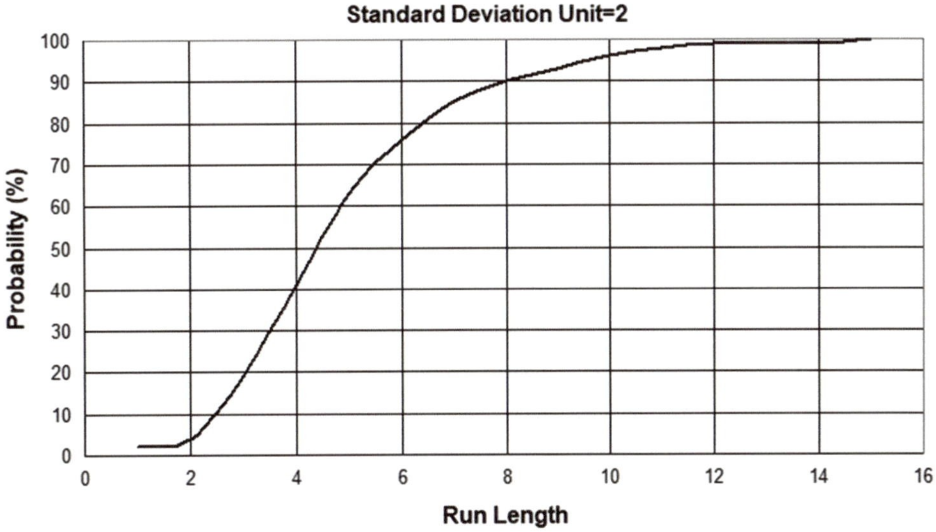

Fig. 7.21 Run length distribution with an increase of mean concentrations of 2 standard deviation units

Table 7.16 Average run lengths using different control limits for specific conductance (SC)

Standard deviation unit	0	0.01	0.1	0.5	1	2	3	4	5
Scenario I	5072.83	4745.17	3889.55	518.4	34.4	5.4	2.8	1.8	1.3
Scenario II	4248.41	3937.05	3627.14	518.4	34.4	5.4	2.8	1.8	1.3

discussed above. Note that in Scenario I, $h = \mathrm{SCL} = 4.5$ and $k = 1$ for all 12 parameters of concern that are subject to the statistical evaluation. In Scenario II, $h = \mathrm{SCL} = 4.5$ and $k = 1$ for 11 detection monitoring analytical parameters, whereas $h = \mathrm{SCL} = 4$ and $k = 0.75$ for SC. Scenario II is proposed for this project.

It is apparent that using lower control limits for SC has improved the ability of the statistical tests to detect potential leakage from the landfill, especially when the leakage is less than 0.1 standard deviation units. When the standard deviation units are equal to or greater than 0.5, the lower control limits for SC still change the run length distribution slightly, but the average run lengths of an out-of-control state remain essentially the same. This is partially caused by the fact that the control chart for Selenium plays the dominant role when the release from the landfill is greater than or equal to 0.5 standard deviation units. Selection of lower control limits for SC can improve the statistical power of the proposed tests.

Although the average in-control run length (standard deviation = 0) decreases substantially from Scenario 1 to Scenario II, the in-control run length for Scenario II is well within the acceptable range (Lucas 1982). Therefore, the lower control limits (i.e., $h=SCL=4$ and $k=0.75$) are recommended for the 12 constituents of concern once the means and standard deviations of the 11 detection monitoring parameters can be estimated from more than 12 background sampling events. If the background data are updated every 2 years by pooling, the lower control limits should be used after the first update.

7.12 Discussions on Using Springs as Monitoring Points

Springs are one of the unique features in karst terranes and are favored as monitoring locations for detection of contaminant release from waste disposal units by many researchers (Quinlan et al. 1991; Ewers 2016). Use of springs as compliance monitoring points shall be based on the conceptual site model of the investigation site and the characteristics of the springs. The following summarizes the main characteristics of springs that shall be considered in their use as monitoring locations:

- **Response to recharge**: Water quality changes with the increase of discharge at the spring. The lag time may vary, depending on the characteristics of the karst aquifer. Knowledge of within-event constituent concentration fluctuations provides limited information to decision-makers regarding control measures that might be required. Sampling under various flow conditions and extensive data analyses are essential to provide interpretations for intelligent decisions. A one-time grab sampling is of little use for estimating the water quality, flow rate, and mass loadings because one instantaneous value is not representative of the average conditions at the spring; even less, the variability of the concentration and mass changes cannot be described. Site-specific sampling and analysis plans may be needed to characterize the water quality of karst aquifers. Due to dynamic responses to recharge events, multiple-parameter, long-term, and high-frequency monitoring may be required. Sampling and analysis plans should be designed to reflect the unique characteristics of the monitoring locations. For example, in some cases, light or dense non-aqueous phase liquids may be observed only during storm events, and other constituents might only be detected on sediment when mobilized. These types of temporal variations are in general not the case in porous media aquifers, and therefore traditional investigation techniques would often not be effective for sites in karst media. Characterization of the natural variations in water quality may require sampling efforts under different flow conditions.
- **Non-repeatability**: Measurements at the karst spring are observational, not experimental because of the continuous movement of groundwater, antecedent condition, rainfall intensity, and its constant reaction with the surrounding environment. Recreating the many natural and anthropogenic influences that affect each measurement is impossible. This is especially true for the karst aquifer under study. The water flow and water quality have a dynamic response to rainfall events. Ancillary and metadata (information about a given data set, including explanatory information and data-quality information) pertinent to the statistical characteristics of the sampled population are investigated using the data from nine sampling events. For reasons that statistical regularity cannot be demonstrated through controlled experiments at the site, ancillary data are important to quantify possibly confounding variables that may preclude meaningful interpretation of data. Any modifications on the land use upstream of the landfill site should be documented.
- **Censored data**: Negative values are not possible for the water quality data at the springs. Limits in methods for sample collection and analysis cause data to be reported as either above or (more typically) below one or more reporting limits, which produces a censored population of data. The effect of censored data can be especially problematic for interpretation of water-quality data (Gibbons and Coleman 2001). Laboratory detection limits change with time and can be dramatically different from laboratory to laboratory and may even be different from method to method within a laboratory. Detection-limit artifacts affect statistical properties of individual data sets. When a data set contains values reported as less than one or more detection limits an overestimation of central-tendency measures and an underestimation of dispersion measures will be caused by truncation of the lower tail of the true population. Because the relative uncertainty in the accuracy and precision of individual values tend to increase as reported concentrations approach the detection limit, the percent error expected for measurements near detection limits is much higher than for values well within the measurement range of the method of analysis.
- **Meaningful statistical outliers**: Valid measurements that are considerably higher or lower than most of the measured population are common among data sets at karst springs. Outliers can be extremely problematic in data interpretation. Outliers can arise from a variety of sources including but not limited to transcription errors, inconsistent sampling procedures, instrument failure, calibration or measurement errors, underestimation of spatial or temporal variability, and other factors. The presence of meaningful high-end outliers (actual but extreme values) contributes to the positive skew and is

a factor producing non-standard distributions. High-end outliers represent times when, for example, regulatory criteria may be exceeded, and the health of the local ecosystems may be affected. Those outliers produce a host of potential problems for interpretation of data sets (Barnett and Lewis 1998). If an outlier is discovered to have a strong influence on the slope of a regression line (the slope or the correlation coefficient changes significantly when the point is omitted), then it must be determined whether the outlier represents extreme values for a single process or if a secondary process is characterized by the outlier. Measurement and documentation of explanatory variables such as precipitation and flow; real-time measures of water-quality characteristics such as SC, pH, temperature, and turbidity, use of ratios between constituents of interest, and results from a comprehensive quality assurance/quality control (QA/QC) program can be used to identify and explain outliers in terms of the potential effect of real physicochemical processes as opposed to the effect of sampling artifacts. Because of the complexity in sampling and analysis in the spring, a strict QA/QC measure is taken to ensure the meaningful outliers are not excluded in the statistical analyses. Elimination of outliers is considered a dangerous and unwarranted practice for the interpretation of water-quality data at the site unless one has substantial objective evidence demonstrating that the outliers are not representative of the population under study. If outliers are not handled in an appropriate manner, unwanted, and potentially unnoticed bias in statistical calculations can occur, which could result in false-positive and/or false-negative detections.

- **Positive skewness**: Data sets that are not symmetrical around mean or median values are typical for spring water quality data sets because the combined effects of a lower bound of zero, censoring, and meaningful outliers tend to produce data sets in which the right tail of the distribution is extended and the left tail truncated.
- **Autocorrelation and independence**: An event is said to be independent of another event when the occurrence of one does not affect the occurrence of another. Spring peak flow-rates separated by a long period of time may be independent, but two peaks close to one another may not be independent. This is true when the recession limb of the first hydrograph at a spring becomes part of the rising limb of the next hydrograph. Natural and anthropogenic effects tend to cause conditions in which consecutive measurements are correlated. The natural and anthropogenic processes controlling groundwater quality and the methods for sampling, processing, and analysis often cause problems with autocorrelation. Autocorrelation is also referred to as serial correlation or correlation—the dependence of residuals in a time sequence because data reflect the effects of preceding conditions. Time-series effect may also occur between subsequent samples within individual sampling events. Autocorrelation can be important because it affects the optimization of regression coefficients, affects estimates of population variance, invalidates results of hypothesis tests, and produces confidence and prediction intervals that are too narrow for the real population being sampled.
- **Interdependence**: Changes in one characteristic of interest such as rainfall intensity, antecedent flow conditions, or temperature cause changes in other characteristics such as measured flows and concentrations. The interdependence makes it very challenging to characterize the natural variations of water quality at karst springs.
- **Temporal variation**: Temporal variation of water quality is a characteristic of many karst springs. Measured water-quality characteristics vary at time scales of hours, days, seasons, years, and even decades because of both natural and anthropogenic influence. Temporal variation may also increase variability in data and affect the comparability of data between sites.

Springs are not isolated hydrogeologic features; they are part of a drainage basin. Use of springs as either monitoring locations or remedial targets shall take their relationship with the entire drainage basin into consideration. One objective of groundwater monitoring programs is to identify releases of contamination as soon as the releases occur. If the spring is too far away from the site, the contaminants may have migrated quite a distance by the time when a spring is contaminated. Under such scenarios, monitoring of springs is still useful but needs to be supplemented with other monitoring efforts immediately downgradient of the disposal facility. Sufficient data should be collected to prove that the springs to be monitored are hydraulically connected to the groundwater system of the investigation site. Some techniques to collect connection data include chemical fingerprinting, tracer testing, isotope analysis, or aquifer testing.

Springs with large drainage basins may contain contamination sources in addition to the disposal facility of interest. The water quality at such springs is impacted by multiple potential contamination sources. An evaluation should be conducted to determine if the chemical constituents can be differentiated from the different sources. Use of springs as monitoring points or remedial action targets shall consider practicability in implementing the program. For example, submerged springs are not readily accessible, and use of the springs requires significant modifications of the discharge areas. Any changes in surface water bodies may require regulatory approval. There is also a possibility of modification of the discharge areas which may result in adverse impacts on aquatic lives.

Use of springs as monitoring points or remedial action targets shall also consider feasibility of developing an effective statistical evaluation plan to evaluate the potential impact of waste disposal units on the groundwater system. Due to the heterogeneity of karst aquifers, intra-locational comparison is generally preferred to inter-locational comparison. Some of the statistical methods typically employed for investigating groundwater in porous media systems are not readily applicable for statistical data analysis of the data collected at springs. A thorough background study is required when a spring is proposed as a remedial action target with an attempt to eliminate the exposure pathway.

7.13 Summaries

The specific procedures discussed in this book may not be directly duplicated; however, the overall technical approaches can shed light on groundwater monitoring programs in other karst areas. Based on our literature review, the monitoring program for this site is the first of its type developed specifically for a karst groundwater setting. This statistical evaluation plan, which used intra-locational statistical comparison, has been developed based on data from one karst spring monitoring location on-site (i.e., QRS). The initial background sample size (n) is 50 for SC and nine for the laboratory analytical parameters, as collected during six storm-pulse sampling events and three 24-h baseflow events. The future pooling of data to background every 2 years (when appropriate pending on the Mann–Kendall testing for trend) would be required prior to the pooling of data into the background data pool (every 2 years) would increase the background size by up to 12. It is also recommended that independence between future sampling and precipitation events be verified at least every 2 years, or prior to the pooling of data to background. Verification of independence can be completed by calculating and evaluating residence times.

The FWC concept, similar to the EMC concept used in stormwater sampling, has been utilized for this project. Single FWC values have been calculated for each parameter of concern for each sampling event using all data points from each of the nine background events. For SC, TDS, chloride, sodium, magnesium, and calcium, FWCs required adjustment to remove dependency on aquifer water levels (i.e., aquifer storage condition), such that the independence assumption was valid. Trend testing was performed on the FWC background data for the 11 detection monitoring parameters subject to the statistical evaluation and on SC using the Mann–Kendall nonparametric test for trend (Kendall 1975; Mann 1945). This testing indicated that, with the exception of Chloride, there is a lack of strong evidence for a positive trend in the 12 constituents of concern.

Combined Shewhart-CUSUM Control Charts have been developed for future statistical testing for this project. These combined control charts are sensitive to both gradual and rapid releases. The statistical power of the proposed control scheme was evaluated using Monte Carlo Simulations. Based on 100 realizations in the simulation, the probability of detecting an increase of two standard deviation units following the first compliance monitoring event is 2%. The probabilities of detecting the same increase after the first and second years of compliance monitoring are 76% and 99%, respectively. For Combined Shewhart-CUSUM Control Charts, the following values should be used:

- For SC: $k=0.75$ and $h=\text{SCL}=4$.
- For the other 11 detection monitoring parameters ($n<12$): $k=1$ and $h=\text{SCL}=4.5$.
- For the other 11 detection monitoring parameters ($n>12$): $k=0.75$ and $h=\text{SCL}=4$.

The range control charts should also be developed and evaluated to check overall variability when sufficient data are available for such statistical tests. The need to initiate compliance monitoring should be closely evaluated if selenium or 3 or more of the other 11 constituents of concern are out of control in future statistical tests.

References

Alt FB (1982) bonferroni inequalities and intervals. In: Kotz S, Johnson N (eds) Encyclopedia of statistical sciences, vol 1. Wiley, New York, pp 294–300

American Society for Testing and Materials (ASTM) (1995) ASTM D 5717-95, Standard guide for design of ground-water monitoring systems in karst and fractured-rock aquifers: 1995 Annual book of standards, Section 7.4 [Construction], Volume 04.09 [Soil and Rock (II)], West Conshohocken, Pennsylvania, pp 440–456; with minor additions in October 1998

ASTM (1998) ASTM D 6312-98, Standard guide for developing appropriate statistical approaches for ground-water detection monitoring programs: 1998 Annual book of standards, West Conshohocken, Pennsylvania, p 1–14

Barfield BJ, Felton GK, Stevens EW, McCann M (2000) A simple model of karst spring flow using modified NRCS procedures. Paper presented at the American Institute of Hydrology Meeting, 4–7 Nov, 2000, Charlotte, NC

Barnett V, Lewis T (1998) Outliers in statistical data: Wiley series in probability and mathematical statistics, 3rd edn. John Wiley & Sons Ltd., West Sussex England, p 584

Bernd MP, Hatzell HH, Crandall CA, Turtora M, Pittman JR, Oaksford ET (1998) Water quality in the Georgia-Florida coastal plain, Georgia and Florida, 1992–1996: U.S. Geological Survey Circular 1151

Berthouex PM, Hunter WG, Pallesen L (1978) Monitoring sewage treatment plants: some quality control aspects. J Qual Control 10(4):139–149

Brooke D, Evans DA (1972) An approach to the probability distribution of CUSUM run length. Biometrika 59:539–549

Burr IW (1967) The effect of non-normality on constants for and R charts. Indus Qual Control 23(11):563–569

Cooke SE, Ahmed SM, MacAlpine NE (2000) Introductory guide to surface water-quality monitoring in agriculture. Conservation and development branch, Alberta Agriculture, Food and Rural Development, Edmonton, Alberta

Currens J, Simmons SA, Pocoke R (1987) Spatial variation in heavy metal deposition rates in urban areas. In: Proceedings of the second international symposium: highway pollution. Sci Tot Environ 59

Danckwerts PV (1953) Continuous flow systems: distribution of residence times. Chem Eng Sci 2:1–13

Davis CB (1998) Power comparisons for control chart and prediction limits procedures. EnviroStat Technical Report 98-1, Environmetrics and Statistics Limited, Henderson, NV

Drever IJ (1997) The geochemistry of natural waters: surface and groundwater environments, Pearson College Division

Dreybrodt W (1988) Processes in karst systems—physics, chemistry, and geology. Springer Series in Physical Environment. Springer-Verlag, Germany

Driscoll ED (1986a) Combined sewer overflow analysis methodology. Prepared for EPA, Municipal Facilities Division, October 1986

Driscoll ED (1986b) Lognormality of point and nonpoint source pollutant concentrations. In: Proceedings of stormwater and water-quality model users group meeting, Orlando, Florida, March 1986. EPA/600/9-86/023, pp 157–176

Driscoll ED, Shelley PE, Strecker EW (1990) Pollutant loadings and impacts from highway stormwater runoff. volume I: design procedure: Federal Highway Administration Final Report FHWA-RD-88-006, p 67

Duke LD (2001) Evaluating critical flow conditions for pollutants in an effluent-dominated, storm-peaking western U.S. stream, California regional water control board, Los Angeles, Draft report: chloride total maximum daily load for the Calleguas creek watershed

Ewers R (2016) On the efficacy of monitoring wells in karstic carbonate aquifers (Chap. 10). In: Caves and karst across time. Geol Soc Spec Pap 516

Felton GK (1994) Hydrologic responses of a karst watershed. Trans ASAE 37(1):L143-150

Fenneman NM (1938) Physiography of the eastern United States. McGraw-Hill, New York

Field MS (1989) The vulnerability of karst aquifers to chemical contamination. In: Moore JE, Zaporozec AA, Csallany SC, Varney TC (eds) Recent advances in ground-water hydrology. American Institute of Hydrology, pp 130–142

Field MS (2002) The QTRACER2 program for tracer-breakthrough curve analysis for tracer tests in karstic aquifers and other hydrologic systems. EPA/600/R-02/001

Ford DC, Williams PW (1989) Karst geomorphology and hydrology. Unwin Hyman, Boston, p 601

Ford DC, Williams PW (2007) Karst hydrology and geomorphology. Wiley, Chichester, UK

Gibbons RD (1994) Statistical methods for groundwater monitoring. John Wiley & Sons Inc., New York

Gibbons RD (1999) Use of combined Shewhart-CUSUM control charts for ground water monitoring applications. Ground Water 37(5):682–691

Gibbons RD, Coleman DE (2001) Statistical methods for detection and quantification of environmental contamination. John Wiley & Sons Inc., New York

Gilbert RO (1987) Statistical methods for environmental pollution monitoring. John Wiley & Sons Inc., New York

Goode DJ (1996) Direct simulation of groundwater age. Water Resour Res 32(2):289–296

Granato GE, Bank FG, Cazenas PA (1998) Data quality objectives and criteria for basic information, acceptable uncertainty, and quality-assurance and quality-control documentation. U.S. Geological Survey Open File Report 99-394, 17 pp

Helsel DR, Hirsch RM (1992) Statistical methods in water resources. U.S. Geological Survey

Helsel DR, Hirsch RM (2002) Chapter 3: Hydrologic analysis and interpretation. In: Statistical methods in water resources, techniques of water-resources investigations of the United States Geological Survey (Book 4). U.S. Geological Survey

Hess JW, White WB (1988) Storm response of the karst carbonate aquifer of south-central Kentucky. J Hydrol 99:235–252

Hillel D (1980) Fundamentals of soil physics. Academic Press, Orlando, Florida, p 413

Hollingsworth E (2009) Karst regions of the world: populating global karst datasets and generating maps to advance the understanding of karst occurrence and protection of karst species and habitats worldwide. MS Thesis, University of Arkansas, Fayetteville, AK

Iglewicz B (1983) In: Hoaglin DC, Mosteller F, Tukey JW (eds) Understanding robust and exploratory data analysis. John Wiley, New York

Interstate Technology & Regulatory Council (2013) Groundwater statistics and monitoring compliance, statistical tools for the project life cycle. GSMC-1. Interstate Technology & Regulatory Council, Groundwater Statistics and Monitoring Compliance Team, Washington, DC, 372 pp

Irish Jr. LB, Barrett ME, Malina Jr. JF, Charbeneau RJ (1998) Use of regression models for analyzing highway storm-water loads. J Environ Eng 124(10):230–254

Kendall MG (1975) Rank correlation methods, 4th edn. Charles Griffin, London

Loftis JC, Porter PS, Settembre G (1987) Statistical analysis of industrial wastewater monitoring data. J WPCF 59(3):145–151

Lounsbury RA, Pettit AJ, Zhou WF, Beck BF (2003) Effective practices for initiating high-frequency, storm-induced sampling of two karst springs. In: Beck BF (ed) Proceedings of the 9th multidisciplinary conference on sinkholes and the engineering and environmental impacts of karst, 6–10 Sept, 2003, Huntsville, Alabama. Published by the American Society of Civil Engineers (ASCE)

Lucas JM (1982) Combined Shewhart-CUSUM quality control scheme. J Qual Technol 14:51–59

Lucas JM (1985) Cumulative Sum (CUSUM) control schemes. Commun Statist Theor Meth 14(11):2689–2704

Lucas JM, Crosier RB (1982) Robust CUSUM: a robustness study for CUSUM quality control schemes. Commun Statist Theor Meth 11(23):2669–2687

Maloszewski P, Zuber A (1991) Influence of matrix diffusion and exchange reactions on radiocarbon ages in fissured carbonated aquifers. Water Resour Res 27(8):1937–1945

Mann HB (1945) Nonparametric tests against trend. Econometrica 13:245–259

Mann CJ (1987) Misuses of linear regression in earth sciences. In: Size, use, and abuse of statistical methods in the earth sciences. International Association for Mathematical Geology Studies in Mathematical Geology No. 1, pp 74–106

McConnell H, Horn JM (1972) Probabilities of surface karst-applications in Mitchell Plain, Indiana. Spatial analysis in geomorphology. Methuen, London, pp 111–133

Millard SP, Neerchal NK (2000) Environmental statistics. CRC Press

Montgomery DC (1997) Introduction to statistical quality control, 3rd edn. John Wiley & Sons, New York

Mull DS, Smoot JL, Liebermann TD (1988) Dye-tracing techniques used to determine ground-water flow in a carbonate aquifer system near Elizabethtown, Kentucky. U.S. Geological Survey. Water Resources Investigation Report 87-4174

Palmer MV, Palmer AN (1975) Landform development in the Mitchell Plain of southern Indiana—origin of a partially karsted plain. Z Geomorph 19(1):1–39

Parkhurst DF (1998) Arithmetic versus geometric means for environmental data. Environmental Science and Technology/News, 92A-98A

Powell RL (1973) Karst areas of Indiana—the south-central karst area (physiography and development). In: Guidebook for the 1973 convention of the national speleological society. Indiana Geological Survey, pp 3–6

Pratt CJ, Elliott G, Fulcher G (1987) Suspended solids discharge from highway gully pots in a residential catchment. In: Proceedings of the second international symposium: highway pollution. Sci Tot Environ 59

Quinlan JF (1990) Special problems of ground-water monitoring in karst terranes. In: Nielsen DM, Johnson AI (eds) Ground water and vadose zone monitoring. American Society for Testing and Materials STP 1053, Philadelphia, Pennsylvania, pp 275–304

Quinlan JF, Alexander EC Jr (1987) How often should samples be taken at relevant locations for reliable monitoring of pollutants from an agricultural, waste disposal, or spill site. In: Beck BF (ed) Proceedings of 2nd multidisciplinary conference on sinkholes and the environmental impacts of karst. Balkema, Rotterdam, The Netherlands, pp 277–286

Quinlan JF, Ewers RO (1989) Subsurface drainage in the Mammoth cave area. In: White WB, White EI (eds) Karst hydrology—concepts from the Mammoth cave are, Van Nostrand Reinhold

Quinlan JF, Smart PL, Schindel GM, Alexander EC Jr, Edwards AJ, Smith AR (1991) Recommended administrative/regulatory definition of karst aquifer, principles for classification of vulnerability of karst aquifers, and determination of optimum sampling frequency at springs. In: Proceedings of the 3rd conference on hydrogeology, ecology, monitoring, and management of ground water in karst terranes, 4–6 Dec, Nashville, Tennessee

Ryan TP (2002) Statistical methods for quality improvement, 2nd edn. John Wiley & Sons, New York, p 446

Schilling EG, Nelson PR (1976) The effect of non-normality on the control limits of charts. J Qual Technol 8(4):183–188

Schindel GM, Quinlan JF, Davies GJ, Ray JA (1996) Guidelines for wellhead and springhead protection area delineation. In: Carbonate rocks. US EPA Region IV Groundwater Protection Branch, 195 pp

Smart CC (1999) Subsidiary conduit system: a hiatus in aquifer monitoring and modeling. In: Palmer AN, Palmer MV, Sasowsky ID (eds) Karst modeling. Karst Water Institute, Special Publication 5, pp 146–157

Starks TH (1988) Evaluation of control chart methodologies for RCRA waste sites; Draft report by Environmental Research Center, University of Nevada, Las Vegas, for exposure assessment research division, Environmental monitoring systems laboratory—Las Vegas, Nevada. EPA Technical Report CR814342-01-3

United States Geological Survey (USGS) (1997) Preliminary estimates of residence times and apparent ages of ground water in the Chesapeake Bay watershed, and water-quality data from a survey of springs. Water Resources Investigations Report 97-4225

U.S. Environmental Protection Agency (USEPA) (1989) Statistical analysis of ground-water monitoring data at RCRA facilities, Interim final guidance. EPA/530-SW-89-026, Office of Solid Waste, U.S. Environmental Protection Agency, Washington, DC

USEPA (1992) Statistical analysis of ground-water monitoring data at RCRA facilities: addendum to interim final guidance. In: Statistical training course for ground-water monitoring data analysis. EPA530-R-93-003

USEPA (2004) Handbook of groundwater protection and cleanup policies for RCRA corrective action for facilities subject to corrective action under Subtitle C of the Resource Conservation and Recovery Act. EPA530-R-04-030, US Environmental Protection Agency, Washington, DC

U.S. Soil Conservation Service (1985) Soil conservation service national engineering handbook, Section 7.4, Hydrology. U.S. Department of Agriculture, Natural Resources Conservation Service, Washington, DC

Vaughan WJ, Russell CS (1983) Monitoring point sources of pollution: answers and more questions from statistical quality control. The Am Statis 37(4):476–487

White WB (1988) Geomorphology and hydrology of karst terrains. Oxford University Press, New York

Zhou WF (2000) Studies of confluent flow in mature karst aquifers using analog models and numerical mixing cell models. In: Sasowsky ID, Wicks CM (eds) Groundwater flow and contaminant transport in carbonate aquifers. A. A. Balkema, Rotterdam, pp 157–168

Zhou W, Beck BF, Stephenson JB (1999) Investigation of groundwater flow in karst areas using component separation of natural potential measurements. Environ Geol 37(1–2):19–25

Zhou W, Beck BF, Stephenson JB (2000) Reliability of dipole-dipole electrical resistivity tomography for defining depth to bedrock in covered karst terranes. Environ Geol 39(7):760–766

Zhou W, Beck BF, Pettit AJ, Stephenson JB (2002) A ground water tracing investigation as an aid of locating ground water monitoring stations on the Mitchell Plain of Southern Indiana. Environ Geol 41:842–851

GPSR Compliance

The European Union's (EU) General Product Safety Regulation (GPSR) is a set of rules that requires consumer products to be safe and our obligations to ensure this.

If you have any concerns about our products, you can contact us on ProductSafety@springernature.com

In case Publisher is established outside the EU, the EU authorized representative is:

Springer Nature Customer Service Center GmbH
Europaplatz 3
69115 Heidelberg, Germany

Batch number: 10391164

Printed by Printforce, the Netherlands